T0349908

Analytic Theory of Subnormal Operators

Analytic Theory of Subnormal Operators

Daoxing Xia

Vanderbilt University, USA

World Scientific

NEW JERSEY • LONDON • SINGAPORE • BEIJING • SHANGHAI • HONG KONG • TAIPEI • CHENNAI

Published by

World Scientific Publishing Co. Pte. Ltd.

5 Toh Tuck Link, Singapore 596224

USA office: 27 Warren Street, Suite 401-402, Hackensack, NJ 07601

UK office: 57 Shelton Street, Covent Garden, London WC2H 9HE

Library of Congress Cataloging-in-Publication Data
Xia, Daoxing, 1930 October–
   [Ci zheng chang suan zi jie xi li lun. English]
   Analytic theory of subnormal operators / by Daoxing Xia (Vanderbilt University, USA).
    pages cm
   Includes bibliographical references and index.
   ISBN 978-9814641333 (hardcover : alk. paper)
   1. Subnormal operators. 2. Operator theory. 3. Mathematics. I. Title.
   QA329.2.X5313 2015
   515'.7246--dc23
                            2014042629

British Library Cataloguing-in-Publication Data
A catalogue record for this book is available from the British Library.

Printed in Singapore

Contents

Preface

The spectral analysis of normal operators was established in the early 20th century. In 1993, Professor I. M. Gelfand delivered the Shanks Lecture at Vanderbilt University. He said "In 1937–1940, I tried to find the infinite dimensional analogue of the Jordan form of an operator; but I did not succeed. On the way, I found the theory of normed rings."

The author thinks that for a long time period, to find the infinite dimensional analogue of the Jordan form shall be one of the central problems of operator theory.

Around the 1960's there appeared L. De Brange's theory, the B. Sz. Nagy-B. Focas model for contractions as well as several models for semi-normal and hypernormal operators established by T. Kato, J. D. Pincus, R. W. Carey-Pincus, P. S. Muhly, and the author. Then some years later the two-dimensional model of hyponormal operators was obtained by M. Martin and M. Putinar. Around the same time another important direction appeared consisting of Pincus' determinant formula, J. W. Helton-R. Howe's trace formula and Carey-Pincus' trace formula of commutators. All of these influenced the author to establish an analytic model for subnormal operators. Subnormal operators are defined to be operators with normal extensions. So they are most closely related to normal operators. The analytic model gives a representation of a subnormal operator as the multiplication by the independent variable of a space of vector-valued functions which are analytic on the resolvent set of its normal extension.

A useful tool in the theory of the analytic model is the mosaic. A mosaic of an operator A is a kind of operator-valued function on some set related to the spectrum of A which is constructed by the operator A and is a complete unitary invariant for the operator. It was firstly introduced by Pincus, Carey and then studied by Carey-Pincus for semi-normal operators

and then by Pincus-Xia for hyponormal operators. The mosaic for a subnormal operator is completely different from the mosaic for a hyponormal
operator, although a subnormal operator is hyponormal. Here the mosaic
is an idempotent operator-valued analytic function on the resolvent set of
the minimal normal extension of a subnormal operator. The value of that
function is the parallel projection to the subspace of the vector value of
these analytic functions in the model Hilbert space. Still, these two kinds
of mosaics have two things in common. Firstly, they are complete unitary
invariants and, secondly, the trace of both mosaics is the Pincus principal
function. Actually, earlier Carey-Pincus discovered that the principal function of a subnormal operator is integer-valued. This gives us some hint that
we should find a mosaic which is idempotent-valued.

Chapter 1 of this monograph contains the basic properties of the analytic model and the mosaic. Besides, we give a trace formula for the trace
of the commutator of the functions of a subnormal operator and its adjoint, expressed as a line integral on the spectrum of the minimal normal
extension.

In Chapter 2, we study the subnormal operators with finite rank self-
commutators. Using quadrature domains in Riemann surfaces, the author
gives the decomposition of the mosaic. By this, the infinite-dimensional
diagonalization of the adjoint of a subnormal operator with finite rank self-
commutator is established. Besides, the trace formula for commutators
is expressed by the multiplicity of the minimal normal extension. This is
related to the principal current studied by Carey-Pincus.

In the middle of the 1980's, important works by J. Conway, R. Curto
and M. Putinar on the subnormal tuples of operators occurred. Under
this circumstance, the author generalized the theory of the analytic model
and the mosaic of a single subnormal operator to the case of subnormal
k-tuples of operators. In this case the mosaic is a family of operator-valued
functions. Some new operator identities were found. Using the mosaic, the
author gives some expression for the products of resolvents of the minimal
normal extension. These are discussed in Chapter 3. The generalization
of the trace formula for the commutator in the single subnormal operator
case to the subnormal tuple case is also included in this chapter.

In Chapter 4 we study the subnormal k-tuple of operators with finite
rank self-commutators. It includes the trace formula of commutators by
Pincus-Xia as well as work by Pincus-D. Zheng. The author studied the
joint point spectrum of the adjoint of the subnormal k-tuple of operators.
This is related to an analytic manifold of complex one dimension, which
could be considered as a generalization of the quadrature domain in a Rie-

mann surface. The author proved that the Pincus principal function in this case is the multiplicity function on the joint point spectrum of the adjoint of a subnormal tuple of operators.

In Chapter 5 we study some more general classes of operators with finite rank self-commutators. This contains part of the theory of the analytic model of hyponormal operators with rank one self-commutators by Pincus, D. Xia and J. Xia, part of the linear analysis of quadrature domains by M. Putinar, and some works on operators with finite rank self-commutator by the author. These are written in a way that is close to the concepts and methods, such as using the mosaic, discussed in the previous chapters. The author tries to show that the ideas and methods here are applicable to more general classes of operators.

According to J. Conway's idea, the combination of the analytic model and M. Putinar's block matrix representation may have some possibility to solve other problems in the theory of subnormal operators as well as hyponormal operators, etc. The block matrix representation of an operator and the commuting tuple of operators are presented in Chapter 1 and Chapter 3, respectively. Some few results in Chapter 1 and Chapter 3 may be also considered as a clue for the application of this kind of combination.

The writing of this monograph is stimulated also by D. Yakubovich and J. Gleason, and joint papers by J. Gleason and C. Rosentrater on this subject. Anyone who knows the theory of functions of one complex variable and spectral analysis of normal operators can read this monograph without difficulty. But the reader who wants to know the whole picture of the theory of subnormal operators, except for this monograph, still needs to read J. Conway's "The Theory of Subnormal Operators".

The author wishes to dedicate this monograph to the memory of his two former advisors: academician Chen Jian Gong (K. K. Chen) who guided the author to study classical analysis, especially the theory of functions of one complex variable and academician Israel M. Gelfand who guided the author to study functional analysis and mathematical physics.

The Chinese version of this monograph has been published by the Higher Education Press, Beijing, China, 2012. In this English edition, the author has added a new section 1.7 and made some small corrections on typos and some others. This edition was completed while the author was visiting the Chen Jiangong Institute for Advance Study, Hangzhou Normal University, Hangzhou, China, during the summer of 2014.

Daoxing Xia

Chapter 1

Subnormal Operators

1.1 Subnormal operators

In this monograph, a Hilbert space means a separable Hilbert space on the complex field $\mathbb{C}$. An operator means a linear bounded operator.

The algebra of all operators on a Hilbert space $\mathcal{H}$ is denoted by $L(\mathcal{H})$. A normal operator N on a Hilbert space $\mathcal{H}$ is an operator satisfying $N^*N = NN^*$, where N^* is the adjoint of N. Let N be a normal operator on the Hilbert space $\mathcal{H}$. Let $\sigma(N)$ be the spectrum of N. Let $\mathcal{B}$ be the family of all Borel sets in $\sigma(N)$. Let $\mathcal{M}$ be the measurable space $(\sigma(N), \mathcal{B})$. It is well-known that there is a projection-valued measure $E(\cdot)$ on $\mathcal{M}$ satisfying the following conditions

(i) $E(\sigma(N)) = I$,
(ii) $(E(\cup_{n=1}^{\infty}B_n)x, y) = \sum_{n=1}^{\infty}(E(B_n)x, y)$ for $x, y \in \mathcal{H}$, if $B_n \in \mathcal{B}$, $n = 1, 2, \ldots$ and $B_m \cap B_n = \emptyset$ for $m \neq n$.
(iii) $N = \int \lambda E(d\lambda)$.

This measure space $(E(\cdot), \mathcal{M})$ is the *spectral measure* for N.

An operator S on a Hilbert space $\mathcal{H}$ is said to be *subnormal* if there is a Hilbert space $\mathcal{K}$ containing $\mathcal{H}$ as a subspace satisfying

$$Sx = Nx \text{ for } x \in \mathcal{H}.$$

This N is said to be a *normal extension* of S and S is a *contraction* of N. Among non-normal operators, the subnormal one is mostly close to the normal operator. The normal extension N is said to be *minimal* if there is no subspace $\mathcal{H}_1 \subset \mathcal{K} \ominus \mathcal{H}$ satisfying $\mathcal{H}_1 \neq \{0\}$ and reducing N. Let m.n.e. be the abbreviation of minimal normal extension.

For any set $\mathcal{F}$ of vectors, let

$$\bigvee \mathcal{F} = \{\sum_{j=1}^{n} c_j x_j : x_j \in \mathcal{F}, c_j \in \mathbb{C}, n = 1, 2, \dots\},$$

be the linear manifold spanned by $\mathcal{F}$. For any set $\mathcal{F}$, let cl$\mathcal{F}$ be the closure of the set $\mathcal{F}$.

Theorem 1.1.1. *Let S be a subnormal operator on a Hilbert space $\mathcal{H}$, then there is an m.n.e. N of S on $\mathcal{K} \supset \mathcal{H}$. This m.n.e. N is unique in the following sense: if there is another m.n.e. N_1 of S on $\mathcal{K}_1 \supset \mathcal{H}$, then there is a unitary operator U from $\mathcal{K}$ to $\mathcal{K}_1$ satisfying $N_1 = UNU^{-1}$ and $Ux = x$ for $x \in \mathcal{H}$.*

Proof. Suppose N_0 is a normal extension of S on $\mathcal{K}_0 \supset \mathcal{H}$. Let

$$\mathcal{F} = \bigvee \{N_0^{*k} x : x \in \mathcal{H}, k = 0, 1, 2 \dots\}$$

and $\mathcal{K} = \text{cl}\mathcal{F}$. It is easy to see that $\mathcal{F}$ is invariant with respect to N_0 and N_0^*. Therefore $\mathcal{K}$ is a subspace of $\mathcal{K}_0$ and $\mathcal{K}$ reduces N_0.

Let us prove that $N \overset{\text{def}}{=} N_0|_{\mathcal{K}}$ is an m.n.e. of S. Suppose there is a subspace $\mathcal{H}_1 \subset \mathcal{K}$ satisfying $\mathcal{H}_1 \perp \mathcal{H}$, $\mathcal{H}_1$ reduces N and $\mathcal{H}_1 \neq \{0\}$. Let $y \in \mathcal{H}_1, y \neq 0$. Then

$$N^k y \perp x \quad \text{for all } x \in \mathcal{H}.$$

Thus $(y, N_0^{*k}x) = (y, N^{*k}x) = (N^k y, x) = 0$ for all $x \in \mathcal{H}$, i.e. $y \perp N_0^{*k}x$, $x \in \mathcal{H}$. Therefore $y \perp \mathcal{K}$. It implies $(y, y) = 0$, which leads to a contradiction. Thus N is an m.n.e. of S, and if N_0 is an m.n.e. of S, then $\mathcal{K} = \mathcal{K}_0$.

Now, suppose N_1 is another m.n.e. of S on $\mathcal{K}_1$. Then

$$\mathcal{K}_1 = \text{cl} \bigvee \{N_1^{*k} x : x \in \mathcal{H}, k = 0, 1, 2 \dots\}.$$

Let U be a mapping satisfying

$$U N_0^{*k} x = N_1^{*k} x, \quad x \in \mathcal{H}.$$

From

$$(N_1^{*k}x, N_1^{*m}y) = (N_1^m x, N_1^k y) = (S^m x, S^k y), \quad x, y \in \mathcal{H},$$

and

$$(N_0^{*k}x, N_0^{*m}y) = (N_0^m x, N_0^k y) = (S^m x, S^k y), \quad x, y \in \mathcal{H},$$

it shows that U is an isometry from $\bigvee\{N_0^{*k}x\}$ to $\bigvee\{N_1^{*k}x\}$. Therefore U extends a unitary operator which is also denoted by U such that $U\mathcal{K}_0 = \mathcal{K}_1$ and $N_1 = UN_0U^{-1}$ which proves the theorem. $\qquad\square$

The m.n.e. of a subnormal operator is a basic and important role in investigating the subnormal operator.

Theorem 1.1.2. *Let S be a subnormal operator on $\mathcal{H}$ with m.n.e. N on $\mathcal{K}$. Then $\sigma(N) \subset \sigma(S)$.*

Proof. Suppose $z \in \rho(S)$. Let $E(\cdot)$ be the spectral measure of N. Define $L = E(B(z, \epsilon))\mathcal{K}$, where $0 < \epsilon < \|(S-z)^{-1}\|^{-1}$ and $B(z, \epsilon) = \{\lambda : |\lambda - z| < \epsilon\} \subset \rho(S)$. For $f \in L$,

$$\|(N-z)^{*k} f\| \le \epsilon^k \|f\|.$$

If $f \in L$ and $h \in \mathcal{H}$, then

$$|(f, h)| = |(f, (S-z)^k (S-z)^{-k} h)| = |(f, (N-z)^k (S-z)^{-k} h)|$$
$$= |((N-z)^{*k} f, (S-z)^{-k} h)| \le (\epsilon \|(S-z)^{-1}\|)^k \|f\| \|h\|.$$

Letting $k \to \infty$, it implies $f \perp h$. That means $L \perp \mathcal{H}$. But L reduces N and N is the m.n.e. of S. Therefore $L = \{0\}$, i.e. $E(B(z, \epsilon)) = 0$ and $z \in \rho(N)$ which proves the theorem. $\qquad\square$

1.2 Block-matrix decomposition of a pure operator

For any two operators A and B, let $[A, B] \overset{\text{def}}{=} AB - BA$ be the *commutator* of A and B. Let T be an operator on a Hilbert space $\mathcal{H}$. Define the *self-commutator* of T:

$$D_T \overset{\text{def}}{=} [T^*, T].$$

It is known that self-commutator is a useful tool in investigating subnormal operators as well as some other classes of operators such as hyponormal operators, etc. In this section, we will study the decomposition of a Hilbert space based on the self commutator of an operator on this Hilbert space.

Define $M_T = \text{cl} D_T \mathcal{H}$ as the *defect* subspace of T. Sometimes, M_T is simply denoted by M. Define

$$K_T \overset{\text{def}}{=} \mathcal{H}_0 = \text{cl} \bigvee_{k=0}^{\infty} T^{*k} M_T,$$

and

$$G_n = \text{cl} \bigvee_{k=0}^{n} T^k \mathcal{H}_0, \quad n \ge 0.$$

Of course, $G_0 = \mathcal{H}_0$. Let $G_\infty = \text{cl} \bigvee_{k=0}^{\infty} T^k \mathcal{H}_0$. An operator T is said to be *pure*, or *completely non-normal*, if there is no proper subspace of $\mathcal{H}$ which reduces T and contains M_T as a subspace.

Proposition 1.2.1. *The subspace G_∞ reduces T, $T|_{G_\infty}$ is pure and the restriction of T on $\mathcal{H} \ominus G_\infty$ is normal.*

Proof. It is obvious that $TG_\infty \subset G_\infty$. By the identity

$$T^*T^k = \sum_{j=0}^{k-1} T^j D_T T^{k-j-1} + T^k T^*, \quad k = 1, 2, \ldots \tag{1.2.1}$$

We have

$$T^*G_n \subset G_n, \quad n = 1, 2, \ldots$$

Hence $T^*G_\infty \subset G_\infty$. Therefore G_∞ reduces T.

If L is any subspace of G_∞ reducing T and $L \supset M_T$, then $L \supset G_\infty$. Thus $L = G_\infty$ which proves that $T|_{G_\infty}$ is pure.

If $x \in \mathcal{H} \ominus G_\infty$, then $[T^*, T]x \in \mathcal{H} \ominus G_\infty$, but $[T^*, T]x \in G_\infty$, i.e. $[T^*, T]x = 0$. Therefore $T|_{\mathcal{H} \ominus G_\infty}$ is normal. $\qquad\square$

Corollary 1.2.2. *The operator T on $\mathcal{H}$ is pure, iff $\mathcal{H} = G_\infty$.*

Since the spectral analysis of normal operators is well studied, therefore we only study the pure operator.

From now on, we assume that T is pure. Define

$$\mathcal{H}_n = G_n \ominus G_{n-1}, \quad n = 1, 2, \ldots$$

It is easy to see that $\mathcal{H}_{n+1} \oplus \mathcal{H}_n = G_{n+1} \ominus G_{n-1}$, for $n > 0$. Let us prove that

$$T\mathcal{H}_n \subset \mathcal{H}_{n+1} \oplus \mathcal{H}_n, \quad n = 0, 1, \ldots \tag{1.2.2}$$

It is obvious that (1.2.2) holds good for $n = 0$. Besides,

$$T\mathcal{H}_n \subset TG_n \subset G_{n+1}.$$

Therefore we only have to prove that if $h \perp G_{n-1}$ then $Th \perp G_{n-1}$. As a matter of fact, from $h \perp G_{n-1}$, we have

$$(h, T^k T^{*j} x) = 0, \quad k = 0, 1, \ldots, n-1, \; j = 0, 1, 2, \ldots, \; x \in M.$$

Therefore $(Th, T^k T^{*j} x) = (h, T^* T^k T^{*j} x)$. By (1.2.1),

$$(Th, T^k T^{*j} x) = \sum_{l=0}^{k-1} (h, T^j D_T x_l) + (h, T^k T^{*j+1} x),$$

where $x_l \in \mathcal{H}$. Thus

$$(Th, T^k T^{*j} x) = 0, \quad k = 0, 1, \ldots, n-1$$

which proves (1.2.2).

Proposition 1.2.3. *If T is pure, then T has the following two-diagonal structure with respect to the orthogonal decomposition $\mathcal{H} = \sum_{n=0}^{\infty} \oplus \mathcal{H}_n$*

$$T = \begin{pmatrix} B_0 & 0 & 0 & 0 & \cdots \\ D_1 & B_1 & 0 & 0 & \cdots \\ 0 & D_2 & B_2 & 0 & \cdots \\ 0 & 0 & D_3 & B_3 & \cdots \\ \vdots & \vdots & \vdots & \vdots & \ddots \end{pmatrix} \tag{1.2.3}$$

where D_l and B_m satisfy

$$[B_0^*, B_0] + D_1^* D_1 = D_T, \quad [B_k^*, B_k] + D_{k+1}^* D_{k+1} = D_k D_k^*, \tag{1.2.4}$$

$k \geq 1$, *and*

$$B_{k+1}^* D_{k+1} = D_{k+1} B_k^*, \quad k \geq 0. \tag{1.2.5}$$

If T is hyponormal (see a paragraph after the proof of Theorem 1.2.5 or Appendix A), i.e. $[T^*, T] \geq 0$, then let $D_T = D_0^2$, $D_0 \geq 0$ and (1.2.4) holds good also for $k = 0$.

Proof. Let P_n be the projection from $\mathcal{H}$ to $\mathcal{H}_n$. Define

$$B_n = P_n T|_{\mathcal{H}_n} \text{ and } D_{n+1} = P_{n+1} T|_{\mathcal{H}_n}, \quad n = 0, 1, 2, \ldots \tag{1.2.6}$$

From (1.2.2) and (1.2.6), it follows (1.2.3). From (1.2.3) and

$$T^* T - T T^* = \begin{pmatrix} D_T & 0 & 0 & \cdots \\ 0 & 0 & 0 & \cdots \\ 0 & 0 & 0 & \cdots \\ \vdots & \vdots & \vdots & \ddots \end{pmatrix}$$

it follows (1.2.4) and (1.2.5). $\qquad\square$

Notice that $T^* \mathcal{H}_0 \subset \mathcal{H}_0$ and $[T^*, T]\mathcal{H}_0 \subset \mathcal{H}_0$. Then we define two operators on $\mathcal{H}_0$ as

$$\Lambda_T \overset{\text{def}}{=} (T^*|_{\mathcal{H}_0})^* \tag{1.2.7}$$

and

$$C_T \overset{\text{def}}{=} [T^*, T]|_{\mathcal{H}_0}. \tag{1.2.8}$$

Theorem 1.2.4. *If T is a pure operator, then T is unitarily equivalent to an operator*

$$\widetilde{T} = \begin{pmatrix} \Lambda_0 & 0 & 0 & 0 & \cdots \\ C_1 & \Lambda_1 & 0 & 0 & \cdots \\ 0 & C_2 & \Lambda_2 & 0 & \cdots \\ 0 & 0 & C_3 & \Lambda_3 & \cdots \\ \vdots & \vdots & \vdots & \vdots & \ddots \end{pmatrix} \tag{1.2.9}$$

on $\widetilde{\mathcal{H}}$ with respect to the orthogonal decomposition $\widetilde{\mathcal{H}} = \sum_{n=0}^{\infty} \oplus \widetilde{\mathcal{H}}_n$, where $\Lambda_0 = \Lambda_T$, and $C_n \geq 0$, $n = 1, 2, \ldots$, $C_1 = (C_T - [\Lambda_0^*, \Lambda_0])^{\frac{1}{2}}$,

$$C_{n+1} = (C_n^2 - [\Lambda_n^*, \Lambda_n])^{\frac{1}{2}}, \quad n = 1, 2, \ldots, \tag{1.2.10}$$

$$\Lambda_{n+1} = C_{n+1}^{-1} \Lambda_n C_{n+1}, \quad n = 0, 1, 2, \ldots \tag{1.2.11}$$

and $\widetilde{\mathcal{H}}_n = cl\ range\ C_n$, $n = 1, 2, \cdots$ satisfying

$$\mathcal{H}_0 = \widetilde{\mathcal{H}}_0 \supset \widetilde{\mathcal{H}}_1 \supset \widetilde{\mathcal{H}}_2 \supset \cdots$$

Proof. From (1.2.3), it is easy to see that

$$TG_n \subset G_n \oplus range\ D_{n+1}.$$

Therefore $G_{n+1} \subset G_n \oplus cl\ range\ D_{n+1}$. Thus

$$cl\ range\ D_{n+1} = \mathcal{H}_{n+1}.$$

Hence there is a unitary operator V_{n+1} from cl range $|D_{n+1}|$ onto $\mathcal{H}_{n+1}$ such that

$$D_{n+1} = V_{n+1}|D_{n+1}|$$

where $|A| \overset{\text{def}}{=} (A^*A)^{\frac{1}{2}}$. Let

$$U \overset{\text{def}}{=} \begin{pmatrix} I & 0 & 0 & \cdots & & 0 & & \cdots \\ 0 & V_1^* & 0 & \cdots & & 0 & & \cdots \\ 0 & 0 & V_1^* V_2^* & \cdots & & 0 & & \cdots \\ \vdots & \vdots & \vdots & \vdots & & \vdots & & \vdots \\ 0 & 0 & 0 & \cdots & V_1^* V_2^* & \cdots V_{n-1}^* & \cdots \\ \vdots & \vdots & \vdots & \vdots & & \vdots & & \ddots \end{pmatrix}$$

be a unitary operator from $\mathcal{H}$ to $\widetilde{\mathcal{H}}$, then $\widetilde{T} = UTU^*$ is of the form (1.2.9), where $\Lambda_0 = B_0$

$$\Lambda_n = V_1^* \cdots V_n^* B_n V_n \cdots V_1, \tag{1.2.12}$$

and

$$C_n = V_1^* \cdots V_{n-1}^* V_n^* D_n V_{n-1} \cdots V_1, \qquad (1.2.13)$$

for $n = 1, 2, \ldots$, where $V_0 = I$. We have to prove that $V_1^* \cdots V_n^*$ is isometric on $\mathcal{H}_n$ by mathematical induction. First, for $n = 1$, V_1^* is a unitary operator from $\mathcal{H}_1$ onto cl range $D_1 \subset \mathcal{H}_0$. Define $\widetilde{\mathcal{H}}_0 = \mathcal{H}_0$. From (1.2.13), it is easy to see that $\widetilde{\mathcal{H}}_n = V_1^* \cdots V_n^* \mathcal{H}_n$, for $n = 1, 2, \ldots$. Suppose $V_1^* \cdots V_n^*$ is isometric on $\mathcal{H}_n$ for $n \geq 1$. Let us prove that $V_1^* \cdots V_n^* V_{n+1}^*$ is isometric on $\mathcal{H}_{n+1}$ and $\widetilde{\mathcal{H}}_{n+1} \subset \widetilde{\mathcal{H}}_n$. From the fact that V_{n+1}^* is a unitary operator from $\mathcal{H}_{n+1}$ onto cl range $|D_{n+1}|$ and $D_{n+1} \mathcal{H}_n \subset \mathcal{H}_{n+1}$, we may conclude that range $|D_{n+1}| \subset \mathcal{H}_n$ and V_{n+1}^* is isometric on $\mathcal{H}_{n+1}$ satisfying cl range $V_{n+1}^* \subset \mathcal{H}_n$. Thus $V_1^* \cdots V_n^* V_{n+1}^*$ is isometric on $\mathcal{H}_{n+1}$ and $\widetilde{\mathcal{H}}_{n+1} \subset \widetilde{\mathcal{H}}_n$. Therefore U is a unitary operator.

From (1.2.4), (1.2.5), (1.2.12) and (1.2.13), it follows (1.2.10) and

$$C_{n+1} \Lambda_{n+1} = \Lambda_n C_{n+1}.$$

For $y \in$ range $\Lambda_n C_{n+1}$, there is a vector $x \in$ range Λ_{n+1} such that

$$C_{n+1} x = y.$$

If there is an x' such that $C_{n+1} x' = y$. Then $x = x'$, since

$$\ker C_{n+1} = (\text{range } C_{n+1})^\perp = \{0\}.$$

We define $x = C_{n+1}^{-1} y$, which proves (1.2.11). $\qquad \square$

To obtain a simple clear complete unitary invariant for an operator is a classical problem in the operator theory. It is well-known that the spectrum and the spectral multiplicities is a complete unitary invariant for a normal operator.

Theorem 1.2.5. *If T is pure, then $\{C_T, \Lambda_T\}$ defined in* (1.2.7) *and* (1.2.8) *is a complete unitary invariant.*

Proof. It is obvious that $\{C_T, \Lambda_T\}$ is a unitary invariant of T. Suppose T and T' are two pure operators on $\mathcal{H}_0$ and $\mathcal{H}_0'$ with corresponding $\{\mathcal{H}_0, C_0, \Lambda_0\}$ and $\{\mathcal{H}_0', C_0', \Lambda_0'\}$ respectively where $C_0 = C_T, \Lambda_0 = \Lambda_T, C_0' = C_{T'}$ and $\Lambda_0' = \Lambda_{T'}$. Suppose there is a unitary operator V from $\mathcal{H}_0$ to $\mathcal{H}_0'$ satisfying

$$C_0' = V C_0 V^{-1} \quad \text{and} \quad \Lambda_0' = V \Lambda_0 V^{-1}.$$

Let

$$\widetilde{T}' = \begin{pmatrix} \Lambda'_0 & 0 & 0 & \cdots \\ C'_1 & \Lambda'_1 & 0 & \cdots \\ 0 & C'_2 & \Lambda'_2 & \cdots \\ \vdots & \vdots & \vdots & \ddots \end{pmatrix}$$

on the corresponding decomposition $\widetilde{\mathcal{H}}' = \sum \oplus \widetilde{\mathcal{H}}'_n$ with $\widetilde{\mathcal{H}}'_n =$ cl range C'_n. Then

$$C'_1 = (C_{T'} - [\Lambda'^*_0, \Lambda'_0])^{\frac{1}{2}} = V(C_T - [\Lambda_0{}^*, \Lambda_0])^{\frac{1}{2}} V^* = V C_1 V^*.$$

Let $V_1 = V|_{\widetilde{\mathcal{H}}_1}$. Then V_1 is a unitary operator from $\widetilde{\mathcal{H}}_1$ to $\widetilde{\mathcal{H}}'_1$, such that

$$C'_1 = V_1 C_1 V_1^*.$$

By (1.2.11), we have $\Lambda'_1 = V_1 \Lambda_1 V_1^*$. By mathematical induction, we may construct unitary operators V_n from $\widetilde{\mathcal{H}}_n$ to $\widetilde{\mathcal{H}}'_n$, $n = 2, \ldots$ such that

$$C'_n = V_n C_n V_n^* \text{ and } \Lambda'_n = V_n \Lambda_n V_n^*, \quad n = 2, 3, \ldots$$

Construct the unitary operator

$$W = \begin{pmatrix} V & 0 & 0 & \cdots \\ 0 & V_1 & 0 & \cdots \\ 0 & 0 & V_2 & \cdots \\ \vdots & \vdots & \vdots & \ddots \end{pmatrix}.$$

Then $\widetilde{T}' = W \widetilde{T} W^*$ which proves the theorem. $\square$

Let H be an operator on a Hilbert space, if $[H^*, H] \geq 0$, i.e.

$$([H^*, H]x, x) \geq 0 \quad \text{for } x \in \mathcal{H}.$$

Then H is said to be *hyponormal*. In §1.3 we will show that a subnormal operator is hyponormal. In Chapter 5, we will give examples to show that hyponormal operator may not be subnormal.

In this monograph, we mainly study subnormal operators, but we also give some theorems on some special classes of hyponormal operators which might help us to understand subnormal operators better.

Proposition 1.2.6. *If H is a pure hyponormal operator on $\mathcal{H}$. Then $\sigma_p(H) = \emptyset$.*

Proof. Suppose $z \in \sigma_p(H)$. From

$$[H^* - \bar{z}, H - z] = [H^*, H] \geq 0$$

we have

$$\|(H^* - \bar{z})f\| \leq \|(H - z)f\|, \quad \text{for } f \in \mathcal{H}.$$

Thus if $f \neq 0$ is an eigenvector of H corresponding to eigenvalue z, then $(H - z)f = (H^* - \bar{z})f = 0$. Therefore $M_f = \{\lambda f : \lambda \in \mathbb{C}\}$ reduces H and $H|_{M_f}$ is $z I_{M_f}$, where I_{M_f} is the identity operator on M_f. Thus H is not pure, which proves the lemma. □

The spectrum of any hyponormal operator H does not have isolated points, since every isolated point in the spectrum is a point spectrum.

1.3 Analytic model for a subnormal operator

Let S be a subnormal operator on a Hilbert space $\mathcal{H}$ with m.n.e. N on $\mathcal{K} \supset \mathcal{H}$. Let $\widetilde{\mathcal{H}} = \mathcal{K} \ominus \mathcal{H}$. Let P and $\widetilde{P}$ be the projections from $\mathcal{K}$ to $\mathcal{H}$ and $\widetilde{\mathcal{H}}$ respectively. Let A be the operator from $\widetilde{\mathcal{H}}$ to $\mathcal{H}$ defined by

$$Ax = PNx, \quad x \in \widetilde{\mathcal{H}}. \tag{1.3.1}$$

Let $\widetilde{S}$ be the operator from $\widetilde{\mathcal{H}}$ to $\widetilde{\mathcal{H}}$ defined by

$$\widetilde{S} \stackrel{\text{def}}{=} N^*|_{\widetilde{\mathcal{H}}}. \tag{1.3.2}$$

The range of $\widetilde{S}$ is in $\widetilde{\mathcal{H}}$, since $\widetilde{\mathcal{H}}$ is an invariant space of N^*. Then we have a block-matric representation of N

$$N = \begin{pmatrix} S & A \\ 0 & \widetilde{S}^* \end{pmatrix}. \tag{1.3.3}$$

The condition $N^*N = NN^*$ is equivalent to the following three identities:

$$[S^*, S] = AA^*, \tag{1.3.4}$$

$$[\widetilde{S}^*, \widetilde{S}] = A^*A \tag{1.3.5}$$

and

$$S^*A = A\widetilde{S}. \tag{1.3.6}$$

From (1.3.4), we know that *subnormal operator is hyponormal.*

The operator $\widetilde{S}$ is also a subnormal operator on $\widetilde{\mathcal{H}}$ with m.n.e. N^* and is called the *dual* of S.

Example. Let $H^2(\mathbb{T})$ be the Hardy space of all analytic functions f on the unit disk satisfying

$$\|f\|^2 = \lim_{r \to 1^-} \frac{1}{2\pi} \int_0^{2\pi} |f(re^{i\theta})|^2 d\theta < \infty.$$

Let $L^2(\mathbb{T})$ be the Hilbert space of all measurable and square integrable functions on $\mathbb{T} = \{e^{i\theta} \in \mathbb{C} : 0 \le \theta \le 2\pi\}$ with

$$(f,g) = \frac{1}{2\pi} \int_0^{2\pi} f(e^{i\theta}) \overline{g(e^{i\theta})} \, d\theta.$$

If we consider $H^2(\mathbb{T})$ as the function space of the boundary value functions on the unit circle of the functions in $H^2(\mathbb{T})$, then $H^2(\mathbb{T})$ is a subspace of $L^2(\mathbb{T})$. Let U_+ be the multiplication operator on $H^2(\mathbb{T})$:

$$(U_+ f)(z) = z f(z), \quad |z| < 1.$$

Then U_+ is a unilateral shift with multiplicity one. Let $\{e^{ni\theta} : n = 0, 1, 2, \ldots\}$ be the orthonormal basis, then

$$U_+ e^{in\theta} = e^{(n+1)i\theta}, \quad n = 0, 1, 2, \ldots$$

The operator $U_+{}^*$ is

$$(U_+{}^* f)(z) = (f(z) - f(0))/z, \quad f \in H^2(\mathbb{T}).$$

Then $U_+{}^* 1 = 0$ and $U_+{}^* e^{ni\theta} = e^{(n-1)i\theta}$, $n = 1, 2, \ldots$. The operator U_+ is subnormal with an m.n.e. N on $L^2(\mathbb{T})$, where

$$(Nf)(e^{i\theta}) = e^{i\theta} f(e^{i\theta}).$$

In this case $\widetilde{\mathcal{H}} = \{\sum_{n=-1}^{-\infty} c_n e^{ni\theta} : \sum |c_n|^2 < +\infty\}$,

$$(Pf)(z) = \frac{1}{2\pi} \int_0^{2\pi} \frac{f(e^{i\theta}) \, d\theta}{1 - ze^{-i\theta}}, \quad f \in L^2(\mathbb{T})$$

and $\widetilde{P} = I - P$. Let $(P_0 f)(z) = \frac{1}{2\pi} \int_0^{2\pi} f(e^{i\theta}) \, d\theta$, $f \in L^2(\mathbb{T})$. Then

$$Af = P_0\big((\cdot)f(\cdot)\big) = c_{-1}$$

where $f(e^{i\theta}) = \sum_{n=-1}^{-\infty} c_n e^{ni\theta}$, and

$$(A^* f)(e^{i\theta}) = P_0(f)e^{-i\theta}, \quad f \in H^2(\mathbb{T}).$$

Let $\widetilde{U}_+$ be the dual operator of U_+. Then

$$(\widetilde{U}_+^* f)(e^{i\theta}) = e^{i\theta} f(e^{i\theta}) - c_{-1}, \quad f \in \widetilde{\mathcal{H}}.$$

It is easy to see that

$$[U_+{}^*, U_+] = P_0 \quad \text{and} \quad [\widetilde{U}_+^*, \widetilde{U}_+] = P_{-1}$$

where $P_{-1}(\sum c_n e^{ni\theta}) = c_{-1}e^{-i\theta}$. Besides, $U_+{}^* A = A\widetilde{U}_+ = 0$.

Lemma 1.3.1. *Let S be a pure normal operator on $\mathcal{H}$. Then*

$$S^*M \subset M, \tag{1.3.7}$$

and

$$\sigma(\Lambda) \subset \sigma(S), \tag{1.3.8}$$

where $M = M_S$ and $\Lambda = \Lambda_S$ defined as in § 1.2.

Proof. From (1.3.4), $M = \text{cl } AA^*\mathcal{H} = \text{cl range } A$. From (1.3.6), we have $S^*AA^* = A\widetilde{S}A^*$. Thus

$$S^*M \subset \text{cl } A\widetilde{S}A^*\mathcal{H} \subset \text{cl range } A = M,$$

which proves (1.3.7).

In order to prove (1.3.8), we only have to prove that

$$\rho(S^*) \subset \rho(\Lambda^*). \tag{1.3.9}$$

Let $\lambda \in \rho(S^*)$, then $\lambda \in \rho(N^*)$ since $\sigma(S) \supset \sigma(N)$, by Theorem 1.1.2. For every $x \in \widetilde{\mathcal{H}}$, let $x_1 = P\big((\lambda - N^*)^{-1}x\big)$ and $x_2 = \widetilde{P}\big((\lambda - N^*)^{-1}x\big)$. Then by (1.3.3)

$$x = (\lambda - N^*)(x_1 + x_2) = (\lambda - S^*)x_1 + (-A^*x_1 + (\lambda - \widetilde{S})x_2).$$

Therefore

$$(\lambda - S^*)x_1 = Px = 0.$$

Hence $x_1 = 0$ by $\lambda \in \rho(S^*)$. Thus $(\lambda - \widetilde{S})x_2 = x$. By (1.3.6),

$$Ax = (\lambda - S^*)Ax_2 = (\lambda - \Lambda^*)Ax_2.$$

Therefore cl range $(\lambda - \Lambda^*) = M$ by $M = \text{cl } A\widetilde{\mathcal{H}}$. On the other hand,

$$\|(\lambda - \Lambda^*)x\| = \|(\lambda - S^*)x\| \geq \|(\lambda - S^*)^{-1}\|^{-1}\|x\|$$

for $x \in M$. Thus $\lambda \in \rho(\Lambda^*)$ which proves (1.3.8). $\qquad\square$

For subnormal operator S, define

$$Q_S(z,w) = (\overline{w} - \Lambda_S{}^*)(z - \Lambda_S) - C_S, \quad z, w \in \mathbb{C} \tag{1.3.10}$$

as $L(M_s)$-valued function of z and w. Sometimes denote $Q_S(z,w)$ simply by $Q(z,w)$.

Define the "*Brodski-Lifshitz kernel*" or "*determining function*"

$$S(z,w) = P_M(\overline{w} - S^*)^{-1}(z - S)^{-1}|_M, \quad z, w \in \rho(S),$$

where P_M is the projection to $M = M_S$. It is not difficult to prove that $\mathcal{H} = \text{cl }\bigvee\{(\lambda - S)^{-1}x : x \in M, \lambda \in \rho(S)\}$. Therefore the determining function $S(\cdot, \cdot)$ determines the inner product of any two vectors in $\mathcal{H}$.

Lemma 1.3.2. *Let S be a pure subnormal operator, then*

$$S(z,w) = Q(z,w)^{-1}, \quad \text{for } z, w \in \rho(S).$$

Proof. Define

$$T(z,w) = P_M (z - S)^{-1} (\overline{w} - S^*)^{-1}|_M.$$

Notice that $[S^*, S] = CP_M$, where $C = C_S$. From the commutation relation, we have

$$[(\overline{w} - S^*)^{-1}, (z - S)^{-1}] = (z - S)^{-1}(\overline{w} - S^*)^{-1} CP_M (\overline{w} - S^*)^{-1}(z - S)^{-1}.$$

Hence

$$S(z,w) - T(z,w) = T(z,w)CS(z,w), \quad z, w \in \rho(S). \tag{1.3.11}$$

If $\overline{\lambda} \in \rho(S^*)$, then $\overline{\lambda} \in \rho(\Lambda^*)$ and

$$(\overline{\lambda} - \Lambda^*)^{-1} x = (\overline{\lambda} - S^*)^{-1} x, \quad x \in M.$$

Therefore

$$(T(z,w)u, v) = ((\overline{w} - S^*)^{-1} u, (\overline{z} - S^*)^{-1} v) = ((\overline{w} - \Lambda^*)^{-1} u, (\overline{z} - \Lambda^*)^{-1} v)$$
$$= ((z - \Lambda)^{-1}(\overline{w} - \Lambda^*)^{-1} u, v)$$

for $u, v \in M$ and $z, w \in \rho(S)$. Thus

$$T(z,w) = (z - \Lambda)^{-1}(\overline{w} - \Lambda^*)^{-1}, \quad z, w \in \rho(S). \tag{1.3.12}$$

From (1.3.11) and (1.3.12), it follows that

$$Q(z,w)S(z,w) = I, \quad z, w \in \rho(S).$$

By the communtation relation, we also have

$$[(\overline{w} - S^*)^{-1}, (z - S)^{-1}] = (\overline{w} - S^*)^{-1}(z - S)^{-1} CP_M (z - S)^{-1}(\overline{w} - S^*)^{-1}.$$

By the same method, we may prove that

$$S(z,w)Q(z,w) = I, \quad z, w \in \rho(S).$$

Hence $Q(z,w)$ is invertible, which proves the lemma. □

Therefore $S(\cdot, \cdot)$ is also a complete unitary invariant for pure subnormal operator S.

Let S be a pure subnormal operator on $\mathcal{H}$ with m.n.e. N on $\mathcal{K} \supset \mathcal{H}$. Let $M = \text{cl } [S^*, S]\mathcal{H}$. Let $(E(\cdot), \mathcal{B})$ be the spectral measure of the operator N (see §1.1). Define

$$e(B) = P_M E(B)|_M, \quad B \in \mathcal{B}. \tag{1.3.13}$$

Then $e(B) \in L(M)$ and $e(B) \geq 0$. Thus $(e(\cdot), \sigma(N), \mathcal{B})$ is a $L(M)$-valued positive measure on $(\sigma(N), \mathcal{B})$ and $e(\sigma(N)) = I$, the identity operator on M.

Let $\mathcal{S}(M, \mathcal{B})$ be the linear space over $\mathbb{C}$ of all M-valued simple functions

$$\sum_{\nu=1}^{n} a_\nu 1_{A_\nu}(\cdot),$$

where n is finite, $a_\nu \in M$, and 1_{A_ν} is the characteristic function of the set $A_\nu \in \mathcal{B}$, with ordinary linear operation. Define the scalar product on $\mathcal{S}(M, \mathcal{B})$ as follows

$$\begin{aligned}
(\sum a_l 1_{A_l}(\cdot), \sum a'_m 1_{A'_l}(\cdot)) &= (\sum_{l,m} e(A_l \cap A'_m) a_l, a'_m) \\
&= \sum_{l,m} (E(A_l) a_l, E(A'_m) a'_m).
\end{aligned}$$

It is obvious that $\mathcal{S}(M, \mathcal{B})$ endowed with $(\cdot, \cdot)$ is an inner product space. Let $L^2(e(\cdot), \mathcal{B})$ or simply $L^2(e)$ be the Hilbert space completion of $\mathcal{S}(M, \mathcal{B})$. It is easy to see that the function $f(\cdot)a \in L^2(e)$, where $f(\cdot)$ is a complex function on $\sigma(N)$, measurable with respect to $\mathcal{B}$, and $a \in M$, iff

$$\int_{\sigma(N)} |f(z)|^2 (e(dz)a, a) < +\infty$$

where $B \mapsto (e(B)a, a)$, $B \in \mathcal{B}$ is a measure on $\mathcal{B}$. Thus for $f(\cdot), g(\cdot) \in L^2(e)$, the inner product is

$$(f(\cdot), g(\cdot)) = \int_{\sigma(N)} (e(du)f(u), g(u)).$$

Let $\varphi(z)$ be a bounded Baire function on $\sigma(N)$. Then the integral $\int \varphi(z) e(dz)$ is defined as $P_M \int \phi(z) E(dz)|_M$. If $T \in L(M)$ then $\int T\phi(z) e(dz)$ is defined as $T \int \phi(z) e(dz)$.

Lemma 1.3.3. *Let S be a pure subnormal operator on $\mathcal{H}$ with m.n.e. N on $\mathcal{K} \supset \mathcal{H}$. Let $e(\cdot)$ be defined in (1.3.13). Then $e(\sigma(N)) = I$,*

$$\int_{\sigma(N)} \frac{u - \Lambda}{u - z} e(du) = 0, \quad z \in \rho(S), \tag{1.3.14}$$

and

$$\int_B Q(u, u) e(du) = \int_B e(du) Q(u, u) = 0, \tag{1.3.15}$$

where $\Lambda = \Lambda_M$, $C = C_M$ and $Q(\cdot, \cdot) = Q_S(\cdot, \cdot)$ are defined in (1.2.7), (1.2.8) and (1.3.10) respectively and B is any Borel set in $\sigma(N)$.

Proof. Denote $\sigma(N)$ by σ. For $x, y \in M$ and $z, w \in \rho(S)$, by Lemma 1.3.1 we have

$$\int_\sigma \frac{(e(du)x, y)}{(\overline{w} - \overline{u})(z - u)} = ((z - N)^{-1}x, (w - N)^{-1}y) = ((z - S)^{-1}x, (w - S)^{-1}y)$$

$$= (S(z, w)x, y) = (Q(z, w)^{-1}x, y)$$

i.e.

$$\int_\sigma \frac{Q(z, w)e(du)}{(\overline{w} - \overline{u})(z - u)} = I.$$

Hence

$$e(\sigma) + \int_\sigma \frac{(\overline{u} - \Lambda^*)e(du)}{\overline{w} - \overline{u}} + \int_\sigma \frac{(u - \Lambda)e(du)}{z - u} + \int_\sigma \frac{Q(u, u)e(du)}{(\overline{w} - \overline{u})(z - u)} = I.$$
$$(1.3.16)$$

It is obvious that $e(\sigma) = I$. Putting $w \to \infty$ in the left-hand side of (1.3.16), we get (1.3.14). Similarly, putting $z \to \infty$, we get

$$\int_\sigma \frac{(\overline{u} - \Lambda^*)e(du)}{\overline{w} - \overline{u}} = 0, \quad w \in \rho(S). \tag{1.3.17}$$

Thus

$$\int_\sigma \frac{Q(u, u)e(du)}{(\overline{w} - \overline{u})(z - u)} = 0, \quad z, w \in \rho(S). \tag{1.3.18}$$

By Stone–Weierstrass Theorem, it is easy to prove that every continuous function on the compact set $\sigma(S)$ can be approximated uniformly by linear combinations of functions $(\overline{w} - \overline{(\cdot)})^{-1}(z - (\cdot))^{-1}$ for $z, w \in \rho(S)$. Thus (1.3.18) implies that

$$\int_\sigma f(u)Q(u, u)e(du) = 0, \tag{1.3.19}$$

for all continuous functions $f(\cdot)$ on $\sigma(S)$. For $a, b \in M$, let

$$\mu_{a,b}(B) \overset{\text{def}}{=} \int_{B \cap \sigma(N)} (Q(u, u)e(du)a, b)$$

for Borel set $B \subset \sigma(S)$. Then $\mu_{a,b}(\cdot)$ is a complex measure on $\sigma(S)$ with finite total variation. Then (1.3.19) imples that

$$\int_{\sigma(S)} f(u)\mu_{a,b}(du) = 0,$$

for all continuous function f on $\sigma(S)$. Thus $\mu_{a,b}(\cdot) = 0$. Hence

$$\int_B Q(u, u)e(du) = 0 \tag{1.3.20}$$

for all $B \in \mathcal{B}$. Taking the adjoint of the both sides of (1.3.20), we prove (1.3.15). $\qquad \square$

Lemma 1.3.3 is a basic lemma for the measure $e(\cdot)$.

Let $R^2(\sigma(S), e)$ be the closure of $\bigvee\{(\lambda - (\cdot))^{-1}\alpha : \lambda \in \rho(S), \alpha \in M\}$ in the $L^2(e)$.

The following theorem gives an analytic model for subnormal operators.

Theorem 1.3.4. *Let S be a pure subnormal operator on a separable Hilbert space $\mathcal{H}$ with m.n.e. N on $\mathcal{K} \supset \mathcal{H}$. Let M, C, Λ, etc. be defined in this section for the operator S. Then there exists an $L(M)$-valued positive measure $e(\cdot)$ on $\sigma(N)$ satisfying $e(\sigma(N)) = I$, (1.3.14) and (1.3.15), and a unitary operator U from $\mathcal{K}$ onto $L^2(e)$ satisfying*

$$U f(N)\alpha = f(\cdot)\alpha \tag{1.3.21}$$

for every bounded Baire function f on $\sigma(N)$ and $\alpha \in M$. The operator U satisfies

$$U\mathcal{H} = R^2(\sigma(S), e), \tag{1.3.22}$$

$$(USU^{-1}f)(u) = uf(u), \quad u \in \sigma(N), \quad f \in R^2(\sigma(N), e), \tag{1.3.23}$$

and

$$(US^*U^{-1}f)(u) = \overline{u}f(u) - (\overline{u} - \Lambda^*)f(\Lambda), \tag{1.3.24}$$

where

$$f(\Lambda) \stackrel{def}{=} \int_{\sigma(N)} e(du)f(u) = U P_M U^{-1} f, \tag{1.3.25}$$

where P_M is the projection to M.

Proof. Let $e(\cdot)$ be the measure in Lemma 1.3.3. Define U by (1.3.21). It is obvious that U is isometric on

$$\{f(N)\alpha : \ f(\cdot) \text{ is a bounded Baire function, } \alpha \in M\}.$$

It extents a unitary operator from $\mathcal{K}$ to $L^2(e)$. We still denote it by U. It is obvious that (1.3.22) and (1.3.23) hold good. For $\alpha, \beta \in M$ and $\lambda \in \rho(S)$, we have

$$(P_M(\lambda - S)^{-1}\alpha, \beta) = (\alpha, (\overline{\lambda} - S^*)^{-1}\beta) = (\alpha, (\overline{\lambda} - \Lambda^*)^{-1}\beta)$$
$$= ((\lambda - \Lambda)^{-1}\alpha, \beta).$$

Therefore $P_M(\lambda - S)^{-1}|_M = (\lambda - \Lambda)^{-1}$ for $\lambda \in \rho(S)$. Thus for any rational function $f(\cdot)$, with poles in $\rho(S)$,

$$P_M f(S)|_M = \int_{\sigma(N)} f(u)e(du) = f(\Lambda). \tag{1.3.26}$$

On the other hand,

$$S^*(\lambda - S)^{-1} - (\lambda - S)^{-1}S^* = (\lambda - S)^{-1}CP_M(\lambda - S)^{-1}, \quad \lambda \in \rho(S).$$

Therefore, for $\alpha \in M$ and $\lambda \in \rho(S)$, by (1.3.26) we have

$$S^*(\lambda - S)^{-1}\alpha = (\lambda - S)^{-1}\Lambda^*\alpha + (\lambda - S)^{-1}C(\lambda - \Lambda)^{-1}\alpha.$$

Thus for any rational function f, with poles in $\rho(S)$, we have

$$(US^*U^{-1}f\alpha)(u)$$
$$= \Lambda^*f(u)\alpha + C(u - \Lambda)^{-1}(f(u) - f(\Lambda))\alpha, \quad u \in \sigma(N). \tag{1.3.27}$$

We have to prove that as a vector in $L^2(e)$, the right-hand side of (1.3) equals to $\overline{u}f(u)\alpha - (\overline{u} - \Lambda^*)f(\Lambda)$. Actually, by (1.3.15), we have

$$e(du)(\overline{u}f(u)\alpha - (\overline{u} - \Lambda^*)f(\Lambda)\alpha - (\Lambda^*f(u)\alpha + C(u - \Lambda)^{-1} \tag{1.3.28}$$
$$(f(u) - f(\Lambda))\alpha)) = e(du)Q(u,u)(u - \Lambda)^{-1}(f(u) - f(\Lambda))\alpha = 0.$$

Therefore (1.3.24) and (1.3.25) hold good if f is in the set

$$\mathcal{F} = \{f(\cdot)\alpha : \; f(\cdot) \text{ is a rational function on } \rho(S), \; \alpha \in M\}.$$

But $U\mathcal{H} = \text{cl} \bigvee \mathcal{F}$, which proves (1.3.24) and (1.3.25). $\qquad\square$

Corollary 1.3.5. *Under the condition of the Theorem 1.3.4,*

$$(UA^*U^{-1}f)(u) = (\overline{u} - \Lambda^*)f(\Lambda), \tag{1.3.29}$$

and

$$(U[S^*, S]U^{-1}f)(u) = Cf(\Lambda), \tag{1.3.30}$$

for $f \in U\mathcal{H}$, and

$$UAU^{-1}f = \int (u - \Lambda)e(du)f(u), \tag{1.3.31}$$

$$(U\widetilde{S}U^{-1}f)(u) = \overline{u}f(u), \tag{1.3.32}$$

$$(U\widetilde{S}^*U^{-1}f)(u) = uf(u) - \int (v - \Lambda)e(dv)f(v), \tag{1.3.33}$$

and

$$([U\widetilde{S}^*U^{-1}, U\widetilde{S}U^{-1}]f)(u) = (\overline{u} - \Lambda^*)\int (v - \Lambda)e(dv)f(v), \tag{1.3.34}$$

for $f \in U\widetilde{\mathcal{H}} = L^2(e) \ominus U\mathcal{H}$, where A and $\widetilde{S}$ are defined by (1.3.1) and (1.3.2). Besides,

$$U\widetilde{\mathcal{H}} = \text{cl} \bigvee\{((\overline{(\cdot)} - \overline{\lambda})^{-1}(\overline{(\cdot)} - \Lambda^*)\alpha, \; \alpha \in M, \; \lambda \in \rho(S)\}. \tag{1.3.35}$$

Proof. By (1.3.3) we have

$$N^* = \begin{pmatrix} S^* & 0 \\ A^* & \widetilde{S} \end{pmatrix}$$

and hence for $x \in \mathcal{H}$, $A^*x = N^*x - S^*x$. It is obvious that

$$(UN^*U^{-1}f)(u) = \overline{u}f(u). \tag{1.3.36}$$

From (1.3.24), (1.3.36) and $UA^*U^{-1}f = UN^*U^{-1}f - US^*U^{-1}f$, it follows (1.3.29). From

$$\begin{aligned}
(UAU^{-1}f, g) = (f, UA^*U^{-1}g) &= \int (e(du)f(u), (\overline{u} - \Lambda^*)g(\Lambda)) \\
&= (\int (u - \Lambda)e(du)f(u), g(\Lambda)) \\
&= \int (e(dv) \int (u - \Lambda)e(du)f(u), g(v))
\end{aligned}$$

for $f \in L^2(e) \ominus U\mathcal{H}$ and $g \in U\mathcal{H}$, it follows (1.3.31). The identity (1.3.30) is a direct consequence of (1.3.15), (1.3.23) and (1.3.24).

The formulas (1.3.2) and (1.3.36) imply (1.3.32). The identity

$$\widetilde{S}^*x = Nx - Ax, \quad \text{for } x \in \widetilde{\mathcal{H}}$$

and (1.3.31) imply (1.3.33).

From (1.3.32), (1.3.33) and the fact that

$$\int (\overline{u} - \Lambda^*)(u - \Lambda)e(du)f(u) = C \int e(du)f(u) = 0,$$

for $f \in U\widetilde{\mathcal{H}}$, it follows (1.3.34).

In order to prove (1.3.35), firstly we have to prove that

$$(\overline{\lambda} - \overline{(\cdot)})^{-1}(\overline{(\cdot)} - \Lambda^*)\alpha \in U\widetilde{\mathcal{H}}, \tag{1.3.37}$$

for $\lambda \in \rho(S)$, $\alpha \in M$. For this, by (1.3.22), we only have to prove that

$$((\mu - (\cdot))^{-1}\beta, (\overline{\lambda} - \overline{(\cdot)})^{-1}(\overline{(\cdot)} - \Lambda^*)\alpha) = 0, \quad \text{for } \beta \in M, \ \mu \in \rho(S). \tag{1.3.38}$$

But, by (1.3.14) we have

$$\int \frac{(u - \Lambda)e(du)}{(u - \lambda)(u - \mu)} = \frac{1}{\lambda - \mu}(\int \frac{u - \Lambda}{u - \lambda}e(du) - \int \frac{u - \Lambda}{u - \mu}e(du)) = 0,$$

which proves (1.3.38) and hence (1.3.37).

Thus we only have to prove that if $f_0 \in U\widetilde{\mathcal{H}}$ and

$$f_0 \perp (\overline{(\cdot)} - \overline{\lambda})^{-1}(\overline{(\cdot)} - \Lambda^*)\alpha, \quad \alpha \in M, \ \lambda \in \rho(S) \tag{1.3.39}$$

then $f_0 = 0$. From (1.3.39), we have

$$\int \frac{(u - \Lambda)e(du)}{u - \lambda} f_0(u) = 0, \quad \lambda \in \rho(S). \tag{1.3.40}$$

Let $F = \bigvee\{\overline{(\cdot)}^m (\cdot)^n f_0 : m, n = 0, 1, 2, \ldots\}$. We have to prove that $F \subset \cup\widetilde{\mathcal{H}}$. Then cl F reduces UNU^{-1} and f_0 must be zero since N is an m.n.e. It is obvious that

$$\{\overline{(\cdot)}^m f_0 : m = 0, 1, 2, \ldots\} \subset \cup\widetilde{\mathcal{H}},$$

since $\cup\widetilde{\mathcal{H}}$ is invariant with respect to UN^*U^{-1}. Now let us prove that

$$\int \frac{u - \Lambda}{u - \lambda} e(du)\overline{u}^m f_0(u) = 0, \quad \lambda \in \rho(S), \tag{1.3.41}$$

and

$$\int \frac{e(du)\overline{u}^m f_0(u)}{u - \lambda} = 0, \quad \lambda \in \rho(S), \tag{1.3.42}$$

for $m = 0, 1, 2, \ldots$ by mathematical induction with respect to m.

First, from $\overline{(\cdot)}^m f_0 \in U\widetilde{\mathcal{H}}$ it follows that $\int e(du)\overline{u}^m f_0(u) = 0$. From

$$\int \frac{u - \Lambda}{u - \lambda} e(du)\overline{u}^m f_0(u) = \int e(du)\overline{u}^m f_0(u) + (\lambda - \Lambda) \int \frac{e(du)\overline{u}^m}{u - \lambda} f_0(u),$$

and $\lambda \in \rho(\Lambda)$ it follows that (1.3.41) and (1.3.42) are equivalent.

For $m = 0$, (1.3.40) implies (1.3.41) and (1.3.42). Suppose (1.3.41) and (1.3.42) hold good for $m = n$, $n \geq 0$. Let us prove that they also hold good for $m = n + 1$. By (1.3.15), we have

$$\int \frac{(u - \Lambda)e(du)\overline{u}^{n+1}}{u - \lambda} f_0(u) = \int \frac{Q(u, u)e(du)\overline{u}^n}{u - \lambda} f_0(u) + C \int \frac{e(du)\overline{u}^n}{u - \lambda} f_0(u)$$

$$+ \Lambda^* \int \frac{(u - \Lambda)e(du)\overline{u}^n}{u - \lambda} f_0(u) = 0$$

which proves (1.3.41) and (1.3.42) for $m = n + 1$ and hence all m.

For fixed m, $m = 0, 1, 2, \ldots$, let us prove that

$$(\cdot)^n \overline{(\cdot)}^m f_0 \in U\widetilde{\mathcal{H}}, \quad n = 0, 1, 2, \ldots \tag{1.3.43}$$

by mathematical induction with respect to n. We already know that (1.3.43) holds good for $n = 0$. Suppose that is true for $n = k$, $k \geq 0$. In order to prove that $UNU^{-1}(\cdot)^k\overline{(\cdot)}^m f_0 \in U\widetilde{\mathcal{H}}$, we only have to prove that $UAU^{-1}(\cdot)^k\overline{(\cdot)}^m f_0 = 0$, that is

$$\int (u - \Lambda)e(du)u^k\overline{u}^m f_0 = 0, \tag{1.3.44}$$

by (1.3.31). But (1.3.44) is a consequence of (1.3.41), which proves the corollary. $\qquad\square$

Notice that there is no point $a \in \partial\sigma(S)$ satisfying $e(\{a\}) \neq 0$ for pure subnormal operator S, since if $a \in \partial\sigma(S)$ and $e(\{a\}) \neq 0$, it is easy to prove that a is in the point spectrum of S. But pure subnormal operator does not have point spectrum, since subnormal operator is hyponormal.

The following theorem is the converse of Theorem 1.3.4.

Theorem 1.3.6. *Let M be a Hilbert space. Let $e(\cdot)$ be a $L(M)$-valued positive measure on a compact set $\sigma \subset \mathbb{C}$, satisfying $e(\sigma) = I$,*

$$\int_\sigma \frac{(u - \Lambda)e(du)}{u - z} = 0, \qquad (1.3.45)$$

for z in the unbounded component D of $\mathbb{C} \setminus \sigma$ and

$$\int_B e(du)Q(u, u) = 0 \qquad (1.3.46)$$

for every Borel set $B \subset \sigma$, where

$$Q(z, w) = (\overline{w} - \Lambda^*)(z - \Lambda) - C,$$

$\Lambda, C \in L(M)$ and C is positive.

Let $\mathcal{H}_1 = \bigvee\{(\lambda - (\cdot))^{-1}\alpha, \ \lambda \in D, \ \alpha \in M\}$, *and* $\mathcal{H} = \mathrm{cl}\,\mathcal{H}_1$. *Define*

$$(Sf)(u) = uf(u), \quad u \in \sigma, \qquad (1.3.47)$$

for $f \in \mathcal{H}$. Then S is a subnormal operator on $\mathcal{H}$ with the normal extension

$$(Nf)(\cdot) = (\cdot)f(\cdot), \quad f \in L^2(e) \qquad (1.3.48)$$

and

$$(S^*f)(u) = \overline{u}f(u) - (\overline{u} - \Lambda^*)f(\Lambda), \quad u \in \sigma \qquad (1.3.49)$$

for $f \in \mathcal{H}$ where

$$f(\Lambda) = \int_\sigma e(du)f(u). \qquad (1.3.50)$$

Proof. It is obvious that the operator S is subnormal, since it has a normal extension N defined in (1.3.48).

By (1.3.45),

$$I + (z - \Lambda)\int \frac{e(du)}{u - z} = 0, \quad \text{for } z \in D,$$

i.e. $\int_\sigma e(du)(z - u)^{-1} = (z - \Lambda)^{-1}$. Therefore (1.3.50) holds good for $f \in \mathcal{H}_1$.

To verify (1.3.49), note that

$$\int_\sigma (u - \Lambda)e(du)g(u) = 0$$

for $g \in \mathcal{H}_1$ by (1.3.45). Hence

$$\int_\sigma (e(du)(\overline{u} - \Lambda^*)x, g(u)) = (x, \int_\sigma (u - \Lambda)e(du)g(u)) = 0,$$

for $x \in M$, $g \in \mathcal{H}_1$. Therefore

$$(\overline{(\cdot)}f(\cdot) - (\overline{(\cdot)} - \Lambda^*)f(\Lambda), g(\cdot))$$
$$= \int (e(du)f(u), ug(u)) - \int (e(du)(\overline{u} - \Lambda^*)f(\Lambda), g(u))$$
$$= (f, Sg).$$

Hence $S^*f = P(\overline{(\cdot)}f(\cdot) - (\overline{(\cdot)} - \Lambda^*)f(\Lambda))$ for $f \in \mathcal{H}$, where P is the projection from $L^2(e)$ to $\mathcal{H}$. Similar to (1.3.28), we have that as a vector in $L^2(e)$

$$\overline{(\cdot)}f(\cdot) - (\overline{(\cdot)} - \Lambda^*)f(\Lambda) = \Lambda^*f(\cdot) + C((\cdot) - \Lambda)^{-1}(f(\cdot) - f(\Lambda)),$$

for $f \in \mathcal{H}_1$, where $f(\Lambda) = \sum(\lambda_j - \Lambda)^{-1}\alpha_j$ for $f(\cdot) = \sum(\lambda_j - (\cdot))^{-1}\alpha_j$, $\lambda_j \in D$ and $\alpha_j \in M$. Therefore

$$\overline{(\cdot)}f(\cdot) - (\overline{(\cdot)} - \Lambda^*)f(\Lambda) \in \mathcal{H}, \quad \text{for } f \in \mathcal{H}_1,$$

which proves (1.3.49) for $f \in \mathcal{H}_1$. Then (1.3.49) and (1.3.50) also hold good for $f \in \mathcal{H}$, since $\mathcal{H}_1$ is dense in $\mathcal{H}$.

$\square$

## 1.4	Mosaic

For a pure subnormal operator S on $\mathcal{H}$ with m.n.e. N on $\mathcal{K}$, let M be the defect space of S. Define an $L(M)$-valued analytic function

$$\mu(z) = P_M(N - SP_M)(N - z)^{-1}|_M \text{ for } z \in \rho(N), \tag{1.4.1}$$

where P_M is the projection from $\mathcal{K}$ to M. This function $\mu(\cdot)$ is said to be the *mosaic* of S. This function is an important tool in the study of subnormal operators. The meaning of this function can be explained in Lemma 1.4.1.

It is easy to see that for $\alpha, \beta \in M$, we have

$$
\begin{aligned}
(\mu(z)\alpha, \beta) &= (N(N-z)^{-1}\alpha, \beta) - (SP_M(N-z)^{-1}\alpha, \beta) \\
&= \int \frac{u}{u-z}(e(du)\alpha, \beta) - ((N-z)^{-1}\alpha, \Lambda^*\beta) \\
&= \int \frac{u}{u-z}(e(du)\alpha, \beta) - \int \frac{1}{u-z}(e(du)\alpha, \Lambda^*\beta).
\end{aligned}
$$

Thus

$$
\mu(z) = \int_{\sigma(N)} \frac{u-\Lambda}{u-z} e(du), \quad z \in \rho(N). \tag{1.4.2}
$$

From (1.3.14),

$$
\mu(z) = 0, \quad z \in \rho(S). \tag{1.4.3}
$$

Lemma 1.4.1. *Let* $z \in \rho(N)$, *then for every* $f \in \mathcal{H}$, *there is a unique vector in* M *which is denoted by* $f(z)$ *and is said to be the value* f *at* z *such that*

$$
(N-z)^{-1}(f - f(z)) \in \mathcal{H} \quad \text{and} \quad (N-z)^{-1}f(z) \in \widetilde{\mathcal{H}}. \tag{1.4.4}
$$

The value of a vector possesses the following properties:

(i) $(\alpha f + \beta g)(z) = \alpha f(z) + \beta g(z)$, *for* $f, g \in \mathcal{H}$ *and* $\alpha, \beta \in \mathbb{C}$,
(ii) $((\lambda - S)^{-1}f)(z) = (\lambda - z)^{-1}f(z)$, *for* $f \in \mathcal{H}$ *and* $\lambda \in \rho(S)$,
(iii) if $\{f_n\} \subset \mathcal{H}, f \in \mathcal{H}$ *and* $\|f_n - f\| \to 0$ *as* $n \to \infty$, *then*

$$
\lim_{n \to \infty} \|f_n(z) - f(z)\| = 0,
$$

(iv) $f(z) = \mu(z)f(z)$, *for* $z \in \rho(N)$ *and* $f(z) = \mu(z) = 0$, *for* $z \in \rho(S)$,
(v) $f(z)$ *is an analytic function of* $z \in \rho(N)$, *and*

$$
f(z) = \int \frac{u-\Lambda}{u-z} e(du)f(u), \quad \text{for } f \in \mathcal{H}, \tag{1.4.5}
$$

and

(vi) if $\mu(z) = 0$, *then* $z \in \rho(S)$.

Proof. Let

$$
f_1 = P(N-z)^{-1}f \quad \text{and} \quad f_2 = \widetilde{P}(N-z)^{-1}f
$$

where P and $\widetilde{P}$ are the projections from $\mathcal{K}$ to $\mathcal{H}$ and $\widetilde{\mathcal{H}}$ respectively.

Then by (1.3.3), we have

$$
f = (N-z)f_1 + (N-z)f_2 = ((S-z)f_1 + Af_2) + (\widetilde{S}^* - z)f_2.
$$

Hence $f = (S - z)f_1 + Af_2$ and $(\widetilde{S}^* - z)f_2 = 0$. From (1.3.23), (1.3.31) and (1.3.33), it follows that

$$f(u) = (u - z)f_1(u) + \int (v - \Lambda)e(dv)f_2(v), \qquad (1.4.6)$$

and

$$(u - z)f_2(u) - \int (v - \Lambda)e(dv)f_2(v) = 0. \qquad (1.4.7)$$

Denote $f(z) = \int(v - \Lambda)e(dv)f_2(v)$. From (1.4.6) and (1.4.7), it follows (1.4.4).

Suppose there is a vector $a \in M$ satisfying

$$(N - z)^{-1}(f - a) \in \mathcal{H} \quad \text{and} \quad (N - z)^{-1}a \in \mathcal{H}. \qquad (1.4.8)$$

From (1.4.4) and (1.4.8), it follows that

$$(N - z)^{-1}(f(z) - a) \in \mathcal{H} \cap \widetilde{\mathcal{H}}.$$

Thus $f(z) - a = 0$ which proves the uniqueness of $f(z)$ satisfying (1.4.4).

From (1.4.7), we have $f_2(u) = (u - z)^{-1}f(z)$. Thus

$$f(z) = \int (u - \Lambda)e(du)f_2(u) = \int \frac{u - \Lambda}{u - z}e(du)f(z)$$

which proves (iv). For any $b \in M$, we have

$$\left(\int \frac{u - \Lambda}{u - z}e(du)(f(u) - f(z)), b\right) = \left(\frac{f(\cdot) - f(z)}{(\cdot) - z}, (\overline{(\cdot)} - \Lambda^*)b\right) = 0,$$

since (1.4.4) and $(\overline{(\cdot)} - \Lambda^*)b \in \widetilde{\mathcal{H}}$ by (1.3.35). Therefore

$$\int \frac{u - \Lambda}{u - z}e(du)(f(u) - f(z)) = 0,$$

which proves (1.4.5) by (iv).

To prove (ii), by (iv) we have

$$((\lambda - S)^{-1}f)(z) = \int \frac{(u - \Lambda)e(du)}{(u - z)(\lambda - u)}$$

$$= \frac{1}{\lambda - z}\int \left(\frac{u - \Lambda}{u - z} - \frac{u - \Lambda}{u - \lambda}\right)e(du)f(u)$$

$$= \frac{1}{\lambda - z}(f(z) - f(\lambda)), \qquad (1.4.9)$$

for $\lambda \in \rho(S)$. However, from (iv), $f(\lambda) = 0$. Thus (1.4.9) implies (ii).

To prove (vi), suppose $\mu(z) = 0$. From (1.4.4), $(N - z)^{-1}f \in \mathcal{H}$ for $f \in \mathcal{H}$. Let $g = (N - z)^{-1}f$, then

$$(S - z)g = (N - z)g = f.$$

Thus range$(S - z) = \mathcal{H}$. However ker$(S - z) = \{0\}$, since S is pure hyponormal and it has no eigenvalue. Therefore $z \in \rho(S)$. $\qquad \square$

Corollary 1.4.2. *If $g(\cdot)$ is a rational function with poles in $\rho(S), a \in M$ and $f = g(S)a$. Then*

$$f(z) = g(z)\mu(z)a, \quad z \in \rho(S).$$

Proof. Without loss of generality, we may assume that $g(u) = (u-\lambda)^{-1}, \lambda \in \rho(S)$. Then

$$
\begin{aligned}
f(z) &= \int_{\sigma(N)} \frac{(u - \Lambda)e(du)a}{(u - z)(u - \lambda)} \\
&= \frac{1}{z - \lambda}\left(\int_{\sigma(N)} \frac{(u - \Lambda)e(du)a}{u - z} - \int_{\sigma(N)} \frac{(u - \Lambda)e(du)a}{u - \lambda}\right) \\
&= \frac{1}{z - \lambda}(\mu(z)a - \mu(\lambda)a) = \frac{1}{z - \lambda}\mu(z)a,
\end{aligned}
$$

since $\mu(\lambda) = 0$ for $\lambda \in \rho(S)$.

$\square$

From (1.4.5), for $f \in \mathcal{H}$,

$$f(z) = P_M(N - SP_M)(N - z)^{-1}f|_M, \quad z \in \rho(N). \tag{1.4.10}$$

We may extend the definition of $f(z)$ to the vector $f \in \mathcal{K}$ by (1.4.10) or by (1.4.5) in the analytic model. Then $f(z)$ in the general case of $f \in \mathcal{K}$ still possesses all the properties (i), (ii), (iii) and (v).

For $z \in \rho(N)$, denote

$$M_z = \{b \in M : (N - z)^{-1}b \in \widetilde{\mathcal{H}}\} \quad \text{and} \quad \widetilde{M}_z = \{b \in M : (N - z)^{-1}b \in \mathcal{H}\}.$$

Theorem 1.4.3. *Let $\mu(\cdot)$ be the mosaic for a pure subnormal operator S on $\mathcal{H}$ with m.n.e. N on $\mathcal{H} \oplus \widetilde{\mathcal{H}}$ and $z \in \rho(N)$. Then*

(i) $\mu(z) = \mu(z)^2$,
(ii) $\mu(z)M = M_z$,
(iii) $(I - \mu(z))M = \widetilde{M}_z$,
(iv) $M = M_z \dotplus \widetilde{M}_z$, where $\dotplus$ means direct sum,
(v) for every vector $a \in M$, $\mu(z)a$ is the value of a at z,
(vi) $M_z = \{f(z) : f \in \mathcal{H}\}$,
(vii) $\widetilde{M}_z = \{f(z) : f \in \widetilde{\mathcal{H}}\}$,
(viii) $(Pf)(z) = \mu(z)f(z)$, for $f \in \mathcal{K}$.

Proof. For $a \in M$, let $a(z)$ be the value of the vector a at $z \in \rho(N)$. Then from (1.4.4), $a(z)$ is the unique vector satisfying

$$a(z) \in M_z \quad \text{and} \quad a - a(z) \in \widetilde{M}_z.$$

That implies (iv) and mapping $a \mapsto a(z)$ and $a \mapsto a - a(z)$ are parallel projections from M to M_z and $\widetilde{M}_z$ respectively.

From (iv) of Lemma 1.4.1, we have

$$a(z) = \mu(z)a(z), \quad \text{for } a \in M.$$

From $a - a(z) \in \widetilde{M}_z$, we have

$$(\mu(z)(a - a(z)), b) = (\frac{a - a(z)}{(\cdot) - z}, \overline{((\cdot)} - \Lambda^*)b) = 0$$

for $b \in M$. Hence

$$\mu(z)a = \mu(z)a(z) = a(z).$$

Therefore (i), (ii) and (iii) are proved.

(v) is a special case of Corollary 1.4.2. (vi) is an immediate consequence of (ii), (v) and (iv) of Lemma 1.4.1. We will prove (vii) after Theorem 1.4.5. From (vii), $(f - Pf)(z) \in (I - \mu(z))M$ for $f \in \mathcal{K}$. Thus $\mu(z)(f - Pf)(z) = 0$. From (i), (ii) and (vi), $\mu(z)(Pf)(z) = (Pf)(z)$, which proves (viii).   □

For $z \in \rho(N)$, let us introduce another inner product $(\cdot, \cdot)_z$ on M as

$$(a, b)_z \overset{\text{def}}{=} \int_{\sigma(N)} \frac{(e(du)a, b)}{|u - z|^2}.$$

It is obvious that there are non-zero $C_1(z) = \max_{u \in \sigma(N)} |u - z|^{-2}$ and $C_2(z) = \min_{u \in \sigma(N)} |u - z|^{-2}$ such that

$$C_2(z)(a, a) \leq (a, a)_z \leq C_1(z)(a, a).$$

Therefore $\| \cdot \|$ and $\| \cdot \|_z$ are topologically equivalent on M.

Corollary 1.4.4. *Suppose $z \in \rho(N)$, then*

$$M = M_z \oplus \widetilde{M}_z$$

where the orthogonal decomposition $\oplus$ is with respect to $(\cdot, \cdot)_z$.

Proof. If $a \in M_z$ and $b \in \widetilde{M}_z$, then $((\cdot) - z)^{-1}a \in \mathcal{H}$ and $((\cdot) - z)^{-1}b \in \widetilde{\mathcal{H}}$. Therefore

$$(a, b)_z = (((\cdot) - z)^{-1}a, ((\cdot) - z)^{-1}b) = 0.$$

□

Theorem 1.4.5. *Let S be a pure subnormal operator on $\mathcal{H}$ with m.n.e. N on $\mathcal{K} \supset \mathcal{H}$. Let $z \in \sigma(S) \cap \rho(N)$. Then $\bar{z} \in \sigma_p(S^*)$ and a vector f satisfying $S^*f = \bar{z}f$ iff*

$$f = (N^* - \bar{z})^{-1}(N^* - \Lambda^*)\mu(z)^*P_M f. \tag{1.4.11}$$

Proof. We may assume that S is in its analytic model. Then $(S^* - \overline{z})f = 0$ iff

$$(\overline{u} - \overline{z})f(u) = (\overline{u} - \Lambda^*)f(\Lambda)$$

by (1.3.24), i.e., $f(u) = (\overline{u} - \overline{z})^{-1}(\overline{u} - \Lambda^*)f(\Lambda)$. In this case, we have

$$f(\Lambda) = P_M f = \int e(du)f(u) = \int e(du)(\overline{u} - \overline{z})^{-1}(\overline{u} - \Lambda^*)f(\Lambda) = \mu(z)^* f(\Lambda)$$

which proves (1.4.11). By Proposition 1.4.8, for any $a \in M$ and $z \in \sigma(S) \cap \rho(N)$, $(N^* - \overline{z})^{-1}(N^* - \Lambda^*)\mu(z)^* a \in \mathcal{H}$. Therefore $\overline{z} \in \sigma_p(S^*)$. $\qquad \square$

Proof of Theorem 1.4.3, (vii).
For $z \in \rho(N)$, we have

$$\int \frac{e(du)}{u - z}(\overline{u} - \Lambda^*)(z - \Lambda) = \int \frac{e(du)}{u - z}(\overline{u} - \Lambda^*)(u - \Lambda - (u - z))$$

$$= \int \frac{e(du)}{u - z}C, \qquad (1.4.12)$$

since $\int e(du)(\overline{u} - \Lambda^*) = 0$. For $a \in CM$, let $b = (z - \Lambda)C^{-1}a$. Then by (1.4.12)

$$\int \frac{u - \Lambda}{u - z}e(du)(\overline{u} - \Lambda^*)b = (z - \Lambda)\int \frac{e(du)(\overline{u} - \Lambda^*)(z - \Lambda)C^{-1}a}{u - z}$$

$$= (z - \Lambda)\int \frac{e(du)}{u - z}a = (\mu(z) - I)a.$$

Let $f(u) = (\overline{u} - \Lambda^*)b \in \widetilde{\mathcal{H}}$, then $f(z) = (\mu(z) - I)a$. Denote $M_0 = \{f(z) : f \in \widetilde{\mathcal{H}}\}$. Then

$$(I - \mu(z))CM \subset M_0.$$

But M_0 is closed. Therefore $\widetilde{M_z} \subset M_0$. For $w \in \rho(S)$, let

$$f(u) = \frac{\overline{u} - \Lambda^*}{\overline{u} - \overline{w}}a, \quad a \in M.$$

Then

$$\left(f(\cdot), \frac{\overline{(\cdot)} - \Lambda^*}{\overline{(\cdot)} - \overline{z}}\mu(z)^* b\right) = 0, \quad b \in M, \quad z \in \rho(S),$$

since $f(\cdot) \in \widetilde{\mathcal{H}}$ and $(\overline{(\cdot)} - \overline{z})^{-1}(\overline{(\cdot)} - \Lambda^*)\mu(z)^* b \in \mathcal{H}$. Therefore

$$\mu(z)f(z) = 0,$$

i.e., $f(z) \in (I - \mu(z))M = \widetilde{M_z}$, which proves $\widetilde{M_z} = M_0$.

Definition. Let B be an operator on a Hilbert space $\mathcal{H}$. If $\{f_j : j \in J\}$ is a set of vectors on $\mathcal{H}$ satisfying the condition that

$$\text{cl} \bigvee \{(\lambda - B)^{-1} f_j : j \in J, \lambda \in \rho(B)\} = \mathcal{H},$$

then $\{f_j : j \in J\}$ is said to be a *set of generators* for B. The smallest cardinal number of the set of generators for B is said to be the *multiplicity* of B. If there is a single set $\{f\}$ of generator for B, then B is said to be *cyclic* and the vector f is said to be a *cyclic vector* for B.

Proposition 1.4.6. *Let S be a pure subnormal operator on $\mathcal{H}$ with m.n.e. N. If the multiplicity of S is n, then rank $\mu(z) \le n$ for all $z \in \rho(N)$, where $\mu(\cdot)$ is the mosaic of S.*

Proof. Let us use the analytic model of S and $\{f_j : j \in J\}$ be the set of generators for B. Let

$$M(z) = \text{cl} \bigvee \{f_j(z) : j \in J\}, \quad z \in \rho(N),$$

where $f_j(z)$ is the value of the vector f_j at z. Then

$$\dim M(z) \le n. \tag{1.4.13}$$

By Lemma 1.4.1, (i) and (ii), if $h \in \bigvee \{(\lambda_j - S)^{-1} f_j : j \in J, \lambda_j \in \rho(N)\}$, then

$$h(z) \in M(z). \tag{1.4.14}$$

For $f \in \mathcal{H}$, there exists a sequence $\{h_n\}$, where $h_n \in \bigvee \{(\lambda_j - S)^{-1} f_j : j \in J, \lambda_j \in \rho(N)\}$ such that $\|h_n - f\| \to 0$ as $n \to \infty$. By (1.4.14), $h_n(z) \in M(z)$. From Lemma 1.4.1 (iii), it follows that $\lim_{n \to \infty} \|h_n(z) - f(z)\| = 0$. Hence $f(z) \in M(z)$. Thus $M_z \subset M(z)$. From (1.4.13), we have $\dim M_z \le n$, which proves the proposition. $\qquad \square$

Lemma 1.4.7. *Let $z \in \rho(N)$, then for every $f \in \widetilde{\mathcal{H}}$, there is a unique vector $m_f \in M$ such that*

$$(N^* - \overline{z})^{-1}(N^* - \Lambda^*) m_f \in \mathcal{H} \quad and \quad (N^* - \overline{z})^{-1}(f - (N^* - \Lambda^*)m_f) \in \widetilde{\mathcal{H}}. \tag{1.4.15}$$

The proof of this lemma is similar to that of Lemma 1.4.1. We only point out the vector m_f in (1.4.15) is $P_M(N^* - \overline{z})^{-1} f$, and omit the details of the proof.

Let

$$M_z^* \overset{\text{def}}{=} \{a \in M : (N^* - \overline{z})^{-1}(N^* - \Lambda^*) a \in \mathcal{H}\}$$

and

$$\widetilde{M_z}^* \overset{\text{def}}{=} \{a \in M : (N^* - \overline{z})^{-1}(N^* - \Lambda^*) a \in \widetilde{\mathcal{H}}\}.$$

Proposition 1.4.8. *Let $z \in \rho(N)$, then operators $\mu(z)^*$ and $I - \mu(z)^*$ are parallel projections from M to M_z^* and $\widetilde{M_z}^*$ respectively.*

Proof. For $a \in M$, let $f(\cdot) \stackrel{\text{def}}{=} (\overline{(\cdot)} - \Lambda^*)a \in \widetilde{\mathcal{H}}$. Then

$$m_f = P_M \frac{\overline{(\cdot)} - \Lambda^*}{\overline{(\cdot)} - \overline{z}} a = \mu(z)^* a.$$

By Lemma 1.4.7, $\mu(z)^* M \subset M_z^*$. On the other hand, for $a \in M_z^*$, from $(\overline{(\cdot)} - \Lambda^*)(\overline{(\cdot)} - \overline{z})^{-1} a \in \mathcal{H}$, it follows

$$\int (u - \Lambda) e(du) \frac{\overline{u} - \Lambda^*}{\overline{u} - \overline{z}} a = 0, \tag{1.4.16}$$

since for any $g(\cdot) \in \mathcal{H}$, $\int (u - \Lambda) e(du) g(\cdot) = 0$ by the fact that $(g(\cdot), (\overline{(\cdot)} - \Lambda^*) b) = 0$ for every $b \in M$. Then

$$\begin{aligned} C\mu(z)^* a &= \int (\overline{u} - \Lambda^*)(u - \Lambda) e(du) \frac{\overline{u} - \Lambda^*}{\overline{u} - \overline{z}} a \\ &= \int (u - \Lambda) e(du)(\overline{u} - \Lambda^*) a + (\overline{z} - \Lambda^*) \int (u - \Lambda) e(du) \frac{\overline{u} - \Lambda^*}{\overline{u} - \overline{z}} a. \end{aligned} \tag{1.4.17}$$

However

$$\int (u - \Lambda) e(du)(\overline{u} - \Lambda^*) = \int \overline{u}(u - \Lambda) e(du) = \int (\overline{u} - \Lambda^*)(u - \Lambda) e(du) = C \tag{1.4.18}$$

since $\int (u - \Lambda) e(du) = 0$. From (1.4.16) and (1.4.17) it follows that

$$C\mu(z)^* a = Ca.$$

But C is invertible, therefore $a = \mu(z)^* a$. Thus

$$M_z^* = \mu(z)^* M.$$

For $a \in M$, by (1.4.15) and letting $f(\cdot) = (\overline{(\cdot)} - \Lambda^*)a$, we have

$$\frac{\overline{(\cdot)} - \Lambda^*}{\overline{(\cdot)} - \overline{z}} (I - \mu(z)^*) a = \frac{f(\cdot) - (\overline{(\cdot)} - \Lambda^*) m_f}{\overline{(\cdot)} - \overline{z}} \in \widetilde{\mathcal{H}}.$$

Hence $(I - \mu(z)^*) M \subset \widetilde{M_z}^*$. If $a \in \widetilde{M_z}^*$, then for $b \in M$,

$$\left(\frac{\overline{(\cdot)} - \Lambda^*}{\overline{(\cdot)} - \overline{z}} a, b \right) = 0.$$

Therefore $\mu(z)^* a = 0$, i.e., $a = (I - \mu(z)^*)a$, which proves $\widetilde{M_z}^* \subset (I - \mu(z)^*) M$. Thus $(I - \mu(z)^*) M = \widetilde{M_z}^*$ which proves the lemma. $\quad\square$

For $z \in \rho(N)$, let us introduce an inner product $[\cdot, \cdot]_z$ on M as follows

$$[a, b]_z = \int \frac{((u - \Lambda)e(du)(\overline{u} - \Lambda^*)a, b)}{|u - z|^2}. \tag{1.4.19}$$

Corollary 1.4.9. *If $z \in \rho(N)$, then*

$$M = M_z^* \oplus \widetilde{M}_z^*$$

where $\oplus$ is with respect to the inner product $[\cdot, \cdot]_z$ defined by (1.4.19).

The mosaic $\mu(\cdot)$ is a unitary invariant, i.e., if S and S_1 are pure subnormal operators on $\mathcal{H}$ and $\mathcal{H}_1$ with m.n.e. N and N_1 on $\mathcal{K}$ and $\mathcal{K}_1$ respectively and there is a unitary operator U from $\mathcal{K}$ onto $\mathcal{K}_1$ such that

$$S_1 = USU^{-1} \quad \text{and} \quad N_1 = UNU^{-1}, \tag{1.4.20}$$

then the unitary operator $V = U|_M$ where $M = \mathrm{cl}[S^*, S]\mathcal{H}$, from M onto $M_1 = \mathrm{cl}[S_1^*, S_1]\mathcal{H}_1$ satisfies

$$\mu_1(z) = V\mu(z)V^{-1}, \quad \text{for} \ \ z \in \rho(N) = \rho(N_1), \tag{1.4.21}$$

where $\mu(\cdot)$ and $\mu_1(\cdot)$ are the mosaics for S and S_1 respectively.

Under certain conditions, the converse of the above statement is also true, i.e. mosaic $\mu(\cdot)$ is a complete unitary invariant in the following sense.

Theorem 1.4.10. *Let S and S_1 be two pure subnormal operators on $\mathcal{H}$ and $\mathcal{H}_1$ with m.n.e. N and N_1 on $\mathcal{K}$ and $\mathcal{K}_1$ respectively. Let $\mu(\cdot)$ and $\mu_1(\cdot)$ be the mosaics of S and S_1 respectively. If $\sigma(N) = \sigma(N_1)$ has no interior point, and there is a unitary operator V from $M = \mathrm{cl}[S^*, S]\mathcal{H}$ onto $M_1 = \mathrm{cl}[S^*, S]\mathcal{H}_1$ such that (1.4.21) holds good, then there is a unitary operator U from $\mathcal{K}$ to $\mathcal{K}_1$ such that (1.4.20) holds good.*

Proof. For simplicity, we only consider the analytic model of S and S_1. Assume that the operators and operator-valued measures in Theorem 1.3.4 corresponding to S and S_1 are $C, \Lambda, e(\cdot)$ and $C_1, \Lambda_1, e_1(\cdot)$ respectively. From (1.4.21) it follows

$$\int_{\sigma(N_1)} \frac{(u - \Lambda_1)e_1(du)}{u - z} = V \int_{\sigma(N)} \frac{(u - \Lambda)e(du)}{u - z} V^{-1}, \quad z \in \rho(N). \tag{1.4.22}$$

Since $\sigma(N)$ has no interior point, every continuous $f(\cdot)$ on $\sigma(N)$ may be approximated by functions in

$$\bigvee \{\frac{1}{(\cdot) - z}, \ z \in \rho(N)\}.$$

Thus (1.4.22) implies that

$$\int_{\sigma(N)} f(u)(u - \Lambda_1)e_1(du) = V \int_{\sigma(N)} f(u)(u - \Lambda)e(du)V^{-1}$$

for every continuous function $f(\cdot)$ on $\sigma(N)$. Hence

$$\int_{\sigma(N)} \frac{(u - \Lambda_1)e_1(du)}{(u - z)(\overline{u} - \overline{w})} = V \int_{\sigma(N)} \frac{(u - \Lambda)e(du)}{(u - z)(\overline{u} - \overline{w})}V^{-1}, \quad \text{for } z, w \in \rho(N).$$

$$(1.4.23)$$

However, by Lemma 1.3.2 we have

$$\int_{\sigma(N)} \frac{(u - \Lambda)e(du)}{(u - z)(\overline{u} - \overline{w})} = \int_{\sigma(N)} \frac{e(du)}{\overline{u} - \overline{w}} + (z - \Lambda)S(z, w)$$

$$= (\Lambda^* - \overline{w})^{-1} + (z - \Lambda)\big((\overline{w} - \Lambda^*)(z - \Lambda) - C\big)^{-1}$$

$$= \sum_{n=1}^{\infty} \frac{(C(z - \Lambda)^{-1} + \Lambda^*)^n - \Lambda^{*n}}{\overline{w}^{n+1}}$$

for sufficiently large $|w|$. Therefore (1.4.23) implies that

$$C_1(z - \Lambda_1)^{-1} = VCV^{-1}(z - V\Lambda V^{-1})^{-1}$$

for $z \in \rho(N)$. Multiplying both sides of (1.4.23) by z and letting $z \to \infty$, we have $C_1 = VCV^{-1}$ and then we can also prove that

$$\Lambda_1 = V\Lambda V^{-1}.$$

By Theorem 1.2.5 we may prove that S and S_1 are unitarily equivalent. Or, by (1.4.23) again, we may prove that $e_1(\cdot) = Ve(\cdot)V^{-1}$ which proves the theorem. □

For a pure subnormal operator S with m.n.e. N in the analytic model

$$S(z, w) = \int_{\sigma(N)} \frac{e(du)}{(z - u)(\overline{w} - \overline{u})}, \quad z, w \in \rho(S),$$

$$(1.4.24)$$

where $S(\cdot, \cdot)$ is in Lemma 1.3.2. Now let us extend the definition of $S(\cdot, \cdot)$ to

$$S(z, w) \stackrel{\text{def}}{=} P_M(\overline{w} - N^*)^{-1}(z - N)^{-1}|_M$$

$$= \int_{\sigma(N)} \frac{e(du)}{(z - u)(\overline{w} - u)}, \quad \text{for } z, w \in \rho(N).$$

The following theorem is a kind of generalization of Lemma 1.3.2.

Theorem 1.4.11. *If $z, w \in \rho(N)$ and $Q(z, w)$ (defined in (1.3.10)) is invertible, then*

$$S(z, w) = Q(z, w)^{-1} - Q(z, w)^{-1}\mu(z) - \mu(w)^*Q(z, w)^{-1}.$$

$$(1.4.25)$$

Proof. By a simple calculation, it is easy to see that

$$Q(z,w)S(z,w) = \int_{\sigma(N)} \frac{(\overline{w} - \Lambda^*)(z - \Lambda) - (\overline{u} - \Lambda^*)(u - \Lambda)}{(\overline{u} - \overline{w})(u - z)} e(du)$$

$$= I - \mu(z) - \int_{\sigma(N)} \frac{\overline{u} - \Lambda^*}{\overline{u} - \overline{w}} e(du). \qquad (1.4.26)$$

Since $(\overline{u} - \Lambda^*)(u - \Lambda)e(du) = Ce(du)$ by (1.3.15) and

$$(\overline{w} - \Lambda^*)(z - \Lambda) - (\overline{u} - \Lambda^*)(u - \Lambda) = (\overline{w} - \overline{u})(z - \Lambda) - (\overline{u} - \Lambda^*)(u - z).$$

On the other hand, $\mu(w)^* = \int e(du)\frac{\overline{u}-\Lambda^*}{\overline{u}-\overline{w}}$ and

$$\int \frac{\overline{u} - \Lambda^*}{\overline{u} - \overline{w}} e(du)(\overline{w} - \Lambda^*) = \int \frac{(\overline{u} - \overline{w}) + (\overline{w} - \Lambda^*)}{\overline{u} - \overline{w}} e(du)(\overline{w} - \Lambda^*)$$

$$= (\overline{w} - \Lambda^*) + (\overline{w} - \Lambda^*) \int \frac{e(du)}{\overline{u} - \overline{w}}(\overline{w} - \Lambda^*)$$

$$= (\overline{w} - \Lambda^*) \int \frac{e(du)(\overline{u} - \Lambda^*)}{\overline{u} - \overline{w}} = (\overline{w} - \Lambda^*)\mu(w)^*.$$

$$(1.4.27)$$

Besides,

$$\int \frac{\overline{u} - \Lambda^*}{\overline{u} - \overline{w}} e(du)\big((\overline{w} - \Lambda^*)\Lambda + C\big)$$

$$= \int \frac{\overline{u} - \Lambda^*}{\overline{u} - \overline{w}} e(du)\big((\overline{w} - \overline{u})\Lambda + (\overline{u} - \Lambda^*)u\big), \qquad (1.4.28)$$

and

$$\big((\overline{w} - \Lambda^*)\Lambda + C\big) \int e(du)\frac{(\overline{u} - \Lambda^*)}{\overline{u} - \overline{w}}$$

$$= \int \big((\overline{w} - \overline{u})\Lambda + (\overline{u} - \Lambda^*)u\big)e(du)\frac{(\overline{u} - \Lambda^*)}{\overline{u} - \overline{w}}. \qquad (1.4.29)$$

However $\int(\overline{u} - \Lambda^*)e(du) = \int e(du)(\overline{u} - \Lambda^*) = 0$. From (1.4.27), (1.4.28) and (1.4.29) it follows

$$\int \frac{(\overline{u} - \Lambda^*)}{\overline{u} - \overline{w}} e(du)Q(z,w) = Q(z,w)\mu(w)^*. \qquad (1.4.30)$$

From (1.4.26), and (1.4.30), it follows

$$Q(z,w)S(z,w)Q(z,w) = (I - \mu(z))Q(z,w) - Q(z,w)\mu(w)^*,$$

which proves (1.4.25).   □

Let

$$R_S(z) = R(z) \overset{\text{def}}{=} C(z - \Lambda)^{-1} + \Lambda^* \quad \text{for } z \in \rho(\Lambda). \qquad (1.4.31)$$

Theorem 1.4.12. *Let S be a pure subnormal operator. If $z \in \rho(\Lambda) \cap \rho(N)$, then*

$$[\mu(z), R(z)] = 0. \tag{1.4.32}$$

Proof. We only have to prove that

$$\left(C + \Lambda^*(z - \Lambda)\right) \int \frac{e(du)}{u - z} = (z - \Lambda) \int \frac{e(du)}{u - z} R(z), \tag{1.4.33}$$

since $\mu(z) = I + (z - \Lambda) \int \frac{e(du)}{u - z}$. It is easy to see that left-hand side of (1.4.33) equals to

$$\int (u - z)^{-1}\left((\overline{u} - \Lambda^*)(u - \Lambda) + \Lambda^*(z - \Lambda)\right)e(du)$$
$$= \int (u - z)^{-1}\overline{u}(u - \Lambda)e(du) - \Lambda^* = (z - \Lambda)\int \overline{u}(u - z)^{-1}e(du),$$

since $\int \overline{u}e(du) = \Lambda^*$. On the other hand, from (1.4.12) we have

$$\int \overline{u}(u - z)^{-1}e(du) = \int e(du)(u - z)^{-1}R(z),$$

which proves (1.4.33) and hence the theorem. $\qquad \square$

Proposition 1.4.13. *If $f \in \mathcal{H}$, then for $z \in \rho(N) \cap \rho(\Lambda)$*

$$(S^*f)(z) = R(z)(f(z) - \mu(z)f(\Lambda)) + \mu(z)\Lambda^*f(\Lambda). \tag{1.4.34}$$

If $f \in \widetilde{\mathcal{H}}$, then

$$(\widetilde{S}f)(z) = R(z)f(z), \quad z \in \rho(N) \cap \rho(\Lambda) \tag{1.4.35}$$

and

$$(\widetilde{S^*}f)(z) = zf(z) - (I - \mu(z)) \lim_{\zeta \to \infty} \zeta f(\zeta), \quad z \in \rho(N). \tag{1.4.36}$$

Proof. By (1.3.24), we have

$$(S^*f)(z) = \int_{\sigma(N)} (u - \Lambda)e(du)(\overline{u}(f(u) - f(\Lambda)) + \Lambda^*f(\Lambda))(u - z)^{-1}$$
$$= \int_{\sigma(N)} (\overline{u} - \Lambda^*)(u - \Lambda)e(du)(f(u) - f(\Lambda))(u - z)^{-1}$$
$$+ \Lambda^* \int_{\sigma(N)} (u - \Lambda)e(du)(f(u) - f(\Lambda))(u - z)^{-1} + \mu(z)\Lambda^*f(\Lambda).$$
$$\tag{1.4.37}$$

From (1.3.15), (1.4.37) and

$$(u - z)^{-1} = (z - \Lambda)^{-1}((u - \Lambda)(u - z)^{-1} - 1), \tag{1.4.38}$$

it follows (1.4.34).

To prove (1.4.35), notice that by (1.3.15) and (1.3.32), we have

$$(\widetilde{S}f)(z) = \int (u - \Lambda)e(du)\overline{u}f(u)(u - z)^{-1}$$

$$= \int (\overline{u} - \Lambda^*)(u - \Lambda)e(du)f(u)(u - z)^{-1} + \Lambda^* f(z)$$

$$= C \int e(du)f(u)(u - z)^{-1} + \Lambda^* f(z). \qquad (1.4.39)$$

By the definition of $f(z)$,

$$\int e(du)f(u)(u - z)^{-1} = (z - \Lambda)^{-1}f(z), \quad z \in \rho(N) \cap \rho(\Lambda) \qquad (1.4.40)$$

since $\int e(du)f(u) = 0$ for $f \in \widetilde{\mathcal{H}}$. From (1.4.39) and (1.4.40), it follows (1.4.35).

By (1.3.33), we have

$$(\widetilde{S}^*f)(z) = \int (u - \Lambda)e(du)uf(u)(u - z)^{-1} + \mu(z)\eta, \quad f \in \widetilde{\mathcal{H}}, \qquad (1.4.41)$$

where $\eta = -\int (v - \Lambda)e(du)f(v) = \lim_{\zeta \to \infty} \zeta f(\zeta)$. On the other hand, it is obvious that

$$\int (u - \Lambda)e(du)uf(u)(u - z)^{-1} = zf(z) - \eta. \qquad (1.4.42)$$

From (1.4.41) and (1.4.42) it follows (1.4.36). $\qquad\qquad\qquad\square$

1.5   Dual operator and some kind of spectra of pure subnormal operators

Suppose S is a pure subnormal operator on $\mathcal{H}$ with m.n.e. N on $\mathcal{K} \supset \mathcal{H}$. Similar to (1.3.3), let us rewrite $\mathcal{K} = \widetilde{\mathcal{H}} \oplus \mathcal{H}$, then

$$N^* = \begin{pmatrix} \widetilde{S} & A^* \\ 0 & S^* \end{pmatrix}.$$

Therefore the dual operator $\widetilde{S}$ is subnormal on $\widetilde{\mathcal{H}}$ with normal extension N^*.

Let us study $\widetilde{S}$ as in the analytic model of S. From (1.4.18), we have $\|(N^* - \Lambda^*)a\| = \|C^{1/2}a\|$ for $a \in M$. Hence the operator Ω defined by

$$\Omega a = (N^* - \Lambda^*)C^{-1/2}a, \quad a \in C^{1/2}M$$

is an isometry and it extends a unitary operator from M to $\mathrm{cl}[\widetilde{S}^*, \widetilde{S}]\widetilde{\mathcal{H}}$ by (1.3.34). Denote

$$\widetilde{M} \stackrel{\text{def}}{=} \mathrm{cl}[\widetilde{S}^*, \widetilde{S}]\widetilde{\mathcal{H}}.$$

Then

$$\Omega M = \widetilde{M}. \qquad (1.5.1)$$

Define

$$\widetilde{C} \stackrel{\text{def}}{=} C_{\widetilde{S}} \stackrel{\text{def}}{=} [\widetilde{S}^*, \widetilde{S}]|_{\widetilde{M}}.$$

From (1.3.34) and (1.4.18), it follows that

$$[\widetilde{S}^*, \widetilde{S}]\Omega a = \Omega C a, \quad \text{for } a \in M. \qquad (1.5.2)$$

Therefore N^* is an m.n.e. of $\widetilde{S}$ and

$$\widetilde{C} = \Omega C \Omega^{-1}. \qquad (1.5.3)$$

Let $\widetilde{\Lambda} = (\widetilde{S}^*|_{\widetilde{M}})^*$. Then by (1.3.33), for $a \in C^{1/2}M$, we have

$$\begin{aligned}
\widetilde{S}^*|_{\widetilde{M}} \Omega a &= \widetilde{S}^*(\overline{(\cdot)} - \Lambda^*)C^{-1/2}a \\
&= (\cdot)(\overline{(\cdot)} - \Lambda^*)C^{-1/2}a - \int (v - \Lambda)e(dv)(\overline{v} - \Lambda^*)C^{-1/2}a \\
&= (\overline{(\cdot)} - \Lambda^*)((\cdot) - \Lambda)C^{-1/2}a + (\overline{(\cdot)} - \Lambda^*)\Lambda C^{-1/2}a - C^{1/2}a \\
&= \Omega C^{1/2}\Lambda C^{-1/2}a,
\end{aligned}$$

as vectors in $L^2(e)$, since $e(du)(\overline{u} - \Lambda^*)(u - \Lambda)b = e(du)Cb$ for $b \in M$. Therefore $C^{1/2}\Lambda C^{-1/2}$ is a bounded operator on $C^{1/2}M$. It extends a bounded operator on M. Thus $\widetilde{S}^*|_M = \Omega C^{1/2}\Lambda C^{-1/2}\Omega^{-1}$ and $C^{1/2}\Omega^{-1}\widetilde{\Lambda} = \Lambda^* C^{1/2}\Omega^{-1}$. Hence the range of $\Lambda^*C^{1/2}$ is in the range of $C^{1/2}$ and then $C^{-1/2}\Lambda^*C^{1/2}$ makes sense. It is a bounded operator. Therefore

$$\widetilde{\Lambda} = \Omega C^{-1/2}\Lambda^* C^{1/2}\Omega^{-1}. \qquad (1.5.4)$$

Let $P_{\widetilde{M}}$ be the projection from $\mathcal{K}$ to $\widetilde{M}$ and

$$\widetilde{S}(z, w) \stackrel{\text{def}}{=} P_{\widetilde{M}}(\overline{w} - \widetilde{S}^*)^{-1}(z - \widetilde{S})^{-1}|_{\widetilde{M}} \quad \text{for } z, w \in \rho(\widetilde{S}).$$

Let

$$\widetilde{Q}(z, w) \stackrel{\text{def}}{=} (\overline{w} - \widetilde{\Lambda}^*)(z - \widetilde{\Lambda}) - \widetilde{C}.$$

Applying Lemma 1.3.2 to $\widetilde{S}$, we have

$$\widetilde{S}(z, w) = \widetilde{Q}(z, w)^{-1}, \quad z, w \in \rho(\widetilde{S}).$$

Let $\widetilde{e}(A) \stackrel{\text{def}}{=} P_{\widetilde{M}} E(\overline{A})|_{\widetilde{M}}$ where $\overline{A} = \{\overline{z} : z \in A\}$, A is a Borel set in $\sigma(N^*)$, and $E(\cdot)$ is the spectral measure of N. Then

$$\widetilde{e}(A) = \Omega e(\overline{A}) \Omega^{-1}.$$

Let $\widetilde{U}$ be the extension of the mapping

$$(N^* - \lambda)^{-1} a \mapsto ((\cdot) - \lambda)^{-1} a, \quad a \in \widetilde{M}.$$

Thus $\widetilde{U}$ is a unitary operator from $\mathcal{K}$ onto $L^2(\widetilde{e})$. In this case, we have

$$(\widetilde{U}\widetilde{S}\widetilde{U}^{-1} f)(u) = u f(u), \quad f \in \widetilde{U}\mathcal{H}, u \in \sigma(N^*). \tag{1.5.5}$$

We will have the similar identities related to $\widetilde{S}, N^*, \ldots$ as those for $S, N, \ldots$ in Section 1.3. Then we have a similar analytic model for $\widetilde{S}$ as in (1.5.5).

Therefore for any f in $\mathcal{K}$ and $z \in \rho(N^*)$, we may define the value $\widetilde{f}(z)$ as

$$\widetilde{f}(z) = \int_{\sigma(N^*)} \frac{u - \widetilde{\Lambda}}{u - z} \widetilde{e}(du) f(u).$$

Let

$$\widetilde{\mu}(z) = \int_{\sigma(N^*)} \frac{u - \widetilde{\Lambda}}{u - z} \widetilde{e}(du)$$

be the mosaic of the operator $\widetilde{S}$.

Theorem 1.5.1. *For* $z \in \rho(N^*)$,

$$\widetilde{\mu}(z) = \Omega C^{1/2} \mu(\overline{z})^* C^{-1/2} \Omega^{-1}. \tag{1.5.6}$$

Proof. For $a, b \in C^{1/2} M$, we have

$$\left(\int_{\sigma(N^*)} \frac{u - \widetilde{\Lambda}}{u - z} \widetilde{e}(du) \Omega a, \Omega b \right) = (a, b) + \left(\int_{\sigma(N^*)} \frac{\widetilde{e}(du)}{u - z} \Omega a, \Omega f \right) \tag{1.5.7}$$

where $f = (\overline{z} - C^{1/2} \Lambda C^{-1/2}) b \in C^{1/2} M$, since $(\overline{z} - \widetilde{\Lambda})^* \Omega b = \Omega f$ by (1.5.4).

On the other hand

$$\int_{\sigma(N^*)} \left(\frac{\widetilde{e}(du)}{u-z}\Omega a, \Omega f\right) = \left((N^* - z)^{-1}(N^* - \Lambda^*)C^{-1/2}a, (N^* - \Lambda^*)C^{-1/2}f\right)$$

$$= \int_{\sigma(N)} \frac{\left((u-\Lambda)e(du)(\overline{u} - \Lambda^*)C^{-1/2}a, C^{-1/2}f\right)}{\overline{u} - z}$$

$$= \int_{\sigma(N)} \frac{\left(e(du)\left((\overline{u} - \Lambda^*)(u - \Lambda) + (\overline{u} - \Lambda^*)\Lambda\right)C^{-1/2}a, C^{-1/2}f\right)}{\overline{u} - z}$$

$$\quad - \left(\Lambda\mu(\overline{z})^*C^{-1/2}a, C^{-1/2}f\right)$$

$$= \int_{\sigma(N)} \frac{\left(e(du)C^{1/2}a, C^{-1/2}f\right)}{\overline{u} - z} + \left((\mu(\overline{z})^*\Lambda - \Lambda\mu(\overline{z})^*)C^{-1/2}a, C^{-1/2}f\right)$$

$$= \left(((\mu(\overline{z})^* - I)(z - \Lambda^*)^{-1}C + \mu(\overline{z})^*\Lambda - \Lambda\mu(\overline{z})^*)C^{-1/2}a, C^{-1/2}f\right)$$

$$= \left((z - \Lambda^*)^{-1}C(\mu(\overline{z})^* - I)C^{-1/2}a, C^{-1/2}f\right)$$

$$= \left(C^{1/2}(\mu(\overline{z})^* - I)C^{-1/2}a, b\right), \tag{1.5.8}$$

since $\mu(\overline{z})^*((z - \Lambda^*)^{-1}C^{-1} + \Lambda) = ((z - \Lambda^*)C^{-1} + \Lambda)\mu(\overline{z})^*$ and

$$C^{-1/2}f = (\overline{z} - \Lambda)C^{-1/2}b.$$

From (1.5.7) and (1.5.8), it follows (1.5.6). $\qquad\square$

Corollary 1.5.2. *If $f = g(N^*)\Omega a$ for $a \in M$, where $g(\cdot)$ is a rational function with poles in $\rho(\widetilde{S})$. Then*

$$\widetilde{f}(z) = g(z)\widetilde{\mu}(z)a.$$

Similar to (1.4.5) we may define another M-valued analytic function $f^*(\cdot)$ on $\rho(N^*)$ by

$$f^*(z) = \int_{\sigma(N)} \frac{e(du)f(u)}{\overline{u} - z}, \quad z \in \rho(N^*)$$

for $f \in \mathcal{K}$, in the contest of the analytic model for S, i.e.

$$f^*(z) = P_M(N^* - z)^{-1}f, \quad z \in \rho(N^*).$$

This $f^*(\cdot)$ is a dual representation of the vector f.

Theorem 1.5.3. *If $f \in \widetilde{\mathcal{H}}$, then*

$$f^*(z) \in \mu(\overline{z})^*M, \quad \text{for } z \in \rho(N^*), \tag{1.5.9}$$

$$((\widetilde{S} - \lambda)^{-1}f)^*(z) = (z - \lambda)^{-1}f^*(z), \quad \text{for } z \in \rho(N^*), \tag{1.5.10}$$

and $\lambda \in \rho(\widetilde{S})$, *and*

$$(\widetilde{S}^* f)^*(z) = R(\overline{z})^* f^*(z) + \mu(\overline{z})^*(z - \Lambda^*)^{-1} \lim_{\xi \to \infty} \xi f(\xi) \qquad (1.5.11)$$

for $z \in \rho(N^*) \setminus \sigma(\Lambda^*)$, *where* $R(\cdot)$ *is defined in* (1.4.31).
   If $f \in \mathcal{H}$, *then*

$$f^*(z) \in (I - \mu(\overline{z})^*)M, \quad for \ z \in \rho(N^*), \qquad (1.5.12)$$

$$(S^* f)^*(z) = z f^*(z) + (I - \mu(\overline{z})^*) f(\Lambda), \quad for \ z \in \rho(N^*), \qquad (1.5.13)$$

$$((S^* - \lambda)^{-1} f)^*(z) = (z - \lambda)^{-1} f^*(z) - (z - \lambda)^{-1}(I - \mu^*(\overline{z})) f^*(\lambda), \ (1.5.14)$$

for $z \in \rho(N^*), \lambda \in \rho(S^*)$ *but* $\lambda \neq z$. *Besides*,

$$(S f)^*(z) = R(\overline{z})^* f^*(z), \quad for \ z \in \rho(N^*) \setminus \sigma(\Lambda^*). \qquad (1.5.15)$$

Proof. Let

$$f_{\xi,a}(u) \overset{\text{def}}{=} \frac{\overline{u} - \Lambda^*}{\overline{u} - \xi} a, \ a \in M, \xi \in \rho(\widetilde{S}) = \rho(S^*), \qquad (1.5.16)$$

then

$$f^*_{\xi,a}(z) = \int_{\sigma(N)} e(du)\left(\frac{\overline{u} - \Lambda^*}{\overline{u} - z} - \frac{\overline{u} - \Lambda^*}{\overline{u} - \xi}\right)\frac{a}{z - \xi} = \frac{1}{z - \xi}(\mu(\overline{z})^* - \mu(\overline{\xi})^*)a$$

$$= \mu(\overline{z})^*(z - \xi)^{-1} a \in \mu(\overline{z})^* M, \qquad (1.5.17)$$

which proves (1.5.9). *From*

$$(\widetilde{S} - \lambda)^{-1} f_{\xi,a} = \frac{(\overline{u} - \Lambda^*)a}{(\overline{u} - \lambda)(\overline{u} - \xi)} = \frac{1}{\lambda - \xi}\left(\frac{(\overline{u} - \Lambda^*)a}{\overline{u} - \lambda} - \frac{(\overline{u} - \Lambda^*)a}{\overline{u} - \xi}\right)$$

and (1.5.17) *we have*

$$((\widetilde{S} - \lambda)^{-1} f_{\xi,a})^*(z) = \frac{1}{\lambda - \xi}\mu(\overline{z})^*\left(\frac{1}{z - \lambda} - \frac{1}{z - \xi}\right)a$$

$$= \frac{1}{z - \lambda}\mu(\overline{z})^*(z - \xi)^{-1} a$$

which proves (1.5.10).
   For $f \in \mathcal{K}$,

$$\int_{\sigma(N)} (v - \Lambda)e(dv)f(v) = -\lim_{\xi \to \infty} \xi \int_{\sigma(N)} \frac{v - \Lambda}{v - \xi}e(du)f(v)$$

$$= -\lim_{\xi \to \infty} \xi f(\xi).$$

Therefore (1.3.33) may be rewritten as

$$(\widetilde{S}^* f)(u) = uf(u) + \lim_{\xi \to \infty} \xi f(\xi), \quad f \in \widetilde{\mathcal{H}}$$

in the analytic model of S. Thus

$$
\begin{aligned}
(\widetilde{S}^* f)^*(z) &= \int \frac{e(du)}{\overline{u} - z} (uf(u) + \lim_{\xi \to \infty} f(\xi)) \\
&= \int \frac{(u - \Lambda)e(du)f(u)}{\overline{u} - z} + \Lambda f^*(z) + \int \frac{e(du)}{\overline{u} - z} \lim_{\xi \to \infty} \xi f(\xi).
\end{aligned}
$$
$$(1.5.18)$$

On the other hand

$$\int \frac{e(du)}{\overline{u} - z} = (I - \mu(\overline{z})^*)(\Lambda^* - z)^{-1} \quad \text{for } z \in \rho(\Lambda^*) \cap \rho(N^*), \qquad (1.5.19)$$

and

$$
\begin{aligned}
Cf^*(z) &= \int \frac{(\overline{u} - \Lambda^*)(u - \Lambda)}{\overline{u} - z} e(du)f(u) \\
&= (z - \Lambda^*) \int \frac{u - \Lambda}{\overline{u} - z} e(du)f(u) + \int (u - \Lambda)e(du)f(u) \\
&= (z - \Lambda^*) \int \frac{u - \Lambda}{\overline{u} - z} e(du)f(u) - \lim_{\xi \to \infty} \xi f(\xi). \qquad (1.5.20)
\end{aligned}
$$

From (1.5.18), (1.5.19) and (1.5.20), it follows (1.5.11).

From Lemma 1.4.1

$$\frac{\mu(\overline{z})}{(\cdot) - \overline{z}} b \in \widetilde{\mathcal{H}} \quad \text{for } z \in \rho(N^*), b \in M.$$

Therefore

$$\int \left(e(du)f(u), \frac{\mu(\overline{z})b}{u - \overline{z}} \right) = (f, \frac{\mu(\overline{z})b}{(\cdot) - \overline{z}}) = 0, \quad f \in \mathcal{H}, z \in \rho(N^*),$$

which proves (1.5.12). By a direct calculation,

$$
\begin{aligned}
(S^* f)^*(z) &= \int \frac{e(du)}{\overline{u} - z} \left(\overline{u}(f(u) - f(\Lambda)) + \Lambda^* f(\Lambda) \right) \\
&= \int e(du)(f(u) - f(\Lambda)) + z \int \frac{e(du)}{\overline{u} - z} f(u) + \int \frac{e(du)}{\overline{u} - z}(-z + \Lambda^*)f(\Lambda)
\end{aligned}
$$

which is equivalent to (1.5.13). To prove (1.5.15), notice that

$$(Sf)^*(z) = \int \frac{e(du)uf(u)}{\overline{u} - z} = \int \frac{(u - \Lambda)e(du)f(u)}{\overline{u} - z} + \Lambda f^*(z). \qquad (1.5.21)$$

On the other hand, for $f \in \mathcal{H}$, $\lim \xi f(\xi) = -\int (u - \Lambda)e(du)f(u) = 0$. From (1.5.20) and (1.5.21), it follows (1.5.15).

To prove (1.5.14), let $g = (S^* - \lambda)^{-1}f$, i.e., $f = (S^* - \lambda)g$. Applying (1.5.13) to g, we have

$$f^*(z) = ((S^* - \lambda)g)^*(z) = (z - \lambda)g^*(z) + (I - \mu(\overline{z})^*)g(\Lambda),$$

i.e.

$$((S^* - \lambda)^{-1}f)^*(z) = \frac{1}{z - \lambda}(f^*(z) - (I - \mu(\overline{z})^*)g(\Lambda)). \qquad (1.5.22)$$

But

$$f(u) = (\overline{u} - \lambda)g(u) - (\overline{u} - \Lambda^*)g(\Lambda).$$

Therefore

$$\frac{f(u)}{\overline{u} - \lambda} = g(u) - \frac{\overline{u} - \Lambda^*}{\overline{u} - \lambda}g(\Lambda).$$

Thus

$$f^*(\lambda) = \int \frac{e(du)}{u - \lambda}f(u) = g(\Lambda) - \mu(\overline{\lambda})^*g(\Lambda).$$

However $\mu(\overline{\lambda})^* = 0$ for $\lambda \in \rho(S^*)$. We have $g(\Lambda) = f^*(\lambda)$ which proves (1.5.14) by (1.5.22). □

For a pure subnormal operator S on $\mathcal{H}$ and $\widetilde{S}$ defined in (1.3.2), let

$$\tau_p(\widetilde{S}) = \{z \in \mathbb{C} : (\widetilde{S} - z)\widetilde{\mathcal{H}} \cap [\widetilde{S}^*, \widetilde{S}]^{1/2}\widetilde{\mathcal{H}} \neq \{0\}\}.$$

Let $\sigma_p(A)$ be the point spectrum of A.

Theorem 1.5.4. *Let S be a pure subnormal operator on $\mathcal{H}$ with m.n.e. N on $\mathcal{K} \supset \mathcal{H}$. Let $\mu(z), z \in \rho(N)$ be the mosaic of S. Then*

$$\sigma_p(S^*) \cap \rho(N^*) = \{\overline{z} \in \rho(N^*), \mu(z) \neq 0\}. \qquad (1.5.23)$$

For $\overline{z} \in \sigma_p(S^*) \cap \rho(N^*)$, let $R_z^* = \ker(S^* - \overline{z})$ be the eigenspace of S^* corresponding to the eigenvalue $\overline{z}$, then

$$(S - z)\mathcal{H} = R_z^{*\perp}. \qquad (1.5.24)$$

Moreover,

$$\tau_p(\widetilde{S}) \cap \rho(N) = \{z \in \rho(N), \mu(z) \neq I\}. \qquad (1.5.25)$$

Let

$$\widetilde{R}_z^* \stackrel{def}{=} (N^* - \overline{z})^{-1}(N^* - \Lambda^*)(I - \mu(z)^*)M. \qquad (1.5.26)$$

Then

$$\widetilde{R}_z^* = (\widetilde{S} - \overline{z})^{-1}[\widetilde{S}^*, \widetilde{S}]^{1/2}\widetilde{\mathcal{H}} \qquad (1.5.27)$$

and

$$R_z^* \oplus \widetilde{R}_z^* = (N^* - z)^{-1}(N^* - \Lambda^*)M. \qquad (1.5.28)$$

Proof. Let $\overline{z} \in \sigma_p(S^*) \cap \rho(N^*)$, from Lemma 1.4.1 (vi), we have $\mu(z) \neq 0$ and

$$R_z^* = (N^* - \overline{z})^{-1}(N^* - \Lambda^*)\mu(z)^* M. \qquad (1.5.29)$$

It is obvious that $R_z^{*\perp} = \mathrm{cl}(S - z)\mathcal{H}$. Suppose $h \in \mathrm{cl}(S - z)\mathcal{H}$. Then from $\mu(z)h(z) = h(z)$, it follows that

$$\|h(z)\|^2 = \big(h(z), \mu(z)^*\alpha\big) = \int \big(e(du)h(u), (\overline{u} - \overline{z})^{-1}(\overline{u} - \Lambda^*)\mu(z)^*\alpha\big) = 0,$$

where $\alpha = h(z)$. By (1.4.4), $f(\cdot) = h(\cdot)\big((\cdot) - z\big)^{-1} \in \mathcal{H}$. Thus $h \in (S - z)\mathcal{H}$ which proves (1.5.24).

If $z \in \rho(N)$ and $\mu(z) \neq 0$, then R_z^* in (1.5.29) is not $\{0\}$. Therefore $\overline{z} \in \sigma_P(S^*) \cap \rho(N^*)$ which proves (1.5.23).

From (1.5.2) and (1.5.3), it follows that

$$[\widetilde{S}^*, \widetilde{S}]^{1/2}\widetilde{\mathcal{H}} = \widetilde{C}^{1/2}\widetilde{M} = \Omega C^{1/2}M.$$

Thus

$$[\widetilde{S}^*, \widetilde{S}]^{1/2}\widetilde{\mathcal{H}} = (N^* - \Lambda^*)M. \qquad (1.5.30)$$

If $\overline{z} \in \tau_p(\widetilde{S}) \cap \rho(N)$ and $f \in \widetilde{\mathcal{H}}$ satisfy $(\widetilde{S} - \overline{z})f \in [\widetilde{S}^*, \widetilde{S}]^{1/2}\widetilde{\mathcal{H}}$, then there is a vector $a \in M, a \neq 0$ such that

$$(\widetilde{S} - \overline{z})f = (N^* - \Lambda^*)a, \qquad (1.5.31)$$

by (1.5.30). However $\widetilde{S}f = N^*f$. Therefore

$$f = (N^* - \overline{z})^{-1}(N^* - \Lambda^*)a.$$

Thus $a \in \widetilde{M}_z^* = (I - \mu(z)^*)M$. It means $\mu(z) \neq I$. Hence

$$\tau_p(\widetilde{S}) \cap \rho(N^*) \subset \{\overline{z} \in \rho(N^*) : \mu(z) \neq I\}. \qquad (1.5.32)$$

Besides, if $z \in \rho(N), \mu(z) \neq I$, then there is a non-zero vector $a \in (I - \mu(z)^*)M = \widetilde{M}_z^*$ such that the vector f defined in (1.5.31) is not zero and $f \in \widetilde{\mathcal{H}}$. Therefore

$$(\widetilde{S} - \overline{z})f = (N^* - \Lambda^*)a \in (\widetilde{S} - \overline{z})\widetilde{\mathcal{H}} \cap [\widetilde{S}^*, \widetilde{S}]^{1/2}\widetilde{\mathcal{H}}.$$

Therefore the opposite inclusion relation for (1.5.32) holds good, which proves (1.5.25).

It is obvious that

$$R_z^* + \widetilde{R_z}^* = (N^* - \overline{z})^{-1}[\widetilde{S}^*, \widetilde{S}]^{1/2}\widetilde{\mathcal{H}},$$

since $M = M_z^* + \widetilde{M}_z^*$. The orthogonality $R_z^* \perp \widetilde{R_z}^*$ comes from $R_z^* \subset \mathcal{H}$ and $\widetilde{R_z}^* \subset \overline{\mathcal{H}}$. Thus (1.5.28) is proved. $\qquad \square$

For a pure subnormal operator S, let
$$\nu_p(S) = \{z \in \mathbb{C} : (S - z)\mathcal{H} \cap M_S \neq \{0\}\}.$$
For a set $F \subset \mathbb{C}$, let $F^* = \{\overline{z} \in \mathbb{C} : z \in F\}$. Similar to the proof of the Theorem 1.5.4, we may prove the following:

Theorem 1.5.5. *Under the condition of Theorem 1.5.4,*
$$\sigma_p(\widetilde{S}^*) \cap \rho(N) = \sigma_p(S^*)^* \cap \rho(N), \tag{1.5.33}$$
and
$$\nu_p(S) \cap \rho(N) = \tau_p(\widetilde{S})^* \cap \rho(N) = \{z \in \rho(N) : \mu(z) \neq I\}. \tag{1.5.34}$$
Let $R_z \overset{def}{=} (S - z)^{-1}M$ and $\widetilde{R}_z \overset{def}{=} \ker(\widetilde{S} - z)$, then
$$R_z = (N - z)^{-1}(I - \mu(z))M \tag{1.5.35}$$
$$\widetilde{R}_z = (N - z)^{-1}\mu(z)M$$
and
$$R_z \oplus \widetilde{R}_z = (N - z)^{-1}M. \tag{1.5.36}$$

Corollary 1.5.6. *Under the condition for Theorem 1.5.4, for $z \in \sigma(S) \cap \rho(N)$,*
$$\dim(\ker(S^* - \overline{z})) = \operatorname{rank}(\mu(z)), \tag{1.5.37}$$
and
$$\dim(\ker(\widetilde{S}^* - z)) = \operatorname{rank}(I - \mu(z)). \tag{1.5.38}$$
Hence
$$\dim(\ker(S^* - \overline{z})) + \dim(\ker(\widetilde{S}^* - z)) = \dim M.$$

Proof. Applying (1.5.23) to operator $\widetilde{S}^*$, we have
$$\sigma_p(\widetilde{S}^*) \cap \rho(N) = \{z \in \rho(N) : \widetilde{\mu}(z) \neq 0\}$$
$$= \{z \in \rho(N) : \mu(\overline{z})^* \neq 0\} \tag{1.5.39}$$
by (1.5.6). From (1.5.23) and (1.5.28), it follows (1.5.33).

Suppose $z \in \nu_p(S) \cap \rho(N)$, there is a vector $f \in \mathcal{H}$ such that
$$(S - z)f = a$$
for some non-zero $a \in M$, therefore $a \in \widetilde{M}_z = (I - \mu(z))M$ by Theorem 1.4.3. Thus $\mu(z) \neq I$, i.e.
$$\nu_p(S) \cap \rho(N) \subset \{z \in \rho(N) : \mu(z) \neq I\}.$$
The opposite inclusion relation can be proved similarly, which proves (1.5.34) by (1.5.25). The rest of this theorem can be proved similarly. $\quad\square$

Theorem 1.5.7. *Let S be a pure subnormal operator with m.n.e. N and dual operator $\widetilde{S}$. Then*

$$\rho(S) = \{z \in \rho(N) : \mu(z) = 0\}, \tag{1.5.40}$$

$$\sigma(S) = \sigma(N) \cup \sigma_p(S^*)^*, \tag{1.5.41}$$

and

$$\sigma(\widetilde{S}^*) = \sigma(S). \tag{1.5.42}$$

Proof. (1.5.40) is the consequence of Lemma 1.4.1 (iii) and (iv).

It is obvious that

$$\sigma_p(S^*)^* \subset \sigma(S).$$

From Theorem 1.1.2, we have

$$\sigma(N) \cup \sigma_p(S^*)^* \subset \sigma(S).$$

If $z \in \sigma(S) \cap \rho(N)$, then $\mu(z) \neq 0$ and $z \in \rho(N) \cap \sigma_p(S^*)^*$. Hence

$$\sigma(S) \subset \sigma(N) \cup \sigma_p(S^*)^*,$$

which proves (1.5.41).

Finally, by the equality (1.5.41) for $\widetilde{S}$ and (1.5.33) we have

$$\begin{aligned}
\sigma(\widetilde{S}) &= \sigma(N^*) \cup \left(\sigma_p(\widetilde{S}^*)^* \cap \rho(N^*)\right) \\
&= \sigma(N^*) \cup \left(\sigma_p(S^*)^* \cap \rho(N)\right)^* \\
&= \left(\sigma(N) \cup \sigma(S)\right)^* = \sigma(S)^*.
\end{aligned}$$

$\square$

The following theorem is a well-known theorem on the spectral of subnormal operators. It has been proved without the using of mosaic.

Theorem 1.5.8. *Let S be a pure subnormal operator. Then the boundary of $\sigma(S)$ must be a subset of $\sigma(N)$.*

Proof. Suppose on contrary, there is a boundary point z of $\sigma(S)$ in $\rho(N)$. Then $\mu(\cdot)$ is regular at z. There is a sequence $\{z_n\} \subset \rho(S)$ such that $z_n \to z$. From $\mu(z_n) = 0$, we have $\mu(z) = 0$. Therefore $z \in \rho(S)$ by (1.5.40) which contradicts the fact that z is a boundary point of $\sigma(S)$, and which proves the theorem. $\square$

1.6   Subnormal operators with compact self-commutator

Let $L_0(\mathcal{H})$ be the ideal of all compact operators on the Hilbert space $\mathcal{H}$. In this section, S denotes a pure subnormal operator on $\mathcal{H}$ with m.n.e. N on $\mathcal{K} \supset \mathcal{H}$. Let $\mu(z), z \in \rho(N)$ be the mosaic of the operator S.

Theorem 1.6.1. *If $[S^*, S] \in L_0(\mathcal{H})$, then*

$$\operatorname{rank} \mu(z) < +\infty, \quad z \in \rho(N).$$

Proof. Let $C = [S^*, S]|_M, M = \operatorname{cl}([S^*, S]\mathcal{H})$. We only have to consider the case that $\dim M = \infty$. Since $C \in L_0(M)$, there is a sequence of projections $\{P_n\}$ on $L(M)$ for C such that $\operatorname{rank} P_n < +\infty$,

$$0 \le P_1 \le P_2 \le \cdots \le P_n \le \cdots$$

and $\lim_{n \to \infty} \lambda_n = 0$ where $\lambda_n = \|C^{1/2}(I - P_n)\|$. From (1.4.18), we have

$$\int \left(e(du)(\overline{u} - \Lambda^*)x, (\overline{u} - \Lambda^*)x\right) = \|C^{1/2}x\|^2 \le \lambda_n^2 \|x\|^2, \quad \text{for } x \perp P_n M.$$

By Cauchy inequality

$$
\begin{aligned}
|(\mu(z)^* x, y)|^2 &= |\int \left(e(du)(\overline{u} - \Lambda^*)x, (u - z)^{-1}y\right)|^2 \\
&\le \int \left(e(du)(\overline{u} - \Lambda^*)x, (\overline{u} - \Lambda^*)x\right) \int (e(du)y, y)|u - z|^{-2} \\
&\le (\operatorname{dist}(z, \sigma(N)))^{-2} \lambda_n^2 \|x\|^2 \|y\|^2,
\end{aligned}
$$

for $y \in M, x \perp P_n M$ and $z \in \rho(N)$. Hence

$$\|\mu(z)^* x\| \le (\operatorname{dist}(z, \sigma(N)))^{-1} \lambda_n \|x\|, \quad \text{for } x \perp P_n M,$$

which implies that

$$\lim_{n \to \infty} \|\mu(z)^* - \mu(z)^* P_n\| = 0, \quad \text{for } z \in \rho(N).$$

Therefore $\mu(z)^*$ and hence $\mu(z)$ are compact. However $\mu(z)^2 = \mu(z)$. Let F be any bounded closed set of $M_z = \mu(z)M$. Then F must be compact, since the restriction of $\mu(z)$ on F is identity and $\mu(z)$ is compact. Therefore M_z must be finite dimensional, which proves the theorem. $\qquad\square$

Let A be an operator on a Hilbert space $\mathcal{H}$ and $z \in \mathbb{C}$. If

$$\operatorname{range}(A - z) \quad \text{and} \quad \operatorname{range}(A^* - \overline{z})$$

are closed, $\dim(\ker(A - z)) < \infty$ and $\dim(\ker(A^* - \overline{z})) < \infty$. Then z is said to be a *Fredholm point* of A. If zero is a Fredholm point of A, then

$$\operatorname{index} A \stackrel{\text{def}}{=} \dim(\ker(A)) - \dim(\ker(A^*)).$$

Lemma 1.6.2. *Let S be a pure subnormal operator in $\mathcal{H}$ with m.n.e. N, and $z \in \rho(N)$. Then*

$$(S^* - \bar{z})\mathcal{H} = \mathcal{H}.$$

Proof. Without loss of generality we may assume that $z = 0$, since $S - z$ is also a pure subnormal operator. Then $\ker(S) = \{0\}$ implies that

$$\text{cl range } S^* = \mathcal{H}. \tag{1.6.1}$$

For $h \in \mathcal{H}$, by (1.6.1), there is a sequence $\{h_n\} \subset S^*\mathcal{H}$ such that

$$\lim_{n \to \infty} \|h_n - h\| = 0. \tag{1.6.2}$$

Let $f_n \in \mathcal{H}$ satisfying $h_n = S^* f_n$. Then in the analytic model by (1.3.24), we have

$$h_n(u) = \bar{u} f_n(u) - (\bar{u} - \Lambda^*) f_n(\Lambda), \quad u \in \sigma(N).$$

Therefore the dual representation $h_n^*(0)$ of $h_n(\cdot)$ is

$$h_n^*(0) = \int \frac{e(du) h_n(u)}{\bar{u}} = (I - \mu(0)^*) f_n(\Lambda).$$

Define $q_n(u) = f_n(u) - (\bar{u} - \Lambda^*)\bar{u}^{-1}\mu(0)^* f_n(\Lambda)$. Then $q_n \in \mathcal{H}$, since $f_n \in \mathcal{H}$ and $(\bar{u} - \Lambda^*)\bar{u}^{-1}\mu(0)^* f_n(\Lambda)$ is an eigenvector of S^* corresponding to the eigenvalue 0. Besides

$$q_n(\Lambda) = (I - \mu(0)^*) f_n(\Lambda) = h_n^*(0).$$

Therefore

$$h_n = S^* q_n. \tag{1.6.3}$$

It is easy to see that

$$q_n(u) = (h_n(u) + (\bar{u} - \Lambda^*) h_n^*(0))\bar{u}^{-1}, \quad u \in \sigma(N).$$

Let

$$q(u) = (h(u) + (\bar{u} - \Lambda^*) h^*(0))\bar{u}^{-1}, \quad u \in \sigma(N).$$

From (1.6.2) and $0 \in \rho(N)$, we may conclude that $\|h_n^*(0) - h^*(0)\| \to 0$, $q \in \mathcal{H}$ and

$$\lim_{n \to \infty} \|q_n - q\| = 0. \tag{1.6.4}$$

From (1.6.3) and (1.6.4), we have $h = S^* q$. Thus $\mathcal{H} = S^*\mathcal{H}$. $\qquad \square$

Theorem 1.6.3. *Let S be a pure subnormal operator on $\mathcal{H}$ with m.n.e. N. Let $z \in \rho(N)$. Then $\operatorname{rank}\mu(z) < \infty$ iff z is a Fredholm point of S and*

$$\operatorname{rank}\mu(z) = \operatorname{index}(S^* - \overline{z}). \tag{1.6.5}$$

Proof. It is evident that $\ker(S - z) = \{0\}$ by Proposition 1.2.6, since S is pure and hyponormal. By (1.5.24), $\operatorname{range}(S - z)$ is closed. From Lemma 1.6.2, $\operatorname{range}(S^* - \overline{z})$ is also closed. Thus z is a Fredholm point of S iff $\dim \ker(S^* - \overline{z}) < \infty$. From (1.5.37), it follows the theorem. □

Let $L^1(\mathcal{H})$ be the set of all operator A on a Hilbert space $\mathcal{H}$ satisfying the following condition: for every pair of orthonormal basis $\{e_j\}$ and $\{e'_j\}$ of $\mathcal{H}$,

$$\sum_j |(Ae_j, e'_j)| < \infty.$$

For $A \in L^1(M)$, we define the trace of A as

$$\operatorname{tr}(A) \overset{\text{def}}{=} \sum_{j=1}^{\infty}(Ae_j, e_j) \tag{1.6.6}$$

where $\{e_j\}$ is any basis for $\mathcal{H}$. This trace is independent of the choice of the basis $\{e_j\}$. If $A \in L^1(\mathcal{H})$ and $B \in L(\mathcal{H})$ then AB and BA are in $L^1(M)$ and

$$\operatorname{tr}(AB) = \operatorname{tr}(BA). \tag{1.6.7}$$

Lemma 1.6.4. *Let S be a pure subnormal operator on $\mathcal{H}$ with m.n.e. N. Assume that $[S^*, S]|_M \in L^1(M)$. If $z, w, v \in \rho(S)$, then*

$$[(\overline{w} - S^*)^{-1}, (z - S)^{-1}](\overline{v} - S^*)^{-1} \in L^1(\mathcal{H}),$$

and

$$\operatorname{tr}([(\overline{w} - S^*)^{-1}, (z - S)^{-1}](\overline{v} - S^*)^{-1}) = \operatorname{tr}\left(\int \frac{e(du)(\overline{u} - \Lambda^*)}{(\overline{w} - \overline{u})^2(\overline{v} - \overline{u})(z - u)}\right). \tag{1.6.8}$$

Proof. Let $A = \overline{w} - S^*, B = z - S$ and $L = \overline{v} - S^*, C = [S^*, S]|_M$, where $M = \operatorname{cl}[S^*, S]\mathcal{H}$. Then $C \in L^1(M)$ and it is easy to calculate the

$$[A^{-1}, B^{-1}]L^{-1} = A^{-1}B^{-1}CP_M B^{-1}A^{-1}L^{-1} \in L^1(\mathcal{H}).$$

Therefore

$$\operatorname{tr}([A^{-1}, B^{-1}]L^{-1}) = \operatorname{tr}(L^{-1}A^{-2}B^{-1}CP_M B^{-1}).$$

However

$$P_M B^{-1} = (z - \Lambda)^{-1} P_M, \tag{1.6.9}$$

and for $a \in M$

$$P_M L^{-1} A^{-2} B^{-1} a = \int \frac{e(du)a}{(\overline{w} - \overline{u})^2 (\overline{v} - \overline{u})(z - u)}.$$

Therefore

$$\text{tr}([A^{-1}, B^{-1}]L^{-1}) = \text{tr}\left(\int \frac{e(du)C(z - \Lambda)^{-1}}{(\overline{w} - \overline{u})^2 (\overline{v} - \overline{u})(z - u)}\right). \tag{1.6.10}$$

From

$$e(du)C(z - \Lambda)^{-1} = e(du)(\overline{u} - \Lambda^*)(u - \Lambda)(z - \Lambda)^{-1}$$
$$= e(du)(\overline{u} - \Lambda^*) + e(du)(\overline{u} - \Lambda^*)(u - z)(z - \Lambda)^{-1},$$

$$\int \frac{e(du)(\overline{u} - \Lambda^*)}{(\overline{w} - \overline{u})^2 (\overline{v} - \overline{u})} = \frac{d}{dw}\frac{1}{\overline{v} - \overline{w}}(\mu(w)^* - \mu(v)^*) = 0,$$

and (1.6.10), it follows (1.6.8). $\qquad\square$

Lemma 1.6.5. *Under the condition of Lemma 1.6.4, if* $[S^*, S]^{1/2} \in L^1(\mathcal{H})$ *then for every bounded Baire function* f *on* $\sigma(N)$

$$\int f(u)e(du)(\overline{u} - \Lambda^*) \in L^1(\mathcal{H}), \tag{1.6.11}$$

and there is a complex measure $v_1(\cdot)$ *on* $(\sigma(N), \mathcal{B})$ *with finite total variation such that*

$$\text{tr}(\int f(u)e(du)(\overline{u} - \Lambda^*)) = \int f(u)v_1(du). \tag{1.6.12}$$

This measure $v_1(\cdot)$ is formally denoted by $\text{tr}(e(du)(\overline{u} - \Lambda^*))$. Similarly we may define the complex measure

$$\text{tr}((u - \Lambda)e(du)) = \overline{\text{tr}(e(du)(\overline{u} - \Lambda^*))}.$$

Proof. Notice that for $a, b \in M$, we have

$$\int |(e(du)(\overline{u} - \Lambda^*)a, b)| \le \left(\int (u - \Lambda)e(du)(\overline{u} - \Lambda^*)a, a\right)^{1/2} \int (e(du)b, b)^{1/2}$$
$$= \|C^{1/2}a\|\|b\|.$$

Therefore, for orthonormal basis $\{e_i\}, \{e_i'\}$ for M, we have

$$\sum |(\int f(u)e(du)(\overline{u} - \Lambda^*)e_i, e_i')| \le \text{ess sup} |f(\cdot)| \text{tr}(C^{1/2}), \tag{1.6.13}$$

which proves (1.6.11). Define a complex set function

$$v_1(F) = \text{tr}\left(\int_F e(du)(\overline{u} - \Lambda^*)\right)$$

for $F \in \mathcal{B}$, the family of all Borel sets in $\sigma(N)$.

It is easy to see that for any sequence of mutually disjoint Borel sets F_j in $\mathcal{B}$, define

$$f(u) = \sum 1_{F_j}(u) \operatorname{sign} v_1(F_i)$$

where $1_F(\cdot)$ is the characteristic function of the set F and

$$\operatorname{sign} c = \begin{cases} \overline{c}/|c| & \text{if } c \neq 0, \\ 0 & \text{if } c = 0, \end{cases}$$

then

$$\sum_j |v_1(F_j)| = \text{tr}\left(\int f(u)e(du)(\overline{u} - \Lambda^*)\right) \leq \text{tr}(C^{1/2})$$

by (1.6.13) which proves the lemma. □

Let σ be a compact set in $\mathbb{C}$. Let $\mathcal{R}(\sigma)$ be the family of rational functions with poles off σ.

$$\mathcal{R}(\sigma) \overset{\text{def}}{=} \bigvee\{\overline{f(u)}h(u) : f, h \in R(\sigma)\}.$$

For any function

$$f(\overline{u}, u) = \sum_{j=1}^{n} \overline{f_j(u)}h_j(u)$$

where $f_i(\cdot)$ and $h_j(\cdot) \in R(\sigma)$, define the following functional calculus (sum of ordering products):

$$f(S^*, S) = \sum_{j=1}^{n} f_j(S)^* h_j(S).$$

Theorem 1.6.6. *Let S be a pure subnormal operator on a Hilbert space $\mathcal{H}$ with m.n.e. N. Assume that $[S^*, S]^{1/2} \in L^1(\mathcal{H})$. Then there is a complex measure with finite total variation*

$$v(du) \overset{\text{def}}{=} i\,\text{tr}((u - \Lambda)e(du))$$

on $\sigma(N)$ such that for $f, h \in \mathcal{R}(\sigma(N))$ and ordering product $f(S^*, S)$, $h(S^*, S)$

$$i\,\text{tr}\left([f(S^*, S), h(S^*, S)]\right) = \int_{\sigma(N)} f(\overline{u}, u)d_v h(\overline{u}, u), \qquad (1.6.14)$$

where

$$d_v h(u, u) = \frac{\partial h(\overline{u}, u)}{\partial \overline{u}}\overline{v(du)} + \frac{\partial h(\overline{u}, u)}{\partial u}v(du).$$

Proof. Let $f(\cdot), h(\cdot), q(\cdot) \in R(\sigma(S))$. From Lemmas 1.6.4 and 1.6.5, we have

$$\text{tr}(i[f(S)^*, h(S)]q(S)^*) = -\int_{\sigma(N)} \overline{q(u)}h(u)\overline{f'(u)}v(du). \qquad (1.6.15)$$

Taking the conjugate of both sides of (1.6.15), it is easy to verify that for $f, h, q \in R(\sigma(S))$, we have

$$\text{tr}(i[f(S)^*, h(S)]q(S)) = \int q(u)\overline{f(u)}h'(u)v(du). \qquad (1.6.16)$$

Therefore, for $f_1, f_2, h_1, h_2 \in R(\sigma(S))$, letting $f_1(S)^*, f_2(S), h_1(S)^*$ and $h_2(S)$ be A_1, A_2, B_1 and B_2 respectively, we have

$$\begin{aligned}
\text{tr}([A_1 A_2, B_1 B_2]) &= \text{tr}(B_1[A_1, B_2]A_2) + \text{tr}(A_1[A_2, B_1]B_2) \\
&= \text{tr}([B_1 A_1, B_2]A_2 - [B_1, B_2]A_1 A_2)) \\
&\quad + \text{tr}(A_1[A_2 B_2, B_1] - A_1 A_2[B_2, B_1]) \\
&= \text{tr}([A_1 B_1, B_2]A_2) + \text{tr}([A_2 B_2, B_1]A_1), \qquad (1.6.17)
\end{aligned}$$

since $\text{tr}([B_1, B_2]A_1 A_2) + \text{tr}(A_1 A_2[B_2, B_1]) = 0$. From (1.6.15), (1.6.16) and (1.6.17), it is easy to see that (1.6.14) holds good for $f(\overline{u}, u) = \overline{f_1(u)}f_2(u)$ and $h(\overline{u}, u) = \overline{h_1(u)}h_2(u)$. Then the general case may be proved by linear operation. $\qquad \square$

Remark: It is well-known that if $C \in L^1(\mathcal{H})$, then for $A, B \in L(\mathcal{H})$, $AC, BC \in L^1(\mathcal{H})$ and $\text{tr}([A, C]) = 0$. Therefore in Theorem 1.6.6, the function calculus $f(S^*, S)$ and $h(S^*, S)$ are independent to the product ordering. For example, if $f(\overline{u}, u) = \overline{f_1(u)}h_1(u)\overline{f_2(u)}h_2(u)$, then in general

$$f_1(S)^* h_1(S) f_2(S)^* h_2(S) \neq h_1(S)f_1(S)^* h_2(S)f_2(S)^*. \qquad (1.6.18)$$

But in (1.6.14), $f(S^*, S)$ may be replaced by either one of the two operators in (1.6.18).

1.7 Complete unitary invariant for some subnormal operators

In this section, we study two special classes of subnormal operators and design some more clear complete unitary invariant for these operators by analytic mode of subnormal operators. An operator A on a Hilbert Space $\mathcal{H}$ is said to be a unilateral shift with multiplicy n, if there is a subspace $M \subset \mathcal{H}$ with $\dim M = n$ such that $\mathcal{H} = \bigoplus_{m=0}^{\infty} \text{cl}A^n M$.

Lemma 1.7.1. *Let S be a pure subnormal operator on a Hilbert space with m.n.e. N and with defect space M. If $\sigma(N) = \mathbb{T}-$ the unit circle, then S is a unilateral shift with multiplicity dim M.*

Proof. For $n \geq 0$, from (1.3.15), we have

$$
\begin{aligned}
0 &= \int_{\sigma(N)} ((\overline{u} - \Lambda^*)(u - \Lambda) - C)u^n e(du) \\
&= \int_{\mathbb{T}} (\overline{u}u^{n+1} - \overline{u}u^n \Lambda)e(du) - C\Lambda^n,
\end{aligned}
\tag{1.7.1}
$$

since $\int_{\sigma(N)} (u - \Lambda)u^n e(du) = 0$. If $n = 0$, then (1.7.1) becomes

$$
I_M - \Lambda\Lambda^* - C = 0,
\tag{1.7.2}
$$

where I_M is the identity on M. If $n \geq 1$, then (1.7.1) implies

$$
C\Lambda^n = 0.
$$

But C is injective, therefore $\Lambda = 0$. From (1.7.2), we have $C = I_M$.

Let $H^2(\mathbb{T}, M)$ be a Hardy space of M-valued, measurable functions $f(\cdot)$ on $\mathbb{T}$ satisfying

$$
\|f\| \overset{\text{def}}{=} \left(\frac{1}{\pi} \int_{\mathbb{T}} \|f(e^{i\theta})\|^2 d\theta \right)^{1/2}.
$$

Let $U_+^{(M)}$ be the operator

$$
(U_+^{(M)} f)(\cdot) = (\cdot)f(\cdot), \qquad f \in H^2(\mathbb{T}, M).
$$

Then it is easy to see that $U_+^{(M)}$ is a unilateral shift with multiplicity dim M and

$$
C_{U_+^{(M)}} = I_M, \qquad \Lambda_{U_+^{(M)}} = 0.
$$

From Theorem 1.2.5, $\{C_{U_+^{(M)}}, \Lambda_{U_+^{(M)}}\}$ is a complete unitary invariant. But $C_{U_+^{(M)}} = C$ and $\Lambda_{U_+^{(M)}} = \Lambda$, therefore S is unitarily equivalent to $U_+^{(M)}$ i.e. S is also a unilateral shift with multiplicity dim M. $\square$

Let γ be an analytic Jordan curve with interior Domain $\mathcal{D}$. By the theory of conformal mapping, there is a univalent and analytic function $f(\cdot)$ on the neighborhood on the unit disk $\mathbb{D}$ such that $f(\mathbb{D}) = \gamma \cup \mathcal{D}$.

Theorem 1.7.2. *Let S be a pure subnormal operator on a Hilbert space $\mathcal{H}$ with m.n.e. N. Suppose that the spectrum $\sigma(N)$ is an analytic Jordan curve. Let f be any univalent analytic function on a neighborhood of the closed unit disk such that $f(\mathbb{T}) = \sigma(N)$. Then*

$$S = f(S_1),$$

where S_1 is a unilateral shift.

Proof. There exists a positive number $\epsilon > 0$ such that the function $f(\cdot)$ is also univalent and analytic on $\{\zeta \in \mathbb{C} : |\zeta| < 1+\epsilon\}$. Let $\gamma = f(\{\zeta : |\zeta| = r\})$, where $1 < r < 1 + \epsilon$. It is easy to see that $(\lambda - S)^{-1}, \lambda \in \gamma$ is a subnormal operator with m.n.e. $(\lambda - N)^{-1}$. Therefore

$$S_1 \stackrel{\text{def}}{=} f^{-1}(S) = \frac{1}{2\pi i} \int_\gamma (\lambda - S)^{-1} f^{-1}(\lambda) d\lambda$$

is also subnormal with m.n.e.

$$N_1 \stackrel{\text{def}}{=} f^{-1}(N) = \frac{1}{2\pi i} \int_\gamma (\lambda - N)^{-1} f^{-1}(\lambda) d\lambda.$$

Besides, S_1 is also pure. We may prove that $\sigma(N_1) = f^{-1}(\sigma(N))$ is the unit circle. From Lemma 1.7.2, S_1 is a unilateral shift. $\square$

Corollary 1.7.3. *Let S be an irreducible subnormal operator on a Hilbert space $\mathcal{H}$ with m.n.e. N. Suppose $\sigma(N)$ is an analytic Jordan curve. Then $\sigma(N)$ is a complete unitary invariant for the operator S.*

Proof. First, let us continue to apply the notation in the proof of Theorem 1.7.2. We have to prove that $S_1 = f^{-1}(S)$ is irreducible. Suppose that on contrary there are non-trivial subspaces $\mathcal{H}_1$ and $\mathcal{H}_2$ of $\mathcal{H}$ such that $\mathcal{H}_j, j = 1, 2$ reduces S_1 and $\mathcal{H} = \mathcal{H}_1 \bigoplus \mathcal{H}_2$, then from

$$S = \frac{1}{2\pi i} \int_{|s|=r} f(\lambda)(\lambda - S_1)^{-1} d\lambda$$

we may prove that $\mathcal{H}_j$ reduces S. It contradicts to the assumption that S is irreducible. Thus S_1 is a unilateral shift with multiplicity one.

It is obvious that $\sigma(N)$ is a unitary invariant. To prove the completeness, assume that S' is an irreducible subnormal operator on a Hilbert space $\mathcal{H}'$ with m.n.e. N' satisfying $\sigma(N') = \sigma(N)$. Let $S_1' = f^{-1}(S')$. Then S_1' is also a unilateral shift with multiplicity one. Therefore there is a unitary operator U from $\mathcal{H}'$ onto $\mathcal{H}$ such that

$$S_1 = US_1'U^{-1}.$$

Then $f(S_1) = Uf(S_1')U^{-1}$, i.e. S and S' are unitary equivalent. $\square$

Lemma 1.7.4. *Let S be a pure subnormal operator on a Hilbert space $\mathcal{H}$ with m.n.e. N. If $\sigma(N) = \mathbb{T} \cup \{z : |z| = r\}$ where $0 < r < 1$, then*

$$S^*S + r^2(SS^*)^{-1} = 1 + r^2. \tag{1.7.3}$$

Proof. It is easy to see that

$$|u|^2 + r^2|u|^{-2} = 1 + r^2, \quad \text{for} \quad u \in \sigma(N). \tag{1.7.4}$$

From (1.3.5) and (1.7.4), we have

$$(Sf, Sh) + r^2(S^{-1}f, S^{-1}h)$$
$$= \int_{\sigma(N)} (e(du)uf(u), uh(u)) + r^2 \int_{\sigma(N)} (e(du)u^{-1}f(u), u^{-1}h(u))$$
$$= \int_{\sigma(N)} (|u|^2 + r^2|u|^{-2})(ed(u)f(u), h(u))$$
$$= (1 + r^2)(f, h)$$

which proves (1.7.3). $\qquad\qquad\qquad\qquad\qquad\qquad\qquad\qquad\qquad\qquad\square$

Lemma 1.7.5. *Let S be an irreducible subnormal operator on a Hilbert space $\mathcal{H}$ with m.n.e. N. If $\sigma(N) = \mathbb{T} \cup \{z : |z| = r\}$ where $0 < r < 1$. Then there is a number $a \in [0, 1)$ such that*

$$\sigma(S^*S) = \{\frac{r^{2a} + r^{2n+2}}{r^{2a} + r^{2n}} : n = 0, \pm 1, \pm 2, \ldots, \} \cup \{r^2, 1\}$$

and every point in $\sigma(S^*S)$ is a simple eigenvalue of S^*S.

Proof. It is obvious that $\sigma((S^*S)^{1/2}) \subset [r, 1]$. Let $F(\cdot)$ be the spectral measure of the operator $(S^*S)^{1/2}$, it is

$$(S^*S)^{1/2} = \int_r^1 \lambda F(d\lambda). \tag{1.7.5}$$

From (1.7.3) and (1.7.5), we have

$$(SS^*)^{1/2} = \int_r^1 \frac{r}{(1 + r^2 - \lambda^2)^{1/2}} F(d\lambda), \tag{1.7.6}$$

since

$$(SS^*)^{1/2} = \frac{r}{(1 + r^2 - ((S^*S)^{1/2})^2)^{1/2}}. \tag{1.7.7}$$

Let $U(S^*S)^{1/2}$ be the polar decomposition of S, i.e. $S = U(S^*S)^{1/2}$, where U is a unitary operator. Then

$$(SS^*)^{1/2} = U(S^*S)^{1/2}U^{-1}. \tag{1.7.8}$$

Let Z be the set of all integers. For $a \in [0, 1)$, let $\lambda_n(a) = (r^{2a} + r^{2n})^{1/2}(r^{2a} + r^{2n-2})^{-1/2}, n \in Z$. Let

$$\mathcal{B}_a \overset{\text{def}}{=} \{\lambda_n(a) : n \in Z\}.$$

It is easy to see that

$$(r, 1) = \cup\{\mathcal{B}_a : a \in [0, 1)\}$$

and $\mathcal{B}_a \cap \mathcal{B}_b = \phi$, iff $a \neq b$. Define a projection-valued $G(\cdot)$ on the set $[0, 1) \times Z$

$$G(A) \overset{\text{def}}{=} \sum F(\{\lambda_n(a) : (a, n) \in A\}).$$

Then

$$(S^*S)^{1/2} = \int \lambda_n(a)G(d(a, n)) + rF(\{r\}) + F(\{1\}). \tag{1.7.9}$$

From

$$\frac{r}{(1 + r^2 - \lambda_n(a)^2)^{1/2}} = \lambda_{n-1}(a)$$

and (1.7.6), we have

$$(SS^*)^{1/2} = \int \lambda_{n-1}(a)G(d(a, n)) + rF(\{r\}) + F(\{1\}). \tag{1.7.10}$$

From (1.7.7), (1.7.8) and (1.7.9), we have $[F(\{1\}), U] = [F(\{r\}), U] = 0$ and

$$G(d(a, n)) = UG(d(a, n+1))U^{-1}. \tag{1.7.11}$$

For any Baire set $A \subset [0, 1)$, define a projection

$$P(A) = G(A \times Z).$$

From (1.7.11), we have $P(A) = UP(A)U^{-1}$. Therefore

$$P(A)\mathcal{H} = F(\{\lambda_n(a) : a \in A, n \in Z\})\mathcal{H}$$

reduces $(S^*S)^{1/2}$ and U. Hence $P(A)\mathcal{H}$ reduces S. Therefore $P(A)$ is either I or 0.

Let us prove that $F(\{r\})\mathcal{H}$ is $\{0\}$. Suppose on contrary, there is a vector $v \in F(\{r\})\mathcal{H}, v \neq 0$, then

$$(S^*S)^{1/2}v = rv.$$

From $[U, F(\{r\})] = 0$, we may conclude that $Uv = rv$, i.e.

$$Sv = rv.$$

However there is no eigenvector of S. Thus $F(\{r\}) = 0$. Similary $F(\{1\}) = 0$. Thus there must be a number $a \in [0,1)$ such that $P(\{a\}) = I$. That is

$$I = \sum_{n \in Z} F(\{\lambda_n(a)\}).$$

Choose a non-zero vector v of $F(\{\lambda_0(a)\})\mathcal{H}$. Let us prove that

$$U^n v = F(\{\lambda_n(a)\})\mathcal{H} \tag{1.7.12}$$

for $n \in Z$, by mathematical induction. Suppose that (1.7.12) holds good for $m \geq 0$. From (1.7.8) and (1.7.12) we have

$$(SS^*)^{1/2}U^{m+1}v = U(S^*S)^{1/2}U^m v = \lambda_m(a)U^{m+1}a. \tag{1.7.13}$$

From (1.7.6) and (1.7.13), we have

$$U^{m+1}v \in F(\{\lambda_{m+1}(a)\})\mathcal{H}.$$

Thus (1.7.12) holds good for $n = m + 1$. Similarly, we may prove that (1.7.12) is true for $n < 0$. Let $\mathcal{H}_0 = \text{cl}\bigvee\{U^n v : n \in Z\}$, then $\mathcal{H}_0$ is invariant with respect to U, U^{-1} and $(S^*S)^{1/2}$. Therefore $\mathcal{H}_0$ reduces S. Thus $\mathcal{H} = \mathcal{H}_0$ which proves Lemma 1.7.5. $\qquad\square$

Example. Let $a \in [0,1)$, and $0 < r < 1$. Let $\mathcal{H}_{r,a}$ be the Hilbert space of all analytic functions $f(\cdot)$ on $\{z : r < z < 1\}$ with Laurent series

$$f(z) = \sum_{n=-\infty}^{\infty} f_n z^n,$$

satisfying

$$\|f\|_{r,a} \overset{\text{def}}{=} (\sum_{-\infty}^{\infty} |f_n|^2 (r^{2a} + r^{2n}))^{1/2} < +\infty.$$

Let $S_{r,a}$ be the multiplication operator

$$(S_{r,a}f)(z) = zf(z), \quad f \in \mathcal{H}_{r,a}, r < |z| < 1.$$

Then

$$(S_{r,a}^* f)(z) = \sum_{n=-\infty}^{\infty} f_{n+1}\frac{r^{2a} + r^{2n+2}}{r^{2a} + r^{2n}}z^n$$

$$(S_{r,a}^* S_{r,a} f)(z) = \sum_{n=-\infty}^{\infty} f_n \frac{r^{2a} + r^{2n+2}}{r^{2a} + r^{2n}}z^n$$

and

$$(S_{r,a} S_{r,a}^* f)(z) = \sum_{n=-\infty}^{\infty} f_n \frac{r^{2a} + r^{2n}}{r^{2a} + r^{2n-2}}z^n$$

for $f(z) = \sum_{n=-\infty}^{\infty} f_n z^n \in \mathcal{H}_{r,a}$.

Theorem 1.7.6. *Let S be an irreducible subnormal operator on a Hilbert space $\mathcal{H}$ with m.n.e. N. If $\sigma(N) = \mathbb{T} \cup \{z : |z| = r\}$ where $0 < r < 1$. Then there is a number $a \in [0,1)$ and a unitary operator V from $\mathcal{H}$ onto $\mathcal{H}_{r,a}$ such that*

$$VSV^{-1} = S_{r,a}. \tag{1.7.14}$$

Proof. For the subnormal operator S, let us continue the proof of Lemma 1.7.5. We choose $v \in F(\{\lambda_n(a)\})\mathcal{H}$ such that $\|v\| = (r^{2a} + 1)^{1/2}$. Define a linear operator V:

$$VS^n v = z^n$$

from $\mathcal{H}$ to $\mathcal{H}_{r,a}$. Then it is easy to prove that V extends a unitary operator from $\mathcal{H}$ onto $\mathcal{H}_{r,a}$ and (1.7.14) holds good. $\qquad\square$

Let $\mathcal{D}$ be a doubly-connected domain with two boundary Jordan curves γ_1 and γ_0, where γ_0 is in the interior domain of γ_1. Then it is well-known that there is an analytic and univalent function on some ring $\mathcal{D}_r \overset{\text{def}}{=} f^{-1}(\mathcal{D}) = \{z : r < |z| < 1\}$, where $0 < r < 1$ such that

$$f(\mathcal{D}_r) = \mathcal{D}$$

and the boundary value extension of f on $\{z : |z| = R\}$ satisfies $f(\{z : |z| = r\}) = \gamma_0$.

From the Theorem 1.7.6, it is easy to prove the following:

Theorem 1.7.7. *Let S be an irreducible subnormal operator on a Hilbert space $\mathcal{H}$ with m.n.e. N. Suppose $\sigma(N)$ consists of two analytic Jordan curves which is the boundary of a doubly-connected domain $\mathcal{D}$. Let f be an analytic and univalent function on $\mathcal{D}_r = \{z : r < |z| < 1\}$ where $0 < r < 1$ which maps $\mathcal{D}_r$ onto $\mathcal{D}$, then there are a number $a \in [0,1)$ and a unitary operator W from $\mathcal{H}$ onto $\mathcal{H}_{r,a}$ such that*

$$S = f(W^{-1}S_{r,a}W).$$

Therefore the object $\{\sigma(N), a\}$ forms a complete unitary invariant for the operator S in the Theorem 1.7.7.

Chapter 2

Subnormal Operators with Finite Rank Self-Commutators

2.1 Operator with rank one self-commutator

An operator A on a Hilbert space $\mathcal{H}$ is said to be a *unilateral shift with multiplicity one*, if there is a unit vector $e \in \mathcal{H}$ such that $\{A^n e : n = 0, 1, 2, \ldots\}$ is an othonormal basis for $\mathcal{H}$. An example is the following. As in §1.3, suppose $\mathcal{H}$ is the Hardy space $H^2(\mathbb{T})$, e is the constant function 1 and

$$(U_+ f)(z) = z f(z), \quad z \in \mathbb{T}, f \in H^2(\mathbb{T}),$$

then U_+ is a unilateral shift with multiplicity one. Any unilateral shift A with multiplicity one is unitarity equivalent to this canonical example U_+ on $H^2(\mathbb{T})$.

If A is a unilateral shift with multiplicity one on $\mathcal{H}$ and e is the unit vector satisfying that $\{A^n e, n = 0, 1, 2, \cdots\}$ is an orthonormal basis for $\mathcal{H}$, then

$$A^* e_n = e_{n-1}, \quad n = 1, 2, \cdots, \tag{2.1.1}$$

where $e_n = A^{n-1} e, n = 1, 2, \cdots$ and $e_0 = 0$. Actually,

$$(A^* e_n, e_m) = (e_n, e_{m+1}) = \delta_{n,m+1}, \quad n, m = 1, 2, \cdots$$

where $\delta_{n,k}$ is the Kronecker δ. Therefore $A^* e_n$ $(n > 1)$ is orthogonal to every $\{e_k : k \neq n - 1\}$. Thus $A^* e_n = \lambda_n e_{n-1}$ and $(\lambda_n e_{n-1}, e_{n-1}) = \delta_{n,n}$. Hence $\lambda_n = 1$. Besides, $A^* e_1$ is orthogonal to every vector in $\mathcal{H}$ which proves (2.1.1) and $e_0 = 0$.

Theorem 2.1.1. *Let A be a pure operator on a Hilbert space $\mathcal{H}$ with rank one self-commutator, i.e. $\dim[A^*, A]\mathcal{H} = 1$. If $[A^*, A]\mathcal{H}$ is invariant with respect to A^*, then A is a linear combination of the identity operator and a unilateral shift with multiplicity one.*

Proof. Let e be a unit vector in $[A^*, A]\mathcal{H}$. Then e must be an eigenvector of A^* corresponding to the eigenvalue $\bar{\beta}$, since $[A^*, A]\mathcal{H}$ is invariant with respect to A^*. Replacing A with $A - \beta I$, we may assume that

$$A^* e = 0. \tag{2.1.2}$$

Let $[A^*, A]e = \lambda e$, for $\lambda \in \mathbb{C}$. From $[A^*, A]x = \lambda(x, e)e, x \in \mathcal{H}$, we have $\lambda \neq 0$, since A is pure. From (2.1.2) we have

$$\lambda = ([A^*, A]e, e) = \|Ae\|^2. \tag{2.1.3}$$

Thus $\lambda > 0$. Replacing A by $A\lambda^{-\frac{1}{2}}$, we may assume that $[A^*, A]e = e$ and hence

$$[A^*, A]x = (x, e)e, \quad x \in \mathcal{H}. \tag{2.1.4}$$

Denote $e_0 = 0$ and $e_n = A^{n-1}e, n = 1, 2, \ldots$. In order to prove that A is a unilateral shift with multiplicity one, we only have to prove that

$$(e_m, e_n) = \delta_{m,n}. \tag{2.1.5}$$

But for proving (2.1.5), we have to prove (2.1.1), by mathematical induction. (2.1.2) means that (2.1.1) holds good for $n = 1$. Suppose that (2.1.1) holds good for $n = 1, 2, \ldots, m$. Then by (2.1.4), we have

$$A^* e_{m+1} = A^* A^m e = [A^*, A]A^{m-1}e + AA^* e_m = (A^{m-1}e, e)e + Ae_{m-1}. \tag{2.1.6}$$

If $m = 1$, then $A^* e_2 = e = e_1$, since $e_0 = 0$. If $m > 1$, then

$$(A^{m-1}e, e) = (e, A^{*m-1}e) = 0, \quad m = 2, 3, \ldots \tag{2.1.7}$$

by (2.1.2). From $Ae_{m-1} = e_m$, (2.1.6) and (2.1.7) it follows (2.1.1) for $n = m + 1$ and hence for all natural number n.

Now, we have to prove (2.1.5) by induction with respect to m and assume that $n \leq m$. For $m = 1$ and $n = 1$, (2.1.4) holds good since $\|e\| = 1$. Suppose that (2.1.5) holds good for $n \leq m \leq k$, with $k \geq 1$. We have to prove that

$$(e_{k+1}, e_n) = \delta_{k+1,n}, \quad \text{for } n = 1, 2, \ldots, k+1, \tag{2.1.8}$$

where $k \geq 1$. By (2.1.1), we have

$$(e_{k+1}, e_n) = (Ae_k, e_n) = (e_k, A^* e_n) = (e_k, e_{n-1}), \tag{2.1.9}$$

where $n - 1 = 0, 1, 2, \ldots, k$. If $n - 1 = 0$, then (2.1.8) equals to $0 = \delta_{k+1,n}$. If $n - 1 = 1, 2, \ldots, k$, then (2.1.8) equals to $\delta_{k,n-1}$ by hypothesis of the induction, and then it equals $\delta_{k+1,n}$ which proves (2.1.8) and then (2.1.5) for all $m, n = 1, 2, \ldots$. $\qquad \square$

For a subnormal operator S on $\mathcal{H}$, the defect space $[S^*, S]\mathcal{H}$ is invariant with respect to S^*. Thus we have the following:

Corollary 2.1.2. *Every pure subnormal operator with rank one self-commutator is a linear combination of the identity operator and a unilateral shift with multiplicity one.*

Lemma 2.1.3. *Let $T \in L(H^2(\mathbb{T}))$. If $[T, U_+] = 0$, then there is a bounded measurable function $T(\cdot)$ on $\mathbb{T}$ which is the boundary value of a bounded analytic function on the unit disk such that*

$$T = T(U_+), \tag{2.1.10}$$

and

$$\|T\| = \operatorname{ess\,sup}_{u \in \mathbb{T}} |T(u)|. \tag{2.1.11}$$

Proof. Define $T(\cdot) = (T1)(\cdot) \in H^2(\mathbb{T})$. Then for any polynomial $p(u) = \sum_{n=0}^{N} p_n u^n$,

$$Tp(\cdot) = Tp(U_+)1 = p(U_+)T(\cdot) = p(\cdot)T(\cdot), \tag{2.1.12}$$

since $[T, U_+] = 0$. For any positive $\epsilon > 0$, let $F_\epsilon = \{u \in \mathbb{T} : |T(u)| \geq \|T\| + \epsilon\}$. Then

$$m(F_\epsilon)(\|T\| + \epsilon)^{2k} \leq \frac{1}{2\pi} \int_0^{2\pi} |T(e^{i\theta})^k 1_{F_\epsilon}(e^{i\theta})|^2 d\theta \leq \|T^k U_+\|^2 \leq \|T\|^{2k}, \tag{2.1.13}$$

by (2.1.12), where $m(F_\epsilon)$ is the normalized measure of F_ϵ, and $1_{F_\epsilon}(\cdot)$ is the characteristic function of F_ϵ. By (2.1.13), we have $m(F_\epsilon) = 0$, which proves that

$$\operatorname{ess\,sup}_{u \in \mathbb{T}} |T(u)| \leq \|T\|.$$

Similarly we may prove that $\operatorname{ess\,sup}_{u \in \mathbb{T}} |T(u)| \geq \|T\|$, which proves (2.1.11). Therefore $T(\cdot)$ is the bounded value of a bounded analytic function on the unit disk since $T(\cdot) \in H^2(\mathbb{T})$ and (2.1.11). We define

$$T(U_+) = \text{Strong} \lim_{r \to 1^-} T(rU_+).$$

From (2.1.12), it follows (2.1.10). $\qquad\square$

Proposition 2.1.4. *Suppose S is a pure subnormal operator with rank one self-commutator on a Hilbert space $\mathcal{H}$ and $T \in L(\mathcal{H})$ commuting with S. Then there are numbers $\alpha \neq 0, \beta \in \mathbb{C}$ which do not depend on T, and a bounded measurable function $T(\cdot)$ on $\mathbb{T}$ which is the boundary value of a bounded analytic function on the unit disk such that $\alpha S + \beta$ is a unilateral shift with multiplicity one and*

$$T = T(\alpha S + \beta).$$

2.2   Decomposition of the mosaic

Firstly, the following is a preliminary for the rest of this section. Let A be an operator on a Hilbert space $\mathcal{H}$ and $v \in \mathcal{H}$. Let $\rho_v(A)$ be the set of all $\lambda \in \mathbb{C}$ satisfying the condition that there is a unique $f \in \mathcal{H}$ such that

$$(\lambda - A)f = v. \tag{2.2.1}$$

The vector f in (2.2.1) is denoted by $f(\lambda, v)$.

Proposition 2.2.1. *Let S be a pure subnormal operator on a Hilbert space $\mathcal{H}$ with m.n.e. N on $\mathcal{K}$. Let $\lambda \in \rho(\Lambda) \cap \rho(N)$ and $v \in (I - \mu(\lambda))M$. If v is an eigenvector of $R(\lambda) \overset{\text{def}}{=} C(\lambda - \Lambda)^{-1} + \Lambda^*$ corresponding to $\bar{z}$, then*

$$(\lambda - \Lambda)^{-1}v \in \mu(z)^*M. \tag{2.2.2}$$

In order to prove this proposition, we have to prove the following lemmas.

Lemma 2.2.2. *Let S be a pure subnormal operator on $\mathcal{H}$ with m.n.e. N on $\mathcal{K}$. Let $f \in \mathcal{K}$. Then $f \in \mathcal{H}$, iff*

$$f(z) = 0, \quad \text{for } z \in \rho(\mathcal{S}), \tag{2.2.3}$$

where

$$f(z) \overset{\text{def}}{=} \int_{\sigma(N)} \frac{u - \Lambda}{u - z} e(du) f(u).$$

Proof. If $f \in \mathcal{H}$, then by Lemma 1.4.1(iv), $f(z) = \mu(z)f(z) = 0$, for $z \in \rho(S)$. On the other hand, suppose (2.2.3) holds good. Then for $\alpha \in M$ and $z \in \rho(S)$,

$$(f, (N^* - \bar{z})^{-1}(N^* - \Lambda^*)\alpha) = (P_M(N - SP_M)(N - z)^{-1}f, \alpha) = (f(z), \alpha) = 0.$$

By (1.3.35), $f \perp \widetilde{\mathcal{H}}$. Thus $f \in \mathcal{H}$. $\qquad \square$

Lemma 2.2.3. *Under the condition of Lemma 2.2.2, suppose $\lambda \in \rho(N)$ and $v \in M$. Then $\lambda \in \rho_v(S)$ iff $\mu(\lambda)v = 0$.*

Proof. We only have to consider the case that $\lambda \notin \rho(S)$. Since $\sigma_p(S) = \emptyset$, by the definition, $\lambda \in \rho_v(S)$ iff there is a vector $f \in \mathcal{H}$ such that $(\lambda - S)f = v$. But $Sf = Nf$. Therefore $\lambda \in \rho_v(S)$ iff $(\lambda - N)^{-1}v \in \mathcal{H}$. By Lemma 2.2.2, it is equivalent to

$$\int_{\sigma(N)} \frac{(u - \Lambda)\, e(du)v}{(u - z)(\lambda - u)} = 0, \quad \text{for } z \in \rho(S). \tag{2.2.4}$$

However (2.2.4) is equivalent to

$$(\lambda - z)^{-1}(\mu(z) - \mu(\lambda))v = 0, \quad \text{for } z \in \rho(S),$$

since $z \neq \lambda$. Thus (2.2.4) is equivalent to $\mu(\lambda)v = 0$. $\qquad \square$

Lemma 2.2.4. *Under the condition of Lemma 2.2.2, let $v \in M$ and $\lambda \in \rho(N) \cap \rho(\Lambda) \cap \rho_v(S)$. Then*

$$S^* f(\lambda, v) = f(\lambda, R(\lambda)v).$$

Proof. Firstly, $\mu(\lambda)R(\lambda)v = R(\lambda)\mu(\lambda)v = 0$ by Theorem 1.4.12 and Lemma 2.2.3. Thus $\lambda \in \rho_{R(\lambda)v}(S)$ by Lemma 2.2.3. It is easy to see

$$(\lambda - S)S^* f(\lambda, v) = [S^*, S]f(\lambda, v) + S^*(\lambda - S)f(\lambda, v) = CP_M f(\lambda, v) + \Lambda^* v.$$

On the other hand,

$$(P_M f(\lambda, v), (\bar{\lambda} - \Lambda^*)\beta) = (f(\lambda, v), (\bar{\lambda} - S^*)\beta) = (v, \beta) \quad \text{for } \beta \in M,$$

which proves

$$(\lambda - \Lambda) P_M f(\lambda, v) = v. \tag{2.2.5}$$

Therefore $(\lambda - S)S^* f(\lambda, v) = R(\lambda)v$, which proves the lemma. $\qquad\square$

Proof. (of the Proposition 2.2.1.) By Lemma 2.2.3, from $v \in (I - \mu(\lambda))M$, it follows $\lambda \in \rho_v(S)$. If $R(\lambda)v = \bar{z}v$, then $S^* f(\lambda, v) = \bar{z}f(\lambda, v)$, by Lemma 2.2.4. From Theorem 1.4.5, there is a vector $\alpha \in \mu(z)^* M$ such that

$$f(\lambda, v) = (N^* - \bar{z})^{-1}(N^* - \Lambda^*)\alpha.$$

Then $P_M f(\lambda, v) = \mu(z)^* \alpha = \alpha$. From (2.2.5), it follows (2.2.2). $\qquad\square$

Lemma 2.2.5. *Let A be an operator. Suppose there is a parallel projection μ to the eigenspace of A corresponding to the eigenvalue k satisfying*

$$A\mu = \mu A = k\mu. \tag{2.2.6}$$

If $v \neq 0$ is a root vector of A corresponding to k, i.e. $(A - k)^m v = 0$ for some $m > 1$, then v must be an eigenvector of A corresponding to k.

Proof. Suppose $(A - k)^m v = 0$ for some $m > 1$, and v is not an eigenvector of A corresponding to k. Without loss of generality, we may assume that

$$\eta \overset{\text{def}}{=} (A - k)^{m-1}v \neq 0. \tag{2.2.7}$$

Then $(A - k)\eta = 0$, η is an eigenvector of A corresponding to k. Thus $\eta = \mu\eta$. From (2.2.6)

$$\mu(A - k)^{m-1} = 0,$$

since $m - 1 > 0$. Thus $\eta = \mu(A - k)^{m-1}v = 0$ which contradicts to (2.2.7). Therefore v must be an eigenvector, which proves the lemma. $\qquad\square$

From now on, we assume that S is a pure subnormal operator on $\mathcal{H}$ with m.n.e. N and with finite rank of self-commutator, i.e.

$$\dim M < \infty.$$

Define

$$P(z,w) \stackrel{\text{def}}{=} \det Q_S(z,w)$$

where $Q_S(z,w)$ is defined in (1.3.10). If $z \in \rho(\Lambda)$, then

$$P(z,w) = \det(R(z) - \bar{w}) \det(\Lambda - z). \tag{2.2.8}$$

Lemma 2.2.6. *Assume S is a pure subnormal operator on $\mathcal{H}$ with m.n.e. N on $\mathcal{K}$ satisfying $\dim M < \infty$. Then*

$$\sigma(N) \subset \{z \in \mathbb{C} : P(z,z) = 0\}. \tag{2.2.9}$$

Proof. Let $a \in \sigma(N)$. If $Q(a,a)$ is invertible, then there is an $\epsilon > 0$ such that $Q(u,u)^{-1}$ is a continuous function of u on $O_\epsilon(a) = \{u : |u - a| < \epsilon\}$. Thus from (1.3.15) we have

$$0 = \int_{\sigma(N) \cap O_\epsilon(a)} (Q(u,u)e(du)x, Q(u,u)^{-1}y) = (e(\sigma(N) \cap O_\epsilon(a))x, y).$$

Therefore $e(\sigma(N) \cap O_\epsilon(a)) = 0$, which contradicts to $a \in \sigma(N)$. Thus $Q(a,a)$ is not invertible, i.e. $P(a,\bar{a}) = 0$, which proves the lemma. □

The polynomial $P(z,w)$ is Hermitian, i.e. it satisfies

$$P(z,w) = \overline{P(w,z)},$$

and the leading term of $P(z,w)$ is $z^m \bar{w}^m$, where $m = \dim M$. There is a unique factorization

$$P(z,w) = Q(z)\overline{Q(\bar{w})}\Pi_{i=1}^l P_i(z,w)^{k_i},$$

where $Q(\cdot)$ is a polynomial with leading coefficient 1, $P_i(z,w)$ is an irreducible Hermitian polynomial with leading term $z^{m_i} \bar{w}^{m_i}$, $m_i \geq 1$, which can not be factorized as $f(z)\overline{f(\bar{w})}$ and $P_i(z,w) \neq P_j(z,w)$ if $i \neq j$. We will consider the algebraic function $w = f_j(z)$ determined by

$$P_j(z, \overline{f_j(z)}) = 0$$

and it can be considered as the algebraic function defined by

$$P(z, \overline{f_j(z)}) = 0. \tag{2.2.10}$$

We also consider the domains in the Riemann surface defined by the algebraic function $w = f_j(z)$.

From (2.2.9), we have

$$\sigma(N) = \{a_1, \ldots, a_l\} \bigcup_{j=1}^{n} \gamma_j, \qquad (2.2.11)$$

where γ_j, $j = 1, 2, \ldots, n$ are algebraic arcs. Then $\sigma(S) \setminus \bigcup_{j=1}^{n} \gamma_j$ is a union of domains D_k, $k = 1, 2, \ldots, p$ and $\mu(\cdot)$ is a meromorphic function on each D_k with possible poles at some of $\{a_1, \ldots, a_l\}$.

For a point $a \in D_k$, except possible finite set points in D_k such as those zeros of $Q(\cdot)$, those points in $\{a_1, \ldots, a_l\}$ or branch points of the algebraic functions determined by (2.2.10), there exists a neighborhood $O_\epsilon(a) = \{z : |z - a| < \epsilon\}$ for some $\epsilon > 0$ and some single valued algebraic functions $S_l(z)$, $l = 1, 2, \ldots, q$, satisfying

$$P(z, \bar{S}_l(z)) = 0, \quad \text{for } z \in O_\epsilon(a), \qquad (2.2.12)$$

which are all different eigenvalues of $R(z)$, for $z \in O_\epsilon(a)$ by (2.2.8). By Jordan's form, $\sum_l \nu(z, S_l(z)) = I$ and

$$R(z) = \sum R(z)\nu(z, S_l(z)), \qquad (2.2.13)$$

where

$$\nu(z, w) = \frac{1}{2\pi i} \int_{\gamma_{w,\delta}} (\lambda - R(z))^{-1} \, d\lambda, \qquad (2.2.14)$$

where $\gamma_{w,\delta} = \{\lambda : |\lambda - w| = \delta\}$ is a positively oriented circle and $\delta > 0$ is a suitable number such that there is no eigenvalue of $R(z)$ in $\{\lambda : 0 < |\lambda - w| \leq \delta\}$. Then $\nu(z, w)$ is the parallel projection from M to the root space

$$\{v \in M : (R(z) - w)^n v = 0, \text{ for some integer } n \geq 0\}$$

satisfying $\nu(z, w)R(z) = R(z)\nu(z, w)$ and hence

$$\mu(z)\nu(z, w) = \nu(z, w)\mu(z).$$

Besides

$$\mu(z) = \sum \mu(z)\nu(z, S_l(z)). \qquad (2.2.15)$$

Definition. Let $\mathcal{D}$ be a finitely connected domain in a Riemann surface with boundary $\partial \mathcal{D}$ consisting of piecewise smooth, simple closed curves. If there exist a meromorphic function $S(\cdot)$ on $\mathcal{D}$ with continuous boundary value on $\partial \mathcal{D}$ and a bounded analytic function $\psi(\cdot)$ on $\mathcal{D}$ with continuous boundary value on $\partial \mathcal{D}$ satisfying

$$S(\zeta) = \overline{\psi(\zeta)}, \quad \zeta \in \partial \mathcal{D}, \qquad (2.2.16)$$

then $\mathcal{D}$ is said to be a *quadrature domain* in a Riemann surface, $S(\cdot)$ is said to be the *Schwartz function* on $\mathcal{D}$ and $\psi(\cdot)$ is said to be the *projection from $\mathcal{D}$ to $\mathbb{C}$*.

If $\mathcal{D} \subset \mathbb{C}$ and $\psi(\zeta) \equiv \zeta$, $\zeta \in \mathcal{D} \cup \partial\mathcal{D}$, then $\mathcal{D}$ is simply a quadrature domain (see Appendix B).

Theorem 2.2.7. *Suppose S is a pure subnormal operator on $\mathcal{H}$ with m.n.e. N on $\mathcal{K} \supset \mathcal{H}$ and with finite rank self-commutator. Then there is a union $\mathcal{D}$ of quadrature domains in some Riemann surfaces with Schwartz function $S(\cdot)$ and projection $\psi(\cdot)$ on $\mathcal{D}$ satisfying (2.2.16),*

$$P(\psi(\zeta), \bar{S}(\zeta)) = 0, \quad \zeta \in \mathcal{D}, \tag{2.2.17}$$

$$\psi(\mathcal{D} \cup \partial\mathcal{D}) = \sigma(S), \tag{2.2.18}$$

and the difference of $L \stackrel{def}{=} \psi(\partial\mathcal{D})$ with $\sigma(N)$ is a finite set.

There is a decomposition of $\mu(\cdot)$

$$\mu(z) = \sum_{\psi(\zeta)=z} \mu(z, S(\zeta)), \quad z \in \sigma(S) \cap \rho(N), \tag{2.2.19}$$

where

$$\mu(z, w) \stackrel{def}{=} \mu(z)\, \nu(z, w) = \nu(z, w)\, \mu(z) \tag{2.2.20}$$

is a parallel projection from M to a subspace of the eigenspace of $R(z)$ corresponding to the eigenvalue w.

Besides, if G is a domain in $\sigma(S) \setminus L$ with a common boundary arc of $\rho(S)$, then $\psi^{-1}(z)$, $z \in G$ is single-valued. After identifying $\psi^{-1}(z)$ with z for $z \in G$, $\mu(z) = \mu(z, S(z))$, $z \in G$ is the projection to the eigenspace of $R(z)$ corresponding to $S(z)$.

Proof. As we have analyzed in (2.10)–(2.16), it is easy to see that for each D_k there exists a union $\mathcal{D}_k$ of domains in some Riemann surface, meromorphic function $S(\cdot)$ and univalent analytic function $\psi(\cdot)$ on $\mathcal{D}_k$ such that $\mathcal{D}_k = \psi^{-1}(D_k)$,

$$P(\psi(\zeta), S(\zeta)) = 0, \quad \zeta \in \mathcal{D}_k \tag{2.2.21}$$

and

$$\mu(z) = \sum_{\psi(\zeta)=z} \mu(z)\, \nu(z, S(\zeta)), \quad z \in D_k \text{ and } \zeta \in \mathcal{D}_k. \tag{2.2.22}$$

Now, let us prove that

$$\mu(\psi(\zeta), S(\zeta)) \stackrel{def}{=} \mu(\psi(\zeta))\, \nu(\psi(\zeta), S(\zeta)), \quad \text{for } \zeta \in \mathcal{D}_k, \tag{2.2.23}$$

is a parallel projection from M to a subspace of the eigenspace of $R(\psi(\zeta))$ corresponding to $S(\zeta)$, i.e.

$$(R(\psi(\zeta)) - S(\zeta))\,\mu(\psi(\zeta), S(\zeta)) = 0, \quad \zeta \in \mathcal{D}_k. \tag{2.2.24}$$

There is a well-known Plemelj's formula: Let $\gamma : z = z(t)$, $0 \le t \le 1$, be a piecewise smooth simple arc in $\mathbb{C}$, and $\sigma(t)$, $0 \le t \le 1$ be a function of bounded variation. Then for almost all $t_0 \in [0,1]$,

$$\lim_{\substack{z \to z(t_0) \\ \text{from left of } \gamma}} \int_0^1 \frac{d\sigma(t)}{z(t) - z} - \lim_{\substack{z \to z(t_0) \\ \text{from right of } \gamma}} \int_0^1 \frac{d\sigma(t)}{z(t) - z} = 2\pi i\, \frac{\sigma'(t_0)}{z'(t_0)}. \tag{2.2.25}$$

Now firstly, suppose D_k has a boundary arc γ_j in (2.2.11) which is in the boundary $\partial\rho(S)$. By Plemelj's formula, we may extend the definition domain of $\mu(\cdot)$ from D_k to $D_k \cup \gamma_j$ by defining

$$\mu(z) = \lim_{\substack{w \to z \\ w \in D_k}} \mu(w) = 2\pi i (z - \Lambda)\,\frac{e(dz)}{dz} \neq 0$$

for almost $z \in \gamma_j$, where dz depends on the orientation of γ_j, since for z in the right of γ_j, $z \in \rho(S)$, $\mu(z) = 0$. From (1.3.15), we have

$$(R(z) - \bar{z})\,\mu(z) = 2\pi i\, Q(z,z)\, e(dz)/dz = 0, \tag{2.2.26}$$

for almost $z \in \gamma_j$. Let $w = S(z)$ be an algebraic function on $D_k \cup \gamma_j$, satisfying

$$P(z, \overline{S(z)}) = 0 \tag{2.2.27}$$

and

$$S(z) = \bar{z}, \quad z \in \gamma_j. \tag{2.2.28}$$

This $S(\cdot)$ exists, since $P(z,z) = 0$, for $z \in \gamma_j$ by (2.2.9). From (2.2.26), we have

$$(R(z) - S(z))\,\mu(z) = 0, \quad z \in D_k \cup \gamma_j. \tag{2.2.29}$$

Therefore $\mu(z) = \mu(z, S(z))$, where $\mu(z, S(z))$ is defined in (2.2.20). There exist vectors x, $y \in M$ such that the meromorphic function $(\mu(z)x, y)$, $z \in D_k$ is not identically zero. Therefore by (2.2.29)

$$S(z) = (R(z)x, y)/(\mu(z)x, y), \quad z \in D_k \cup \gamma_j$$

is a single-value meromorphic function on $D_k \cup \gamma_j$.

The next step is to prove that

$$\mu(z) = \nu(z, S(z)), \quad z \in D_k \cap \rho(\Lambda), \tag{2.2.30}$$

(where $\nu(z,w)$ is defined in (2.2.14)) is the parallel projection from M to the eigenspace of $R(z)$ corresponding to $S(z)$. Except finite set of points, for every $a \in \gamma_j$, we may choose a small positive number δ such that $O_\delta(a) \stackrel{\text{def}}{=} \{z : |z - a| < \delta\}$ and its boundary are in $\rho(\Lambda) \cap O(\gamma_j)$, where $O(\gamma_j)$ is a neighborhood of γ_j on which the function $S(\cdot)$ has an analytic continuation and the conformal mapping $w = S(z)$ maps $O_\delta(a) \cap \rho(S)$ into $\{\bar{w} : w \in D_k\}$ and it also maps $O_\delta(a) \cap D_k$ into $\{\bar{w} : w \in \rho(S)\}$, since $S(\cdot)$ satisfies (2.2.28). Thus

$$\overline{S(\lambda)} \in D_k, \quad \text{for } \lambda \in O_\delta(a) \cap \rho(S). \tag{2.2.31}$$

Therefore for $\lambda \in O_\delta(a) \cap \rho(S)$, $\mu(\lambda) = 0$ and if $v \in M$ satisfying

$$R(\lambda)v = S(\lambda)v$$

then

$$(\lambda - \Lambda)^{-1}v \in \mu(\overline{S(\lambda)})^* M$$

by (2.2.29) and Proposition 2.2.1. Thus $(\lambda - \Lambda)^{-1}$ is a one to one mapping from $M(\lambda, S(\lambda))$ into $\mu(\overline{S(\lambda)})^* M$, where $M(\lambda, S(\lambda))$ is the eigenspace of $R(\lambda)$ corresponding to $S(\lambda)$. Thus,

$$\dim M(\lambda, S(\lambda)) \leq \dim(\mu(\overline{S(\lambda)})^* M). \tag{2.2.32}$$

However, $\dim M(\lambda, S(\lambda))$ is a continuous function and hence a constant for $\lambda \in O_\delta(a)$ except few points. From (2.2.29)

$$\dim M(z, S(z)) \geq \dim \mu(z)M, \quad z \in O_\delta(a) \cap D_k. \tag{2.2.33}$$

From (2.2.32) and (2.2.33) we have $\dim M(z, S(z)) = \dim \mu(z)M$, which proves (2.2.30) and that $\mu(z)$ is the projection to $M(z, S(z)) = \nu(z, S(z))M$ by Lemma 2.2.5.

Now, suppose D_k is a domain for which (2.2.21) holds good, and there is another D_l with arc γ_j as a part of common boundary of D_k. By Plemelj's formula, the function

$$\widetilde{\mu}(z) \stackrel{\text{def}}{=} \lim_{\substack{w \in D_l \\ w \to z}} \mu(w) - \lim_{\substack{w \in D_k \\ w \to z}} \mu(w) = 2\pi i\,(z - \Lambda)\,\frac{e(dz)}{dz} \neq 0, \tag{2.2.34}$$

satisfies $[R(z), \widetilde{\mu}(z)] = 0$ and

$$(R(z) - \bar{z})\widetilde{\mu}(z) = 0, \tag{2.2.35}$$

for almost $z \in \gamma_j$.

Firstly, let us consider the case that no branch of the algebraic function $S(\psi^{-1}(z))$ for $z \in D_k$ has the boundary value $\bar{z}$ for $z \in \gamma_j$. There exists a domain $\hat{\mathcal{D}}_l$ with boundary arc $\hat{\gamma}$ in some Riemann surface, an algebraic function $\hat{S}(\cdot)$, a projection $\psi(\cdot)$ from $\hat{\mathcal{D}}_l \cup \hat{\gamma}$ to $D_l \cup \gamma_j$ and a projection-valued $\hat{\mu}(\cdot)$ on $\hat{\mathcal{D}}_l \cup \hat{\gamma}$ such that $\psi(\hat{\gamma}) = \gamma_j$, $\tilde{\mu}(\psi(\zeta)) = \hat{\mu}(\zeta)$, for $\zeta \in \hat{\gamma}$ and

$$(R(\psi(\zeta)) - \hat{S}(\zeta))\,\hat{\mu}(\zeta) = 0, \quad \text{for } \zeta \in \hat{\mathcal{D}}_l \cup \hat{\gamma}. \tag{2.2.36}$$

The right-hand side of (2.2.36) actually is an analytic continuation of the right-hand side of (2.2.35). Besides,

$$\hat{S}(\zeta) = \overline{\psi(\zeta)}, \quad \text{for } \zeta \in \hat{\gamma}. \tag{2.2.37}$$

From (2.2.34) it is obvious that $\hat{\mu}(\zeta)$ is a projection to a subspace of the eigenspace of $R(\psi(\zeta))$ corresponding to $\hat{S}(\zeta)$, and

$$\hat{\mu}(\zeta) = \mu(\psi(\zeta))\,\nu(\psi(\zeta), \hat{S}(\zeta)) = \mu(\psi(\zeta), \hat{S}(\zeta)).$$

The domain $\hat{\mathcal{D}}_l$ is a part of $\mathcal{D}_l$. On the other hand, by (2.2.32), we can see that $\mathcal{D}_k$ and $\mathcal{D}_l \setminus \hat{\mathcal{D}}_l$ can be connected by $\psi^{-1}(\gamma_j)$ and $\mu(\psi(\zeta))\,\nu(\psi(\zeta), S(\zeta))$ for $\zeta \in (\mathcal{D}_l \setminus \hat{\mathcal{D}}_l) \cup \psi^{-1}(\gamma)$ is the analytic continuation of $\mu(\psi(\zeta), S(\zeta))$ for $\zeta \in \mathcal{D}_k$. Therefore (2.2.22) holds good for $\zeta \in \mathcal{D}_l \setminus \hat{\mathcal{D}}_l$. Therefore (2.2.22) is proved for $\zeta \in \mathcal{D}_l$.

In the case that there are branches of $S(\psi^{-1}(z))$ for $z \in D_k$ and for $z \in D_l$ which have the same boundary value $\bar{z}$ for $z \in \gamma_j$, by the same method, we may prove that there are subdomains $\hat{\mathcal{D}}_k$ and $\hat{\mathcal{D}}_l$ in $\mathcal{D}_k$ and $\mathcal{D}_l$, respectively such that there are algebraic functions $S_k(\cdot)$, $S_l(\cdot)$, projection-valued functions $\mu_k(\cdot)$ and $\mu_l(\cdot)$ on $\hat{\mathcal{D}}_k \cup \hat{\gamma}$ and $\hat{\mathcal{D}}_l \cup \hat{\gamma}$ (where $\hat{\gamma} \subset \psi^{-1}(\gamma_j)$), respectively satisfying

$$S_k(\zeta) = S_l(\zeta) = \overline{\psi(\zeta)}, \quad \zeta \in \hat{\gamma},$$

$$(R(\psi(\zeta)) - S_k(\zeta))\,\mu_k(\zeta) = 0 \ \text{ for } \zeta \in \hat{\mathcal{D}}_k \cup \hat{\gamma},$$

and

$$(R(\psi(\zeta)) - S_l(\zeta))\,\mu_l(\zeta) = 0 \ \text{ for } \zeta \in \hat{\mathcal{D}}_l \cup \hat{\gamma}.$$

Then $\mu_k(\zeta) = \mu(\psi(\zeta), S_k(\zeta))$ and $\mu_l(\zeta) = \mu(\psi(\zeta), S_l(\zeta))$. Besides, $\mathcal{D}_k \setminus \hat{\mathcal{D}}_k$ and $\mathcal{D}_l \setminus \hat{\mathcal{D}}_l$ can be connected by $\psi^{-1}(\gamma_j)$ and the function $S(\zeta)$ for $\zeta \in \mathcal{D}_l \setminus \hat{\mathcal{D}}_l$ is the analytic continuation of the function $S(\zeta)$ for $\zeta \in \mathcal{D}_k \setminus \hat{\mathcal{D}}_k$, and hence (2.2.22) holds good for $\zeta \in \mathcal{D}_l$, which proves the Theorem. $\square$

Corollary 2.2.8. *Under the condition of Theorem 2.3.7, if $\sigma(N) \setminus \partial\rho(S)$ is a finite set, then $D \overset{def}{=} \sigma(S) \setminus \partial\rho(S)$ is a finite union of quadrature domains (in $\mathbb{C}$), and $\mu(z)$, $z \in D \cap \rho(\Lambda)$ is a parallel projection from M to the eigenspace of $R(z)$ corresponding to the eigenvalue $S(z)$, where $S(\cdot)$ is the Schwartz function of D.*

Corollary 2.2.9. *Under the condition of Theorem 2.3.7, if S is cyclic, then the conclusion of Corollary 2.2.8 holds good.*

Proof. By Proposition 1.4.6, it is easy to see that for $z \in \sigma(S) \cap \rho(N) \cap \rho(\Lambda)$, rank $\mu(z) = 1$. If there is an arc γ in $\sigma(N)$ which is in the common boundary of two domains G_1 and G_2 in $\sigma(S) \cap \rho(N)$. Then by the Plemelj's formula

$$\widetilde{\mu}(z) = \lim_{\substack{w \in G_2 \\ w \to z}} \mu(w) - \lim_{\substack{w \in G_1 \\ w \to z}} \mu(w) = 2\pi i \, (z - \Lambda) \frac{e(dz)}{dz} \neq 0 \qquad (2.2.38)$$

for almost $z \in \gamma$. Let $\mu_j(z)$ be the boundary value of $\mu(\cdot)$ on G_j, $j = 1, 2$, $S_j(z)$ be the eigenvalue of $R(z)$ corresponding to the $\mu_j(z)$ for $z \in G_j \cup \gamma$. From (2.2.35),

$$R(z) \, \widetilde{\mu}(z) = \bar{z} \, \widetilde{\mu}(z)$$

and $R(z)\mu_j(z) = S_j(z)\,\mu_j(z)$, we have

$$\bar{z} \, \widetilde{\mu}(z) = S_2(z) \, \mu_2(z) - S_1(z) \, \mu_1(z).$$

If one of $S_j(\cdot)$, says $S_1(\cdot)$ is not equal to $\bar{z}$ identically, then

$$S_2(z) \, \mu_2(z) = \bar{z} \, \widetilde{\mu}(z) + S_1(z) \, \mu_1(z). \qquad (2.2.39)$$

The range of right-hand side of (2.2.39) is of dimension 2, but the left-hand is of dimension 1. It leads to a contradiction. Therefore $S_j(z) = \bar{z}$ for $z \in \gamma$. Thus $\sigma(S) \setminus \cup \gamma_j = \cup D_k$ and each D_k is a quadrature domain, where $\cup \gamma_j$ is in (2.2.10). The Schwartz function $S(z)$ of D_k is the eigenvalue of $R(z)$ corresponding to $\mu(z)$, which proves the corollary. □

Let D be a finitely connected domain with piecewise smooth non-degenerated boundary curves. Let $\mathcal{M}$ be a finite dimensional inner product space and $v(\cdot)$ be an $L(\mathcal{M})$-valued (matrix valued) meromorphic function on $D \cup \partial D$ with pole off ∂D, satisfying

$$v(z)dz \geq 0, \; z \in \partial D. \qquad (2.2.40)$$

Let $H^2(v)$ be the Hilbert space completion of $\mathcal{M}$-valued analytic functions f on $D \cup \partial D$ satisfying

$$\|f\| \overset{\text{def}}{=} \left(\int_{\partial D} \|f(z)\|_{\mathcal{M}}^2 v(z)dz \right)^{1/2}.$$

Theorem 2.2.10. *Let S be an irreducible subnormal operator on a Hilbert space $\mathcal{H}$ with finite rank. Let N be the m.n.e. of S. Assume that*

$$\sigma_{ess}(N) \subset \partial\sigma(S).$$

Then $\sigma(S) \setminus \partial\sigma(S)$ is a quadrature domain D (in $\mathbb{C}$). There exists a space $H^2(v)$, a finite set $\{a_j\} = \sigma_p(N) \subset D$, a set of $e_j \in L(\mathcal{M})$, satisfying $e_j \geq 0$, and a unitary operator U from $\mathcal{H}$ onto $H^2(v, \mathcal{M}, \{e_j\})$ such that

$$U S U^{-1} = M_z,$$

where M_z is the multiplication operator by the independent variable z, $H^2(v, \mathcal{M}, \{e_j\})$ is closure of

$$\bigvee\{(\lambda - M_z)^{-1}\mathcal{M} : \lambda \in \rho(S)\}$$

in $H^2(v)$ with inner product

$$(f, h) = \int_{\partial D} (f(z), h(z))_{\mathcal{M}} v(z) dz + \sum_j (e_j f(a_j), h(a_j)). \qquad (2.2.41)$$

Proof. By Corollary 2.2.8, D is a quadrature domain. Let

$$v(z) = \frac{1}{2\pi i}(z - \Lambda)^{-1}\mu(z),$$

then $e(dz) = v(z)dz$, $e_j = e(\{a_j\})$, $\mathcal{M} = M$ and (2.2.41) is just

$$(f, g) = \int_{\sigma(N)} (e(dz)f(z), g(z)).$$

$\square$

Corollary 2.2.11. *Let S be a cyclic subnormal operator on a Hilbert space $\mathcal{H}$ with finite rank self-commutator. Let N be its m.n.e.. Then $D = \sigma(S) \setminus \sigma_{ess}(N)$ is a quadrature domain. There are a non-degenerated measurable function $\rho(\cdot)$ on ∂D satisfying $\int \rho(s)ds < \infty$, a finite set $\{a_j\} = \sigma_p(N)$, a set of positive number $\{c_j\}$, and a unitary operator U from $\mathcal{H}$ onto $H^2(\rho)$ such that*

$$U S U^{-1} = M_z,$$

where M_z is the multiplication operator by the independent variable z, but the inner product in $H^2(\rho)$ is modified to

$$(f, h) = \int_{\partial D} f(z)\overline{h(z)}\rho(z)|dz| + \sum_j c_j f(a_j)\overline{h(a_j)}.$$

Furthermore, if there are rational functions $f_j(\cdot)$ with poles in $\rho(S)$ and vectors b_j in the defect space such that

$$\sum_{j=1}^{n} f_j(S)b_j$$

is a cyclic vector, then there is a meromorphic function $w(\cdot)$ on $D \cup \partial D$ with poles off ∂D such that

$$\rho(z) = w(z)S(z)^{-\frac{1}{2}},$$

where $S(z)$ is the Schwartz function of D.

Proof. By Corollary 2.2.9, S satisfies the condition in Theorem 2.2.10. Let f_0 be the cyclic vector. Define

$$\rho(z) = (v(z)f_0(z), f_0(z))\frac{dz}{ds}, \quad z \in \partial D,$$

where v is the function in Theorem 2.2.10, and $c_j = (e_j f_0(a_j), f_0(a_j))$. Then the first half of this corollary is proved.

If $f_0 = \sum_j f_j(S)b_j$, then

$$\rho(z) = (v(z)\sum f_j(z)b_j, \sum f_j(z)b_j)\frac{dz}{ds}.$$

It is easy to see that $dz/ds = S(z)^{-1/2}$. Besides

$$(v(z)\sum f_j(z)b_j, \sum f_j(z)b_j) = \sum_{i,j} \overline{f_i(z)}(v(z)\sum f_j(z)b_j, b_i) \quad (2.2.42)$$

and

$$\overline{f_i(z)} = \overline{(f_i(\overline{S(z)}))}, \quad z \in \partial D.$$

Therefore (2.2.42) is the boundary value of a meromorphic function, which proves the second half of this corollary. $\qquad\square$

2.3   The reproducing kernel Hilbert space model

Let us adopt the notations in Theorem 2.2.7. Let

$$\nu(\zeta) \stackrel{\text{def}}{=} \mu(\zeta, S(\zeta)).$$

Then from (2.2.19) we have the decomposition

$$\mu(z) = \sum_{\psi(\zeta)=z} \nu(\zeta), \quad z \in \sigma(S) \cap \rho(N). \quad (2.3.1)$$

By Theorem 1.4.5 and (2.3.1), a vector f satisfying $S^* f = \bar{z} f$ for $z \in \sigma(S) \cap \rho(N)$, iff f is a sum of vectors of form

$$E_{\zeta,\alpha} \overset{\text{def}}{=} (N^* - \bar{z})^{-1} (N^* - \Lambda^*) \nu(\zeta)^* \alpha, \text{ for } z = \psi(\zeta), \ \zeta \in \mathcal{D}. \tag{2.3.2}$$

Then $S^* E_{\zeta,\alpha} = \overline{\psi(\zeta)} E_{\zeta,\alpha}$. Let

$$\mathcal{H}_0 = \bigvee \{ E_{\zeta,\alpha} : \zeta \in \mathcal{D}, \alpha \in M \}.$$

Lemma 2.3.1. *If S is a pure subnormal operator on $\mathcal{H}$ with finite rank of self-commutator. Then* cl $\mathcal{H}_0 = \mathcal{H}$.

Proof. Let $\mathcal{H}_0^{\perp} = \mathcal{H} \ominus \mathcal{H}_0$. If $f_0 \in \mathcal{H}_0^{\perp}$, then for $z \in \rho(N)$, $\alpha \in M$

$$0 = (f_0, (N^* - \bar{z})^{-1} (N^* - \Lambda^*) \mu(z)^* \alpha) = (f_0(z), \alpha).$$

Therefore $f_0(z) = 0$ where $z \in \rho(N)$. Thus by Plemelj's formula, we have $f_0(u) = 0$, for $u \in \sigma(N)$ except a finite set of points $\{a_1, \ldots, a_n\} \subset \sigma(N)$ where

$$(a_j - \Lambda) e(\{a_j\}) f_0(a_j) = 0.$$

From (1.3.24), we have

$$(S^* f_0)(u) = \bar{u} f_0(u) - (\bar{u} - \Lambda^*) f_0(\Lambda).$$

For any polynomial P, $P(S)^* \mathcal{H}_0 \subset \mathcal{H}_0$. Therefore $P(S) \mathcal{H}_0^{\perp} \subset \mathcal{H}_0^{\perp}$. Thus $P(S) f_0 \in \mathcal{H}_0^{\perp}$. We may choose a P such that

$$P(a_j) = \overline{a_j}, \quad j = 1, 2, \ldots, n.$$

Therefore $\bar{u} f_0(u) = P(u) f_0(u)$, for $u \in \sigma(N)$. We have

$$(S^* f_0 - P(S) f_0)(u) = -(\bar{u} - \Lambda^*) f_0(\Lambda). \tag{2.3.3}$$

The left-hand side of (2.3.3) is in $\mathcal{H}$ and the right-hand side is in $\widetilde{\mathcal{H}}$. Therefore it is zero. Thus $S^* f_0 \in \mathcal{H}_0^{\perp}$. Therefore $\mathcal{H}_0^{\perp}$ reduces S. However, S is pure. Thus $\mathcal{H}_0^{\perp} = \{0\}$ which proves the lemma. $\qquad \square$

For ξ, $\zeta \in \mathcal{D}$, let

$$E(\zeta, \xi) \overset{\text{def}}{=} \nu(\zeta) Q(\overline{S(\xi)}, \overline{S(\zeta)}) \nu(\xi)^* (\psi(\zeta) - \overline{S(\xi)})^{-1} (\overline{\psi(\xi)} - S(\zeta))^{-1}, \tag{2.3.4}$$

where $Q(\cdot, \cdot)$ is the polynomial defined in (1.3.10).

Lemma 2.3.2. *Under the condition of Lemma 2.3.1, for ξ, $\zeta \in \mathcal{D}$ and α, $\beta \in M$,*

$$(E_{\xi,\alpha}, E_{\zeta,\beta}) = (E(\zeta, \xi) \alpha, \beta).$$

Proof. It is obvious that

$$(E_{\xi,\alpha}, E_{\zeta,\beta}) = (\nu(\zeta) \int \frac{(u - \Lambda)e(du)(\bar{u} - \Lambda^*)}{(u - z)(\bar{u} - \bar{w})} \nu(\xi)^* \alpha, \beta) \qquad (2.3.5)$$

where $z = \psi(\zeta)$ and $w = \psi(\xi)$. Let $\lambda = \overline{S(\zeta)}$ and $\mu = \overline{S(\xi)}$, then $[R(z), \nu(\zeta)] = [R(w), \nu(\xi)] = 0$,

$$R(z)\nu(\zeta) = \bar{\lambda}\nu(\zeta) \quad \text{and} \quad R(w)\nu(\xi) = \bar{\mu}\,\nu(\xi). \qquad (2.3.6)$$

Besides, we have

$$(z - \Lambda)Q(z, w)^{-1} = (\bar{w} - R(z))^{-1} \text{ and } Q(z, w)^{-1}(\bar{w} - \Lambda^*) = (z - R(w)^*)^{-1}. \qquad (2.3.7)$$

From (1.4.2) and (1.4.25), it is easy to calculate that

$$\int \frac{(u - \Lambda)e(du)(\bar{u} - \Lambda^*)}{(u - z)(\bar{u} - \bar{w})} = \mu(z) + \mu(w)^* - I + (z - \Lambda)(Q(z, w)^{-1}(\mu(z) - I)$$
$$+ \mu(w)^* Q(z, w)^{-1})(\bar{w} - \Lambda^*). \qquad (2.3.8)$$

From (1.4.32), (2.3.6), (2.3.7) and (2.3.8), we have

$$\nu(\zeta) \int \frac{(u - \Lambda)e(du)(\bar{u} - \Lambda^*)}{(u - z)(\bar{u} - \bar{w})} \nu(\xi)^* = \nu(\zeta)(\Lambda - \mu)\nu(\xi)^*(z - \mu)^{-1}. \quad (2.3.9)$$

However,

$$E(\zeta, \xi) = \frac{\nu(\zeta)(\mu - \Lambda)\nu(\xi)^*}{\mu - z} + \frac{\nu(\zeta)Q(\mu, w)\nu(\xi)^*}{(\mu - z)(\bar{\lambda} - \bar{w})}, \qquad (2.3.10)$$

and $Q(\mu, w)\nu(\xi)^* = (\bar{w} - \Lambda^*)(\mu - R(w)^*)\nu(\xi)^* = 0$ by (2.3.6). From (2.3.5), (2.3.9) and (2.3.10), it follows the Lemma. $\qquad \square$

Define a positive semi-definite kernel on the $\mathcal{D} \times M$ as following

$$K((\xi, \alpha), (\zeta, \beta)) \stackrel{\text{def}}{=} (E(\zeta, \xi)\alpha, \beta). \qquad (2.3.11)$$

Let $K((\xi, \alpha), \cdot)$ be the function on $\mathcal{D} \times M$ defined as

$$(\zeta, \beta) \mapsto K((\xi, \alpha), (\zeta, \beta)).$$

Let $\hat{\mathcal{H}}_0 \stackrel{\text{def}}{=} \bigvee \{K((\xi, \alpha), \cdot) : (\xi, \alpha) \in \mathcal{D} \times M\}$. Define an inner product on $\hat{\mathcal{H}}_0$ as

$$(K((\xi, \alpha), \cdot), K((\zeta, \beta), \cdot)) \stackrel{\text{def}}{=} K((\xi, \alpha), (\zeta, \beta)). \qquad (2.3.12)$$

Define a linear operator V from $\mathcal{H}_0$ to $\hat{\mathcal{H}}_0$ as

$$V E_{\xi,\alpha} = K((\xi, \alpha), \cdot). \qquad (2.3.13)$$

Theorem 2.3.3. *Let S be a pure subnormal operator on $\mathcal{H}$ with finite rank self-commutator. Let V be the extension of (2.3.13) from $\mathcal{H}_0$ to $\mathcal{H}$. Then V is a unitary operator from $\mathcal{H}$ to the reproducing kernel Hilbert space $\hat{\mathcal{H}}$, the completion of $\hat{\mathcal{H}}_0$, satisfying*

$$(Vx)(\zeta, \beta) = (x, E_{\zeta,\beta}), \quad x \in \mathcal{H}, \tag{2.3.14}$$

$$(VSV^{-1}f)(\zeta, \beta) = \psi(\zeta)f(\zeta, \beta), \quad f \in \hat{\mathcal{H}}, \tag{2.3.15}$$

and

$$(VS^*V^{-1})K((\xi, \alpha), \cdot) = \overline{\psi(\xi)}K((\xi, \alpha), \cdot), \ (\xi, \alpha) \in \mathcal{D} \times M. \tag{2.3.16}$$

Proof. By Lemma 2.3.2, (2.3.11) and (2.3.12), (2.3.14) holds good for $x = E_{\xi,\alpha}$, $(\xi, \alpha) \in \mathcal{D} \times M$. By (2.3.12), V is an isometry from $\mathcal{H}_0$ to $\hat{\mathcal{H}}_0$. By Lemma 2.3.1, V extends a unitary operator from $\mathcal{H}$ to $\hat{\mathcal{H}}$ and (2.3.14) holds good for $x \in \mathcal{H}$. If $f = K((\xi, \alpha), \cdot)$ then

$$VSV^{-1}f = VSE_{\xi,\alpha}.$$

Therefore, by (2.3.14)

$$(VSV^{-1}f)(\zeta, \beta) = (SE_{\xi,\alpha}, E_{\zeta,\beta}) = (E_{\xi,\alpha}, S^*E_{\zeta,\beta}) = \psi(\zeta)(E_{\xi,\alpha}, E_{\zeta,\beta})$$

which proves (2.3.15) for $f = K((\xi, \alpha), \cdot)$. By the linearity (2.3.15) holds good for $f \in \mathcal{H}_0$ and hence also for $f \in \mathcal{H}$, since V is unitary and $\mathcal{H}_0$ is dense in $\mathcal{H}$. The equality (2.3.16) is a direct consequence of

$$S^*E_{\xi,\alpha} = \overline{\psi(\xi)}E_{\xi,\alpha},$$

which proves the theorem. $\square$

Theorem 2.3.3 gives an infinite dimensional analogue of the diagonalization for the operator S^*.

2.4 Area integral formula and trace formula

First, let us introduce an area integral formula related to the mosaic.

Theorem 2.4.1. *let S be a pure subnormal operator on a Hilbert space $\mathcal{H}$ with finite rank self-commutator. Let $f(\cdot)$ and $g(\cdot)$ be analytic functions on a neighborhood of $\sigma(S)$. Then*

$$[g(S)^*f(S), S]|_M = \frac{1}{\pi}\iint_{\sigma(S)} \overline{g'(z)}f(z)\mu(z)dA(z), \tag{2.4.1}$$

where $A(\cdot)$ is the Lebesgue planar measure.

In the right hand side of (2.4.1), it is an improper integral. Suppose $\{a_1, \ldots, a_n\}$ is the set of all poles of $\mu(\cdot)$ in $\sigma(S)$. Let

$$O_\epsilon = \bigcup_{j=1}^{n} \{z : |z - a_j| < \epsilon\}.$$

Then the right hand side of (2.4.1) is defined as

$$\lim_{\epsilon \to 0} \frac{1}{\pi} \iint_{\sigma(S) \setminus O_\epsilon} \overline{g'(z)} f(z) \mu(z) dA(z).$$

Proof. We only have to prove the following

$$[(\overline{\xi} - S^*)^{-1}(\lambda - S)^{-1}, S]|_M = \frac{1}{\pi} \iint_{\sigma(S)} \frac{\mu(z) dA(z)}{(\overline{\xi} - \overline{z})^2(\lambda - z)}, \quad \lambda, \xi \in \rho(S).$$
$$(2.4.2)$$

It is easy to see that the left hand side equals to

$$[(\overline{\xi} - S^*)^{-1}, S](\lambda - S)^{-1}|_M = (\overline{\xi} - \Lambda^*)^{-1} C P_M (\overline{\xi} - S^*)^{-1}(\lambda - S)^{-1}|_M$$
$$= (\overline{\xi} - \Lambda^*)^{-1} C (\lambda - \Lambda)^{-1}(\overline{\xi} - R(a))^{-1},$$
$$(2.4.3)$$

since $P_M(\overline{\xi} - S^*)^{-1}(\lambda - S)^{-1}|_M = Q(\lambda, \xi)^{-1} = (\lambda - \Lambda)^{-1}(\overline{\xi} - R(a))^{-1}$ by Lemma 1.3.2.

Let N be the m.n.e. of S and $\sigma_p(N) = \{e_1, \ldots, e_n\}$. It is easy to see that $\sigma_p(N) \subset \sigma(S) \setminus \partial\rho(S)$. Let us adopt the notations in §2.2 and (2.3.1).

First, let us assume that $\sigma_p(N) \cap L = \emptyset$, where $L = \psi(\mathcal{L})$. For any small $\epsilon > 0$, let $O_\epsilon(a_j) = \{z \in \mathbb{C} : |z - a_j| < \epsilon\}$, $\gamma_\epsilon(a_j)$ be the boundary circle of $O_\epsilon(a_j)$ with clock-wise orientation. Let $O_\epsilon = \cup_{j=1}^n O_\epsilon(a_j)$ and $\gamma_\epsilon = \cup_{j=1}^n \gamma_\epsilon(a_j)$. Then

$$\frac{1}{\pi} \iint_{\sigma(S) \setminus O_\epsilon} \frac{\mu(z) dA(z)}{(\overline{\xi} - \overline{z})^2(\lambda - z)} = \frac{1}{2\pi i} \iint_{\sigma(S) \setminus O_\epsilon} d\left(\frac{\mu(z) dz}{(\overline{\xi} - \overline{z})(\lambda - z)}\right)$$
$$= \frac{1}{2\pi i} \iint_{\mathcal{D} \setminus \psi^{-1}(O_\epsilon)} d\left(\frac{\nu(\zeta) d\psi(\zeta)}{(\overline{\xi} - \overline{\psi(\zeta)})(\lambda - \psi(\zeta))}\right)$$
$$= \frac{1}{2\pi i} \int_{\mathcal{L} \cup \psi^{-1}(\gamma_\epsilon)} \frac{\nu(\zeta) d\psi(\zeta)}{(\overline{\xi} - \overline{\psi(\zeta)})(\lambda - \psi(\zeta))}$$
$$= \int_L \frac{(z - \Lambda) e(dz)}{(\overline{\xi} - \overline{z})(\lambda - z)} + \frac{1}{2\pi i} \int_{\gamma_\epsilon} \frac{\mu(z) dz}{(\overline{\xi} - \overline{z})(\lambda - z)}. \quad (2.4.4)$$

Notice that

$$\lim_{\epsilon \to 0} \frac{1}{2\pi i} \int_{\gamma_\epsilon} \frac{\mu(z)dz}{(\bar{\xi} - \bar{z})(\lambda - z)} = \lim_{\epsilon \to 0} \sum_{j=1}^{n} \frac{1}{2\pi i} \int_{\gamma_\epsilon(a_j)} \frac{\frac{a_j - \Lambda_j}{a_j - z}e(\{a_j\})}{(\bar{\xi} - \bar{z})(\lambda - z)}dz$$

$$= \sum_{j=1}^{n} \frac{(a_j - \Lambda_j)e(\{a_j\})}{(\bar{\xi} - \bar{a_j})(\lambda - a_j)}. \qquad (2.4.5)$$

From (2.4.4) and (2.4.5), the right hand side of (2.4.2) equals to

$$\int_{\sigma(N)} \frac{(z - \Lambda)e(dz)}{(\bar{\xi} - \bar{z})(\lambda - z)} = -\int \frac{e(dz)}{\bar{\xi} - \bar{z}} + (\lambda - \Lambda)\int \frac{e(dz)}{(\bar{\xi} - \bar{z})(\lambda - z)}$$

$$= -(\bar{\xi} - \Lambda^*)^{-1} + (\lambda - \Lambda)Q(\lambda, \xi)^{-1}$$

$$= -(\bar{\xi} - \Lambda^*)^{-1} + (\bar{\xi} - R(\lambda))^{-1},$$

which equals to (2.4.3). Thus (2.4.2) is proved under the condition that $\sigma_p(N) \cap L = \emptyset$.

If $\sigma_p(N) \cap L \neq \emptyset$, we may construct a sequence of subnormal operators $\{S_m\}$ with m.n.e. $\{N_m\}$ such that (i) $\sigma(N_m) = L \cup \{a_j^{(m)}\}$ with $a_j^{(m)} \notin L$ and $|a_j^{(m)} - a_j| < \frac{1}{m}$, and (ii) the $e(\cdot)$ measure corresponding to S_m, name it as $e_m(\cdot)$, satisfies $e_m(F) = e(F)$ for $F \subset L$ and $e_m(\{a_j^{(m)}\}) = e(\{a_j\})$. Then for the operator S_m and its mosaic μ_m, the formula (2.4.1) is satisfied. Let $m \to \infty$, then we will obtain (2.4.1) for S. $\qquad\qquad \square$

Corollary 2.4.2. *Let S be a pure subnormal operator with finite rank self-commutator. Then*

$$C = \frac{1}{\pi} \iint_{\sigma(S)} \mu(z)dA(z). \qquad (2.4.6)$$

Proof. In (2.4.1), let $g(z) = z$, $f(z) = 1$, we have (2.4.6). $\qquad\qquad \square$

Now let us study the trace formula (1.6.14) in the special case of

$$\text{rank}[S^*, S] < \infty. \qquad (2.4.7)$$

Theorem 2.4.3. *Let S be a pure subnormal operator with m.n.e. N satisfying (2.4.7). Then $\sigma(N) = L \cup \{a_1, \ldots, a_n\}$, where L is a finite union of algebraic cycles. Let $m(u), u \in L$ be the multiplicity of N at regular point $u \in L$. Then m is piecewise constant on L and for any pair of rational functions $f(\bar{u}, u)$ and $h(\bar{u}, u)$ with poles off $\sigma(S)$,*

$$\text{tr}[f(S^*, S), h(S^*, S)] = \frac{1}{2\pi i} \int_L mf dh. \qquad (2.4.8)$$

Proof. On L, the measure $v(\cdot)$ in Theorem 1.6.6 satisfies the equality

$$v(d\psi(\xi)) = i\operatorname{tr}((\psi(\xi) - \Lambda)e(d\psi(\xi))) = \frac{1}{2\pi}\operatorname{tr}(\nu(\xi)d\psi(\xi)), \quad \xi \in \mathcal{L},$$

by §2.2 and (2.3.1). Let

$$m_1(\xi) \overset{\text{def}}{=} \operatorname{tr}(\nu(\xi)), \quad \xi \in \mathcal{L}.$$

Then $m_1(\xi) = \operatorname{rank}\nu(\xi)$, since $\nu(\xi)^2 = \nu(\xi)$. Except a null set in L,

$$m_1(\xi) = \operatorname{rank}(\frac{e(d\psi(\xi))}{d\psi(\xi)}), \quad \xi \in \mathcal{L}.$$

Thus $m(z) \overset{\text{def}}{=} \sum_{\psi(\xi)=z} m_1(\xi)$ can be proved that is the multiplicity of N at $z \in L$. Therefore the differential $d_v h(\overline{u}, u)$ in Theorem 1.6.6 equals to

$$\frac{m(u)}{2\pi}(\frac{\partial h}{\partial \overline{u}}\overline{du} + \frac{\partial h}{\partial u}du) = \frac{m(u)}{2\pi}dh, \quad \text{for } u \in L.$$

Let

$$\lambda_j = i\operatorname{tr}((a_j - \Lambda)e\{a_j\}).$$

Then from (1.6.14), we have

$$2\pi i\operatorname{tr}[f(S^*, S), h(S^*, S)] = \int_L mfdh + J(f, h), \tag{2.4.9}$$

where

$$J(f, h) = 2\pi i\sum_j f(\overline{a_j}, a_j)(\frac{\partial h}{\partial \overline{u}}(\overline{a_j}, a_j)\overline{\lambda_j} + \frac{\partial h}{\partial u}(\overline{a_j}, a_j)\lambda_j). \tag{2.4.10}$$

The function $m(u)$ is the boundary value of piecewise constant function $\operatorname{tr}(v(\psi(\lambda)))$, $\lambda \in \mathcal{D}$. Therefore $m(\cdot)$ is piecewise constant on L. Let $L = \cup_{j=1}^l \gamma_j$, where γ_j is algebraic arc in L with end points b_j and c_j such that $m(u) = m_j, u \in \gamma_j$. Then

$$\int_L mfdh = \sum_{j=1}^l m_j \int_{\gamma_j} fdh.$$

Thus

$$\int_L mfdh + \int_L mhdf = \sum_{j=1}^l m_j \int_{\gamma_j} (fdh + hdf)$$

$$= \sum_{j=1}^l m_j \int_{\gamma_j} d(fh)$$

$$= \sum_{j=1}^l m_j(f(\overline{b_j}, b_j)h(\overline{b_j}, b_j) - f(\overline{c_j}, c_j)h(\overline{c_j}, c_j)).$$

$$\tag{2.4.11}$$

On the other hand

$$J(h, f) = 2\pi i \sum h(\overline{a_j}, a_j)(\frac{\partial f}{\partial \overline{u}}(\overline{a_j}, a_j)\overline{\lambda_j} + \frac{\partial f}{\partial u}(\overline{a_j}, a_j)\lambda_j). \qquad (2.4.12)$$

However

$$2\pi i \operatorname{tr}[f(S^*, S), g(S^*, S)] + 2\pi i \operatorname{tr}[g(S^*, S), f(S^*, S)] = 0.$$

Therefore $J(f, h) + J(h, f) + \int_L mf dh + \int_L mh df = 0$. By (2.4.10), (2.4.11) and (2.4.12), it is easy to see $\lambda_j = 0, j = 1, \ldots, l$, which proves (2.4.8). $\square$

Corollary 2.4.4. *Under the condition of Theorem 2.4.3. Let $m(z) = \operatorname{tr}(\mu(z)), z \in \rho(N)$. Then for any pair of polynomials f and h*

$$2\pi i \operatorname{tr}[f(S^*, S), h(S^*, S)] = \int_{\sigma(S)} mdf \wedge dh, \qquad (2.4.13)$$

where $df \wedge dh = 2i\frac{\partial(f,h)}{\partial(\overline{z},z)}dA(z)$ is the Cartan's exterior product of the differentials df and dh, and $A(z)$ is the Lebesgue planar measure.

Proof. The set $D = \sigma(S) \cap \rho(N)$ is covered by $\mathcal{D} = \cup_{j=1}^{p}\mathcal{D}_j, \mathcal{D}_j, j = 1, \ldots, p$ are quadrature domains in some Riemann surfaces. For $\xi \in \mathcal{D}$, let $m(\xi) \overset{\text{def}}{=} \operatorname{tr}(\nu(\xi)) = \operatorname{rank}(\nu(\xi))$. Applying Cartan's formula (or Green's formula)

$$\int_{\partial\Omega} \omega = \int_{\Omega} d\omega, \qquad (2.4.14)$$

where ω is a differential form on the boundary of a region Ω, we have

$$\begin{aligned}
\int_L mf dh &= \int_{\mathcal{L}} m(\xi)f(\overline{\psi(\xi)}, \psi(\xi))dh(\overline{\psi(\xi)}, \psi(\xi)) \\
&= \int_{\mathcal{D}} d(m(\xi)f(\overline{\psi(\xi)}, \psi(\xi))dh(\overline{\psi(\xi)}, \psi(\xi))) \\
&= \int_{\mathcal{D}} m(\xi)df(\overline{\psi(\xi)}, \psi(\xi)) \wedge dh(\overline{\psi(\xi)}, \psi(\xi)) \\
&= \int_{\mathcal{D}} m(z)df(\overline{z}, z) \wedge dh(\overline{z}, z)
\end{aligned}$$

where

$$m(z) = \sum_{\psi(\xi)=z} m(\xi) = \sum_{\psi(\xi)=z} \operatorname{tr}(\nu(\xi)) = \operatorname{tr}\mu(z),$$

which proves the corollary. $\square$

The function $m(z)$ in (2.4.13) also equals to the dimension of the eigen space of the operator S^* corresponding to the eigenvalue $\overline{z}$.

Chapter 3

Analytic Model for Subnormal k-Tuple of Operators

3.1 Subnormal k-tuple of operators

Let $\mathbb{T} = (T_1, \ldots, T_k)$ be a k-tuple of operators on a Hilbert space $\mathcal{H}$. If $[T_i, T_j] = 0$, $i, j = 1, 2, \ldots, k$, then $\mathbb{T}$ is said to be a commuting k-tuple of operators. If $[T_i, T_j] = 0$ and $[T_i, {}^* T_j] = 0$, $i, j = 1, 2, \ldots, k$. Then $\mathbb{T}$ is said to be a doubly commuting k-tuple of operators.

Theorem 3.1.1. *(Fuglede-Putnam) Let L and N be normal operators on a Hilbert space $\mathcal{H}$. If $B \in L(\mathcal{H})$ and*

$$BL = NB. \tag{3.1.1}$$

Then

$$BL^* = N^*B. \tag{3.1.2}$$

The original theorem is for unbounded case. Besides, the proof of this theorem is from M. Rosenblum.

Proof. From (3.1.1) it is easy to see that $BL^n = N^n B$ for $n = 0, 1, 2, \ldots$. Therefore, for any complex number z,

$$Be^{i\bar{z}L} = e^{i\bar{z}N} B.$$

That is $B = e^{-i\bar{z}N} Be^{i\bar{z}L}$. Therefore

$$e^{izN^*} Be^{-izL^*} = e^{izN^*} e^{-i\bar{z}N} Be^{i\bar{z}L} e^{-izL^*}.$$

It is easy to see that $e^{izN^*} e^{-i\bar{z}N}$ and $e^{i\bar{z}L} e^{-izL^*}$ are unitary operators. Thus the $L(\mathcal{H})$-valued entire function $z \mapsto e^{izN^*} Be^{-izL^*}$ is bounded by $\|B\|$. By Liouville's theorem, this function is a constant. The value of this function at $z = 0$ is B. Thus

$$e^{izN^*} Be^{-izL^*} = B \text{ for } z \in \mathbb{C}.$$

Upon taking derivative with respect to z, and then setting $z = 0$, we proved (3.1.2). $\qquad\square$

A commuting k-tuple of normal operators is said to be a *normal k-tuple* of operators.

Corollary 3.1.2. *If* $\mathbb{N} = (N_1, \ldots, N_k)$ *is a normal k-tuple of operators, then* $\mathbb{N}$ *is doubly commuting.*

Let F be a compact set in $\mathbb{C}^k$, $\mathcal{B}(F)$ be the σ-algebra of all Borel subsets in F. Suppose $A \mapsto E(A)$, $A \in \mathcal{B}(F)$ be a projection valued measure on $(F, \mathcal{B}(F))$, i.e., for any $A \in \mathcal{B}(F)$, $E(A)$ is a projection on a Hilbert space $\mathcal{K}$, for any multually disadjoint sequence of sets $\{A_j\} \subset \mathcal{B}(F)$

$$E(\bigcup_{j=1}^{\infty} A_j) = \sum_{j=1}^{\infty} E(A_j)$$

and $E(F) = I$. If F is the support of $E(\cdot)$, i.e. there is no open set $O \subset \mathbb{C}^k$ satisfying $O \cap F \neq \emptyset$ and $E(O \cap F) = 0$, then $(F, \mathcal{B}(F), E(\cdot))$, or simply $E(\cdot)$ is said to be a *spectral measure*.

Proposition 3.1.3. *Let* $\mathbb{N} = (N_1, \ldots, N_k)$ *be a normal k-tuple of operators on a Hilbert space* $\mathcal{K}$. *Then there exists a spectral measure* $E(\cdot)$ *such that for any* $(\lambda_1, \ldots, \lambda_k) \in \rho(N_1) \times \cdots \times \rho(N_k)$

$$\prod_j (\lambda_j - N_j)^{-1} = \int \prod_j (\lambda_j - z_j)^{-1} dE(z). \tag{3.1.3}$$

Proof. By the spectral theorem for the N_j, there exists spectral measure $E_j(\cdot)$ on $\sigma(N_j)$ such that

$$(\lambda - N_j)^{-1} = \int_{\sigma(N_j)} (\lambda - z)^{-1} dE_j(z). \tag{3.1.4}$$

By Corollary 3.1.2, $\mathbb{N}$ is doubly commuting. Thus

$$[E_i(A_i), E_j(A_j)] = 0, \text{ for } i, j = 1, 2, \ldots, h$$

where $A_i \subset \sigma(N_i)$, $i = 1, 2, \ldots, k$, are Borel sets in $\mathbb{C}$. Define

$$E(A_1 \times A_2 \times \cdots \times A_K) = \prod_{j=1}^{k} E_j(A_j).$$

Then it is easy to see that $E(\cdot)$ is a projection-valued σ-additive set function on $(\times_j \sigma(N_j), \mathcal{B}(\times_j \sigma(N_j)))$. It is easy to see there exists a support F for this set function. Then the restriction of $E(\cdot)$ on F is the spectral measure of N. From (3.1.4), it follows (3.1.3). $\qquad \square$

This compact set F is said to be the *spectrum* of $\mathbb{N}$ and is denoted by sp($\mathbb{N}$). In many cases, sp($\mathbb{N}$) is a proper subset of $\times_{j=1}^{k} \sigma(N_j)$.

Definition. Let $\mathbb{S} = (S_1, \ldots, S_k)$ be a k-tuple of operators on a Hilbert space $\mathcal{H}$. If there is a normal k-tuple of operators $\mathbb{N} = (N_1, \ldots, N_k)$ on a Hilbert space $\mathcal{K}$ containing $\mathcal{H}$ as a subspace such that

$$S_j = N_j|_{\mathcal{H}}, \quad j = 1, 2, \ldots, k.$$

Then $\mathbb{S}$ is said to be a subnormal k-tuple of operators and $\mathbb{N}$ is said to be a normal extension of $\mathbb{S}$.

It is obvious that if $\mathbb{S} = (S_1, \ldots, S_k)$ is subnormal, then $\mathbb{S}$ is commuting and each S_j is subnormal. But there are some commuting k-tuple of subnormal operators which are not subnormal k-tuple of operators.

Let $\mathbb{S}$ be a subnormal k-tuple of operators on $\mathcal{H}$ with a normal extension $\mathbb{N}$ on $\mathcal{K} \supset \mathcal{H}$. If there is no proper subspace in $\mathcal{K} \ominus \mathcal{H}$ which is reducing $\mathbb{N}$, then $\mathbb{N}$ is said to be a *minimal normal extension*, or simply *m.n.e.*, of $\mathbb{S}$. It is easy to prove that for any subnormal k-tuple of operators, the m.n.e. exists and is essentially unique; i.e. unique under unitary equivalence.

For a k-tuple of operators $\mathbb{A} = (A_1, \ldots, A_k)$ on a Hilbert space $\mathcal{H}$, let

$$\mathrm{sp}_r(\mathbb{A}) \stackrel{\text{def}}{=} \{(\lambda_1, \ldots, \lambda_k) \in \mathbb{C}^k : \sum_{j=1}^{k}(A_j - \lambda_j)(A_j - \lambda_j)^* \text{ is not invertible}\}$$

be the *right spectrum* of $\mathbb{A}$.

Theorem 3.1.4. *Let $\mathbb{S}$ be a subnormal k-tuple of operators on $\mathcal{H}$ with m.n.e. $\mathbb{N}$ on a Hilbert space $\mathcal{K} \supset \mathcal{H}$. Then*

$$\mathrm{sp}(\mathbb{N}) \subset \mathrm{sp}_r(\mathbb{S}). \tag{3.1.5}$$

Proof. Without loss of generality, we only have to prove that if $0 = (0, \ldots, 0) \notin \mathrm{sp}_r(\mathbb{S})$ then $0 \notin \mathrm{sp}(\mathbb{N})$. Suppose $0 \notin \mathrm{sp}_r(\mathbb{S})$ then $\sum_{j=1}^{k} S_j S_j^*$ is invertible. Let $A_j = S_j^*(\sum_l S_l S_l^*)^{-1}$, then $\sum S_j A_j = I$. Hence for every $h \in \mathcal{H}$ there exist $h_j = A_j h$, $j = 1, 2, \ldots, k$ such that

$$\sum_{j=1}^{k} S_j h_j = h, \tag{3.1.6}$$

where $\sum_{j=1}^{k} \|h_j\|^2 \leq (\sum \|A_j\|^2)\|h\|^2$. By changing $\mathbb{S}$ to $c\mathbb{S}$ and $\mathbb{N}$ to $c\mathbb{N}$ for some positive constant c, we may assume $\sum \|A_j\|^2 \leq 1$, i.e.

$$\sum_{j=1}^{k} \|h_j\|^2 \leq \|h\|^2. \tag{3.1.7}$$

Let $E(\cdot)$ be the spectral measure of N. Let
$$\mathcal{L} \overset{\text{def}}{=} E(\{z : |z| \leq 1/2\})\mathcal{K}.$$
Then $\mathcal{L}$ reduces N_j, $j = 1, 2, \ldots, k$. If we prove that $\mathcal{L} \perp \mathcal{H}$, then by the minimality of the extension of N, $\mathcal{L}$ must be $\{0\}$ and 0 is not in the sp(N).

Apply (3.1.6) and (3.1.7) p times, for $h \in \mathcal{H}$, there exists $h_{j_1 \cdots j_p}$ such that

$$\sum_{j_2=1}^{k} S_{j_2} h_{j_1 j_2} = h_{j_1},$$

$$\cdots$$

$$\sum_{j_p=1}^{k} S_{j_p} h_{j_1 j_2 \cdots j_p} = h_{j_1 \cdots j_{p-1}},$$

and

$$\sum_{j_2=1}^{k} \|h_{j_1 j_2}\|^2 \leq \|h_{j_1}\|^2,$$

$$\cdots$$

$$\sum_{j_p=1}^{k} \|h_{j_1 \cdots j_p}\|^2 \leq \|h_{j_1 \cdots j_{p-1}}\|^2.$$

Therefore

$$\sum_{j_1,\ldots,j_p} S_{j_1} S_{j_2} \cdots S_{j_p} h_{j_1 \cdots j_p} = h, \qquad (3.1.8)$$

and

$$\sum_{j_1,\ldots,j_p} \|h_{j_1 \cdots j_p}\|^2 \leq \|h\|^2. \qquad (3.1.9)$$

On the other hand, for any $l \in \mathcal{L}$,
$$\sum_{j_1,\ldots,j_p} \|N_{j_1}^* \cdots N_{j_p}^* l\|^2 = \sum \int |z_{j_1}|^2 \cdots |z_{j_p}|^2 (dE(z)l, l) \leq \frac{1}{2^{2p}}\|l\|^2.$$
Thus, from (3.1.8) and (3.1.9), we have
$$|(l, h)|^2 = |(l, \sum_{j_1,\ldots,j_p} S_{j_1} \cdots S_{j_p} h_{j_1 \cdots j_p})|^2 = |(l, \sum_{j_1,\ldots,j_p} N_{j_1} \cdots N_{j_p} h_{j_1 \cdots j_p})|^2$$

$$= |\sum_{j_1,\ldots,j_p} (N_{j_1}^* N_{j_2}^* \cdots N_{j_p}^* l, h_{j_1 \cdots j_p})|^2$$

$$\leq \sum_{j_1,\ldots,j_p} \|N_{j_1}^* \cdots N_{j_p}^* l\|^2 \sum_{j_1,\ldots,j_p} \|h_{j_1 \cdots j_p}\|^2 \leq \frac{1}{2^{2p}}\|l\|^2\|h\|^2 \quad (3.1.10)$$

for every pair of $l \in \mathcal{L}$ and $h \in \mathcal{H}$. Let $p \to \infty$ in (3.1.10). We have $(l, h) = 0$ for all $l \in \mathcal{L}$ and $h \in \mathcal{H}$. Thus $\mathcal{L} \perp \mathcal{H}$, which leads to $\mathcal{L} = \{0\}$ and thus proves the theorem. $\qquad \square$

3.2 Block-matrix decomposition of a pure *k*-tuple of commuting operators

Let $\mathbb{T} = (T_1, \ldots, T_k)$ be a commuting k-tuple of operators, i.e. a k-tuple of commuting operators on a Hilbert space $\mathcal{H}$. Define

$$D_{lm} = [T_l^*, T_m],$$

and define

$$M_{\mathbb{T}} = \text{cl} \bigvee_{l,m=1}^{k} D_{lm} \mathcal{H} \tag{3.2.1}$$

as the *defect* subspace of $\mathbb{T}$. Sometimes, we simply write $M_{\mathbb{T}}$ as M. Define

$$K \stackrel{\text{def}}{=} K_{\mathbb{T}} \stackrel{\text{def}}{=} \mathcal{H}_0 \stackrel{\text{def}}{=} \text{cl} \bigvee \{ T_1^{*m_1} \cdots T_k^{*m_k} M_{\mathbb{T}} : m_1, \ldots, m_k = 0, 1, 2, \ldots \} \tag{3.2.2}$$

and

$$G_n = \text{cl} \bigvee \{ T_1^{m_1} \cdots T_k^{m_k} \mathcal{H}_0 : m_1 + m_2 + \cdots + m_k \leq n \}. \tag{3.2.3}$$

Of course, $G_0 = \mathcal{H}_0$. Let

$$G_\infty = \text{cl} \bigvee \{ T_1^{m_1} \cdots T_k^{m_k} \mathcal{H}_0 : m_1, \ldots, m_k = 0, 1, 2, \ldots \}. \tag{3.2.4}$$

The operator tuple $\mathbb{T}$ is said to be pure, or completely non-normal, if there is no proper subspace of $\mathcal{H}$ which reduces $\mathbb{T}$ and contains $M_{\mathbb{T}}$ as a subspace. Similar to the proof of Proposition 1.2.1 and Corollary 1.2.2, by

$$T_l^{*} T_m^n = \sum_{j=0}^{n-1} T_l^j D_{lm} T_m^{n-j-1} + T_m^n T_l^*, \quad n = 1, 2, \ldots \tag{3.2.5}$$

we may prove the following:

Proposition 3.2.1. *Let $\mathbb{T}$ be a tuple of commuting operators on $\mathcal{H}$. The subspace G_∞ reduces $\mathbb{T}$, $\mathbb{T}|_{G_\infty}$ is pure and the restriction of $\mathbb{T}$ on $\mathcal{H} \ominus G_\infty$ is normal. The operator tuple $\mathbb{T}$ is pure iff $\mathcal{H} = G_\infty$.*

Since the spectral analysis of the normal tuple of operators is relatively well-studied. Therefore we only study pure commuting operator tuples. From now on, we assume that $\mathbb{T}$ is pure. As in Section 1.2, define

$$\mathcal{H}_n = G_n \ominus G_{n-1}, \quad n = 1, 2, \ldots$$

Then $\mathcal{H}_{n+1} \oplus \mathcal{H}_n = G_{n+1} \ominus G_{n-1}$, for $n > 0$. We have to prove that

$$T_j \mathcal{H}_n \subset \mathcal{H}_{n+1} \oplus \mathcal{H}_n, \quad n = 0, 1, \ldots, \ j = 1, 2, \ldots, k. \tag{3.2.6}$$

It is obvious that (3.2.6) holds good for $n = 0$. Besides,

$$T_j \mathcal{H}_n \subset T_j G_n \subset G_{n+1}.$$

Therefore we have to prove that for $n > 0$ if $h \perp G_{n-1}$ then $T_j h \perp G_{n-1}$. As a matter of fact, from $h \perp G_{n-1}$, we have

$$(h, T_1^{m_1} \cdots T_k^{m_k} T_1^{*n_1} \cdots T_k^{*n_k} x) = 0, \quad m_1 + \cdots + m_k \leq n - 1, \ x \in M. \tag{3.2.7}$$

However

$$(T_j h, T_1^{m_1} \cdots T_k^{m_k} T_1^{*n_1} \cdots T_k^{*n_k} x) = (h, T_j^* T_1^{m_1} \cdots T_k^{m_k} T_1^{*n_1} \cdots T_k^{*n_k} x).$$

By (3.2.5) we have

$$T_j^* T_1^{m_1} \cdots T_k^{m_k} T_1^{*n_1} \cdots T_k^{*n_k} x = T_1^{m_1} T_j^* T_2^{m_2} \cdots T_k^{m_k} T_1^{*n_1} \cdots T_k^{*n_k} x$$
$$+ \sum_{l=0}^{m_1 - 1} T_1^l x_{l1},$$

where $x_{l1} \in M$. Continuing this process, we may prove that

$$T_j^* T_1^{m_1} \cdots T_k^{m_k} T_1^{*n_1} \cdots T_k^{*n_k} x = T_1^{m_1} \cdots T_k^{m_k} T_j^* T_1^{*n_1} \cdots T_k^{*n_k} x$$
$$+ \sum_{l=0}^{m_1 - 1} T_1^l x_{l1} + T_1^{m_1} \sum_{l=0}^{m_2 - 1} T_2^l x_{l2} + \cdots + T_1^{m_1} \cdots T_{k-1}^{m_{k-1}} \sum_{l=0}^{m_k - 1} T_k^l x_{lk},$$

where $x_{lj} \in M$. By (3.2.7), we have

$$(T_j h, T_1^{m_1} \cdots T_k^{m_k} T_1^{*n_1} \cdots T_k^{*n_k} x) = 0, \quad m_1 + \cdots + m_k \leq n - 1, \ x \in M,$$

which proves $T_j h \perp G_{n-1}$ and hence (3.2.6).

Proposition 3.2.2. *Let* $\mathbb{T} = (T_1, \ldots, T_k)$ *be a pure k-tuple of commuting operators, then* T_j, $j = 1, 2, \ldots, k$, *has the following two-diagonal structure with respect to the orthogonal decomposition* $\mathcal{H} = \sum_{n=0}^{\infty} \oplus \mathcal{H}_n$:

$$T_j = \begin{pmatrix} B_{j0} & 0 & 0 & 0 & \cdots \\ E_{j1} & B_{j1} & 0 & 0 & \cdots \\ 0 & E_{j2} & B_{j2} & 0 & \cdots \\ 0 & 0 & E_{j3} & B_{j3} & \cdots \\ \vdots & \vdots & \vdots & \vdots & \ddots \end{pmatrix}, \tag{3.2.8}$$

where B_{jp} *and* E_{jp} *satisfy*

$$[B_{l0}^*, B_{m0}] + E_{l1}^* E_{m1} = C_{lm} \overset{def}{=} D_{lm}|_{\mathcal{H}_0}, \tag{3.2.9}$$

$$[B_{lp}^*, B_{mp}] + E_{l(p+1)}^* E_{m(p+1)} = E_{mp} E_{lp}^*, \quad p \geq 1, \tag{3.2.10}$$

and

$$E_{m(p+1)}^* B_{l(p+1)} = B_{lp} E_{m(p+1)}^*, \quad p \geq 0. \tag{3.2.11}$$

Proof. Let P_p be the projection from $\mathcal{H}$ to $\mathcal{H}_p$, $p = 0, 1, \ldots$. Define

$$B_{lp} = P_p T_l|_{\mathcal{H}_p} \text{ and } E_{l(p+1)} = P_{p+1} T_l|_{\mathcal{H}_p}, \quad p = 0, 1, 2, \ldots \quad (3.2.12)$$

From (3.2.6) and (3.2.12), it follows (3.2.8). From (3.2.8) and

$$[T_l^*, T_m] = \begin{pmatrix} C_{lm} & 0 & 0 & \cdots & 0 \\ 0 & 0 & 0 & \cdots & 0 \\ 0 & 0 & 0 & \cdots & 0 \\ \cdots & \cdots & \cdots & \cdots & \ddots \end{pmatrix},$$

it follows (3.2.9), (3.2.10) and (3.2.11). □

For a Hilbert space H and a natural number k, let H^k be the Hilbert space of all vectors $\left\{ \begin{pmatrix} x_1 \\ \vdots \\ x_k \end{pmatrix} : x_j \in H \right\}$ with inner product

$$\left(\begin{pmatrix} x_1 \\ \vdots \\ x_k \end{pmatrix}, \begin{pmatrix} y_1 \\ \vdots \\ y_k \end{pmatrix} \right) = \sum_l (x_l, y_l).$$

Define the operator $C \stackrel{\text{def}}{=} C_{\mathbb{T}}$ as an operator on $L(\mathcal{H}_0^k)$

$$C_{\mathbb{T}} \stackrel{\text{def}}{=} \begin{pmatrix} D_{11} & D_{12} & \cdots & D_{1k} \\ D_{21} & D_{22} & \cdots & D_{2k} \\ \cdots & \cdots & \cdots & \cdots \\ D_{k1} & D_{k2} & \cdots & D_{kk} \end{pmatrix}, \quad (3.2.13)$$

i.e.

$$C_{\mathbb{T}} \begin{pmatrix} x_1 \\ \vdots \\ x_k \end{pmatrix} = \begin{pmatrix} \sum D_{1m} x_m \\ \vdots \\ \sum D_{km} x_m \end{pmatrix}.$$

Define an operator E_p from $\mathcal{H}_{p-1}^k$ to $\mathcal{H}_p$ as $E_p \stackrel{\text{def}}{=} (E_{1p}, \ldots, E_{kp})$. Therefore

$$E_p \begin{pmatrix} x_1 \\ \vdots \\ x_k \end{pmatrix} = \sum_{l=1}^{k} E_{lp} x_l. \quad (3.2.14)$$

By the method of proving cl(range D_p) = $\mathcal{H}_p$ as in the proof of Theorem 1.2.4, we may also prove

$$\text{cl(range } E_p) = \mathcal{H}_p. \quad (3.2.15)$$

Let H be a Hilbert space. For any operator

$$A = \begin{pmatrix} A_{11} & \cdots & A_{1k} \\ \cdots & \cdots & \cdots \\ A_{k1} & \cdots & A_{kk} \end{pmatrix},$$

on H^k, it means

$$A \begin{pmatrix} x_1 \\ \vdots \\ x_k \end{pmatrix} = \begin{pmatrix} \sum A_{1l} x_l \\ \vdots \\ \sum A_{kl} x_l \end{pmatrix}.$$

Let the transpose of A be

$$A^t = \begin{pmatrix} A_{11} & \cdots & A_{k1} \\ \cdots & \cdots & \cdots \\ A_{1k} & \cdots & A_{kk} \end{pmatrix}.$$

For any operator

$$A = \begin{pmatrix} A_1 \\ \vdots \\ A_k \end{pmatrix}$$

from H to H^k, Ax means $\begin{pmatrix} A_1 x \\ \vdots \\ A_k x \end{pmatrix}$.

Then

$$A^t \begin{pmatrix} x_1 \\ \vdots \\ x_k \end{pmatrix} = \sum_{l=1}^{k} A_l x_l.$$

On the other hand, for an operator

$$A = (A_1 \cdots A_k)$$

from H^k to H, it means

$$A \begin{pmatrix} x_1 \\ \vdots \\ x_k \end{pmatrix} = \sum_{l=1}^{k} A_l x_l.$$

Then A^* and A^t are operators from H to H^k

$$A^* x = \begin{pmatrix} A_1^* x \\ \vdots \\ A_k^* x \end{pmatrix}, \qquad A^t x = \begin{pmatrix} A_1 x \\ \vdots \\ A_k x \end{pmatrix}.$$

For the operators $B_{1p}, \ldots, B_{kp}$ in (3.2.8), Let

$$B_p \overset{\text{def}}{=} (B_{1p}\ B_{2p} \cdots\ B_{kp}).$$

Under these notations, the formulas (3.2.9) and (3.2.10) may be rewritten as

$$B_0^* B_0 - (B_0^t B_0^{*t})^t + E_1^* E_1 = C_{\mathbb{T}}, \tag{3.2.16}$$

and

$$B_p^* B_p - (B_p^t B_p^{*t})^t + E_{p+1}^* E_{p+1} = (E_p^t E_p^{*t})^t. \tag{3.2.17}$$

Let us define the Hilbert space J_p, unitary operator V_p for $p = 0, 1, 2, \ldots$ and the positive operator $d_p \in L^k(J_p)$ for $k = 1, \ldots$ by mathematical induction, where $L^k(J_p)$ is the algebra of all $k \times k$ matrices with entries from $L(J_p)$. Let $J_0 = \mathcal{H}_0$ and $V_0 = I_{J_0}$ the identity operator on J_0. Let V_1 be the unitary operator from $J_1 \overset{\text{def}}{=} \text{cl range} |E_1 V_0|$ onto $\mathcal{H}_1$ and $d_1 \overset{\text{def}}{=} |E_1 V_0|$ satisfying

$$E_1 V_0 = V_1 d_1,$$

where $|A| \overset{\text{def}}{=} (A^* A)^{\frac{1}{2}}$ for an operator A, it is easy to see that

$$J_1 = J_0^k \ominus \ker(d_1).$$

Suppose we have defined J_p, V_p and d_p consequentially satisfying, $d_p = |E_p V_{p-1}|$,

$$E_p V_{p-1} = V_p d_p$$

where

$$E_p V_{p-1} \begin{pmatrix} x_1 \\ \vdots \\ x_k \end{pmatrix} = \sum_{j=1}^{k} E_{jp} V_{p-1} x_j$$

and $J_p = J_{p-1}^k \ominus \ker(d_p)$. Then we define $d_{p+1} = |E_{p+1} V_p|$, V_{p+1} satisfying

$$E_{p+1} V_p = V_{p+1} d_{p+1}$$

and $J_{p+1} = J_p^k \ominus \ker(d_{p+1})$. Thus

$$d_p = V_p^* E_p V_{p-1}, \quad p = 1, 2, \ldots \tag{3.2.18}$$

d_p is a positive operator on J_{p-1}, V_p is a unitary operator from J_{p-1} onto $\mathcal{H}_p$. Define d_{jp}, the operator from J_{p-1} to J_p satisfying

$$d_p \begin{pmatrix} x_1 \\ \vdots \\ x_k \end{pmatrix} = \sum_{j=1}^{k} d_{jp} x_j$$

for $x_j \in J_{p-1}$. Define

$$\Lambda_{jp} = V_p^* B_{jp} V_p, \quad j = 1, 2, \ldots, k, \ p = 0, 1, 2, \ldots \tag{3.2.19}$$

Proposition 3.2.3. *Let* $\mathbb{T} = (T_1, \ldots, T_k)$ *be a pure* k-*tuple of commuting operators on* $\mathcal{H}$. *Then* $\mathbb{T}$ *is unitary equivalent to* $\widetilde{T} = (\widetilde{T_1}, \widetilde{T_2}, \ldots, \widetilde{T_k})$ *which has the following two-diagonal structure with respect to the orthogonal decomposition* $\widetilde{\mathcal{H}} = \sum \oplus J_p$,

$$
\widetilde{T_j} = \begin{pmatrix}
\Lambda_{j0} & 0 & 0 & 0 & \cdots \\
d_{j1} & \Lambda_{j1} & 0 & 0 & \cdots \\
0 & d_{j2} & \Lambda_{j2} & 0 & \cdots \\
0 & 0 & d_{j3} & \Lambda_{j3} & \cdots \\
\vdots & \vdots & \vdots & \vdots & \ddots
\end{pmatrix}
\tag{3.2.20}
$$

where $J_0 = \mathcal{H}_0$, $J_p = J_{p-1}^k \ominus \ker(d_p)$

$$
\Lambda_{j0} = (T_j^*|_{\mathcal{H}_0})^*,
\tag{3.2.21}
$$

$$
d_1 = (C_{\mathbb{T}} - \Lambda_0^* \Lambda_0 + (\Lambda_0^t \Lambda_0^{*t})^t)^{\frac{1}{2}},
\tag{3.2.22}
$$

where $\Lambda_0(x_1, \ldots, x_k) = \sum_{l=1}^k \Lambda_l x_l$,

$$
d_{p+1} = (d_p^t d_p^{*t} - \Lambda_p^* \Lambda_p + (\Lambda_p^t \Lambda_p^{*t})^t)^{\frac{1}{2}}, \quad p \geq 1,
\tag{3.2.23}
$$

and

$$
\Lambda_{j(p+1)} = d_{p+1}^{-1} \begin{pmatrix}
\Lambda_{j,p} & 0 & \cdots & 0 \\
0 & \Lambda_{j,p} & \cdots & 0 \\
\cdots & \cdots & \cdots & \cdots \\
0 & 0 & \cdots & \Lambda_{j,p}
\end{pmatrix} d_{p+1}, \quad p \geq 0,
\tag{3.2.24}
$$

and $\Lambda_p = (\Lambda_{1,p}, \ldots, \Lambda_{n,p})$.

Proof. Let V be the diagonal matrix with diagonal elements $V_0, \ldots, V_p, \ldots$ on $\widetilde{\mathcal{H}} = \sum_j \oplus J_j$. Let $\widetilde{T_j} = V^* T_j V$, then (3.2.8) implies (3.2.20) by (3.2.18) and (3.2.19). Then (3.2.16) implies (3.2.22) and (3.2.17) implies (3.2.23), since d_{p+1} is positive. From (3.2.18), we have

$$
d_{m(p+1)} = V_{p+1}^* E_{m(p+1)} V_p.
$$

Therefore $d_{m(p+1)}^* = V_p^* E_{m(p+1)}^* V_{p+1}$. From (3.2.11) and (3.2.19) we have

$$
d_{m(p+1)}^* \Lambda_{j(p+1)} = \Lambda_{jp} d_{m(p+1)}^*, \quad m = 1, 2, \ldots, k.
$$

Thus $d_{p+1}^* \Lambda_{j(p+1)} = \Lambda_{jp} d_{p+1}^*$. But $d_{p+1}^* = d_{p+1}$, since d_{p+1} is positive. Therefore

$$
d_{p+1} \Lambda_{j(p+1)} = \Lambda_{jp} d_{p+1}.
\tag{3.2.25}
$$

From (3.2.25) and $\ker(d_{p+1}|_{J_{p+1}}) = \{0\}$, it follows (3.2.24). $\qquad\square$

Let $\Lambda_{\mathbb{T}}$ be the k-tuple of operators $(\Lambda_{10}, \ldots, \Lambda_{k0})$, and later in some sections, we denote it as $(\Lambda_1, \ldots, \Lambda_k)$.

Proposition 3.2.4. *Let* $\mathbb{T} = (T_1, \ldots, T_k)$ *be a pure k-tuple of commuting operators on $\mathcal{H}$. Then $\{\Lambda_{\mathbb{T}}, C_{\mathbb{T}}\}$ is a complete unitary invariant for $\mathbb{T}$.*

Proof. It is easy to see that $\{\Lambda_{\mathbb{T}}, C_{\mathbb{T}}\}$ is a unitary invariant. Let $\mathbb{T}' = (T_1', \ldots, T_k')$ be another pure k-tuple of commuting operators on $\mathcal{H}'$, and $\{\Lambda_{\mathbb{T}'}, C_{\mathbb{T}'}\}$ be the corresponding operator tuples and matrix on $\mathcal{H}_0'$. Suppose there is a unitary operator U_0 from $\mathcal{H}_0$ onto $\mathcal{H}_0'$ satisfying

$$\Lambda'_{j0} = U_0 \Lambda_{j0} U_0^*, \quad j = 1, 2, \ldots, k, \tag{3.2.26}$$

and

$$D'_{lm} = U_0 D_{lm} U_0^*, \quad l, m = 1, 2, \ldots, k, \tag{3.2.27}$$

where D'_{lm} is the entry of $C_{\mathbb{T}'}$. Let us consider the case that $\mathbb{T}$ and $\mathbb{T}'$ are in the form (3.2.20). Let $\{\Lambda'_{jp}, d'_{jp}\}$ be the entries of T_j' in the form (3.2.20). Let us define a sequence of unitary operators U_p, $p = 1, 2, \ldots$ from J_p onto J_p' satisfying

$$d'_p = U_p d_p U_p^* \tag{3.2.28}$$

and

$$\Lambda'_{jp} = U_p \Lambda_{jp} U_p^*, \tag{3.2.29}$$

by mathematical induction. From (3.2.26), it follows that (3.2.29) holds good for $p = 0$. We may define a unitary operator U_1 from J_0^k to $J_0'^k$ by

$$U_1 = \begin{pmatrix} U_0 & 0 & \cdots & 0 \\ 0 & U_0 & \cdots & 0 \\ \cdots & \cdots & \cdots & \cdots \\ 0 & 0 & \cdots & U_0 \end{pmatrix}$$

for $(x_1, \ldots, x_k) \in J_0^k$. From (3.2.26), (3.2.27) and (3.2.22) it is easy to see that $U_1 J_1 = J_1'$ and (3.2.28) holds good for $p = 1$ after we redefine U_1 as its restriction on J_1. By (3.2.24), we may prove that (3.2.29) holds good for $p = 1$. Suppose U_p, $p \geq 1$, has been defined as a unitary operator from J_p onto J_p' satisfying (3.2.28) and (3.2.29). Let us define U_{p+1} from J_p^k onto $J_p'^k$ by

$$U_{p+1} = \begin{pmatrix} U_p & 0 & \cdots & 0 \\ 0 & U_p & \cdots & 0 \\ \cdots & \cdots & \cdots & \cdots \\ 0 & 0 & \cdots & U_p \end{pmatrix}.$$

Then by (3.2.23), we may prove that $U_{p+1}J_{p+1} = J'_{p+1}$ and

$$
\begin{aligned}
U_{p+1}d_{p+1}U_p^* &= ((U_p d_p U_{p-1}^*)^t (U_p d_p U_{p-1}^*))^{*t} - (U_p \Lambda_p U_p^*)^* (U_p \Lambda_p U_p^*) \\
&\quad + ((U_p \Lambda_p U_p^*)^t (U_p \hat{\Lambda}_p U_p^*)^*)^{\frac{1}{2}} \\
&= d'_{p+1}.
\end{aligned}
$$

Besides, by (3.2.24), we have

$$
\begin{aligned}
&U_{p+1}\Lambda_{j(p+1)}U_{p+1}^* \\
&= (U_{p+1}d_{p+1}U_{p+1}^*)^{-1} \operatorname{diag}(U_p \Lambda_{jp} U_p^*, \ldots, U_p \Lambda_{jp} U_p^*) U_{p+1} d_{p+1} U_{p+1}^* \\
&= d_{p+1}'^{-1} \operatorname{diag}(\Lambda'_{jp}, \ldots, \Lambda'_{jp}) d'_{p+1}.
\end{aligned}
$$

Thus (3.2.28) and (3.2.29) holds good for $p = 1, 2, \ldots$. Let $U = \operatorname{diag}(U_0, U_1, \ldots, U_p, \ldots)$, i.e.

$$
U = \begin{pmatrix}
U_0 & 0 & \cdots & 0 & \cdots \\
0 & U_1 & \cdots & 0 & \cdots \\
& & \ddots & & \\
0 & \cdots & \cdots & U_p & \cdots \\
\cdots & \cdots & \cdots & \cdots & \ddots
\end{pmatrix}.
$$

Then $T'_j = U T_j U^{-1}$ which proves the proposition. $\qquad\square$

3.3   Some operator identities

In this section, we assume that $\mathbb{T} = (T_1, \ldots, T_k)$ is a pure commuting k-tuple of operators on $\mathcal{H}$. Let K be defined in (3.2.2),

$$
\Lambda_j \stackrel{\text{def}}{=} (T_j^* |_K)^*, \quad j = 1, 2, \ldots, k, \tag{3.3.1}
$$

which is the same Λ_{j0} defined in (3.2.21). Let

$$
\Lambda_{\mathbb{T}} \stackrel{\text{def}}{=} (\Lambda_1, \ldots, \Lambda_k). \tag{3.3.2}
$$

Let

$$
C_{lm} \stackrel{\text{def}}{=} [T_l^*, T_m]|_K, \quad l, m = 1, 2, \ldots, k, \tag{3.3.3}
$$

which is the same operator D_{lm} defined in (3.2.9). In this section, we will establish some operator identities related to Λ_j, C_{lm}, which will be useful in the next sections and for the theory of hyponormal tuples of commuting operators.

Similar to the function $Q(z, w)$ defined in (1.3.10), for $z, w \in \mathbb{C}$, define the $L(K)$-valued functions,

$$Q_{lm}(z, w) \overset{\text{def}}{=} (\bar{w} - \Lambda_l^*)(z - \Lambda_m) - C_{lm}, \tag{3.3.4}$$

and similar to (1.4.31), let us introduce the rational function

$$R_{lm}(z) = C_{lm}(z - \Lambda_m)^{-1} + \Lambda_l^*, \text{ for } z \in \rho(\Lambda_m), \tag{3.3.5}$$

where $l, m = 1, 2, \ldots, k$. It is obvious that

$$Q_{lm}(z, w) = (R_{lm}(z) - \bar{w})(\Lambda_m - z), \text{ for } z \in \rho(\Lambda_m). \tag{3.3.6}$$

Let P_K be the projection from $\mathcal{H}$ to K. Similar to Lemma 1.3.2, we have the following.

Lemma 3.3.1. *If* $w \in \rho(\Lambda_l) \bigcap \rho(T_l)$ *and* $z \in \rho(\Lambda_m) \bigcap \rho(T_m)$, *then* $Q_{lm}(z, w)$ *is invertible and*

$$P_K(T_l^* - \bar{w})^{-1}(T_m - z)^{-1}|_K = Q_{lm}(z, w)^{-1}. \tag{3.3.7}$$

Proof. Let $A \overset{\text{def}}{=} (T_l^* - \bar{w})^{-1}$ and $B \overset{\text{def}}{=} (T_m - z)^{-1}$, then from $[T_l^*, T_m] = C_{lm} P_K$ and

$$[A, B] = AB|_K C_{lm} P_K BA \tag{3.3.8}$$

we have

$$P_K AB|_K = P_K AB|_K C_{lm} P_K BA|_K + P_K BA|_K. \tag{3.3.9}$$

Notice that if $z \in \rho(\Lambda_m) \bigcap \rho(T_m)$, then $(\Lambda_m^* - \bar{z})K = K$. Thus $(T_m^* - \bar{z})K = K$. For $y \in K$, let $u \in K$ satisfying $y = (\Lambda_m^* - \bar{z})u$. Then $(T_m^* - \bar{z})u = y$ and

$$(T_m^* - \bar{z})^{-1}y = (\Lambda_m^* - \bar{z})^{-1}y = u. \tag{3.3.10}$$

Therefore, for any operator $Y \in L(\mathcal{H})$ and $x, y \in K$,

$$\begin{aligned} (BYx, y) &= (Yx, (T_m^* - \bar{z})^{-1}y) \\ &= (Yx, (\Lambda_m^* - \bar{z})^{-1}y) \\ &= ((\Lambda_m - z)^{-1}P_K Yx, y). \end{aligned}$$

Thus

$$P_K(T_m - z)^{-1}Y|_K = (\Lambda_m - z)^{-1}P_K Y|_K. \tag{3.3.11}$$

From (3.3.10), we have

$$(T_l^* - \bar{w})^{-1}|_K = (\Lambda_l^* - \bar{w})^{-1}. \tag{3.3.12}$$

From (3.3.9), (3.3.11) and (3.3.12), it follows

$$P_K AB|_K = P_K AB|_K C_{lm}(\Lambda_m - z)^{-1}(\Lambda_l^* - \bar{w})^{-1} + (\Lambda_m - z)^{-1}(\Lambda_l^* - \bar{w})^{-1}.$$

Thus

$$P_K AB|_K Q_{lm}(z, w) = I_K,$$

where I_K is the identity operator on K. Similarly from the commutator formula

$$[A, B] = BA|_K C_{lm} P_K AB,$$

we have

$$Q_{lm} P_K AB|_K = I_K.$$

Therefore $Q_{lm}(z, w)$ is invertible and (3.3.7) holds good. $\square$

Lemma 3.3.2. *If $z \in \rho(\Lambda_j) \bigcap \rho(T_j)$, then*

$$[R_{pj}(z), R_{qj}(z)] = 0, \quad p, q = 1, 2, \ldots, k. \tag{3.3.13}$$

Furthermore, if $w_n \in \rho(\Lambda_{m_n}) \bigcap \rho(T_{m_n})$, $1 \leq m_n \leq k$, then

$$P_K \prod_{n=1}^{l} (T_{m_n}^* - \bar{w}_n)^{-1}(T_j - z)^{-1}|_K = (\Lambda_j - z)^{-1} \prod_{n=1}^{l} (R_{m_n j}(z) - \bar{w}_n)^{-1},$$

$$\tag{3.3.14}$$

and

$$P_K (T_j^* - \bar{z})^{-1} \prod_{n=1}^{l} (T_{m_n} - w_n)^{-1}|_K$$

$$= \Big(\prod_{n=1}^{l} (R_{m_n j}(z)^* - w_n)^{-1} \Big)(\Lambda_j^* - \bar{z})^{-1}$$

$$= \Big(\prod_{n=1, n \neq t}^{l} [(\Lambda_{m_n} - w_n)^{-1}(R_{j m_n}(w_n) - \bar{z})^{-1} C_{j m_n}(\Lambda_{m_n} - w_n)^{-1}$$

$$+ (\Lambda_{m_n} - w_n)^{-1}] \Big)(\Lambda_{m_t} - w_t)^{-1}(R_{j m_t}(w_t) - \bar{z})^{-1}.$$

$$\tag{3.3.15}$$

Proof. Denote $A_n = (T_{m_n}^* - \bar{w}_n)^{-1}$ and $B = (T_j - z)^{-1}$. By (3.3.8), we have

$$P_K (\prod_{n=1}^{l} A_n) B|_K = P_K (\prod_{n=1}^{l} A_n) B|_K C_{lj} P_K BA_l|_K + P_K (\prod_{n=1}^{l-1} A_n) BA_l|_K.$$

Thus by (3.3.7), we have

$$P_K(\prod_{n=1}^{l} A_n)B|_K\big(I - C_{lj}(\Lambda_j - z)^{-1}(\Lambda_{m_l}^* - \bar{w}_l)^{-1}\big)$$

$$= P_K(\prod_{n=1}^{l-1} A_n)B|_K(\Lambda_{m_l}^* - \bar{w}_l)^{-1}$$

or

$$P_K(\prod_{n=1}^{l} A_n)B|_K = P_K(\prod_{n=1}^{l-1} A_n)B|_K(R_{m_l j}(z) - \bar{w}_l)^{-1}. \qquad (3.3.16)$$

By mathematical induction with respect to l, and (3.3.7), we have

$$P_k(\prod_{n=1}^{l} A_n)B|_K = Q_{1j}(z, w_1)^{-1}(R_{m_2 j}(z) - \overline{w_2})^{-1} \cdots (R_{m_l j}(z) - \overline{w_j})^{-1}.$$

$$(3.3.17)$$

By means of

$$Q_{lm}(z, w) = (R_{lm}(z) - \bar{w})(\Lambda_m - z),$$

we may prove that

$$P_K(\prod_{n=1}^{l} A_n)B|_K = (\Lambda_j - z_j)^{-1}(R_{m_1 j}(z) - \bar{w}_1)^{-1} \cdots (R_{m_l j}(z) - \bar{w}_l)^{-1}.$$

$$(3.3.18)$$

In the case of $m_1 = p$ and $m_2 = q$ in (3.3.8), we have

$$P_K(T_p^* - \bar{w}_1)^{-1}(T_q^* - \bar{w}_2)^{-1}(T_j - z)^{-1}|_K$$
$$= (\Lambda_j - z_j)^{-1}(R_{pj}(z) - \bar{w}_1)^{-1}(R_{qj}(z) - \bar{w}_2)^{-1}. \qquad (3.3.19)$$

In (3.3.19) exchanging p and q, w_1 and w_2, we have (3.3.13) since

$$[(T_p^* - \bar{w}_1)^{-1}, (T_q^* - \bar{w}_2)^{-1}] = 0. \qquad (3.3.20)$$

Therefore (3.3.18) implies (3.3.14). Taking adjoint of the both sides of (3.3.14), we have (3.3.15). $\qquad \square$

Notice that in (3.3.15), it contains an identity:

$$(\Lambda_{m_n} - w_n)^{-1}(R_{jm_n}(w_n) - \bar{z})^{-1}C_{jm_n}(\Lambda_{m_n} - w_n)^{-1} + (\Lambda_{m_n} - w_n)^{-1}$$
$$= (R_{jm_n}(z)^* - w_n)^{-1}.$$

This can be proved easily. It is also useful in (3.3.24).

In the case of $z_j \in \rho(\Lambda_j)$, let us introduce some $n \times n$ matrix

$$R_{m,l_1,\ldots,l_n}(z_1,\ldots,z_n) = (a_{ij}), \quad i, j = 1,\ldots,n.$$

In the matrix (a_{ij}), $a_{ij} = 0$ for $i > j, a_{ii} = R_{ml_i}(z_i)$

$$a_{ij} = -C_{ml_i} \prod_{p=i}^{j} (\Lambda_{l_p} - z_p)^{-1}, \qquad i < j.$$

$R_{m,l_1,\ldots,l_n}(z_1, z_2, \ldots, z_n)$ also can be denoted by $R_{m,L_n}(z)$, where L_n stands for the ordered set of integers $\{l_1, \ldots, l_n\}$ satisfying $1 \le l_j \le k$, and $z = (z_1, \ldots, z_n)$. It is easy to see that R_{m,L_n} is invertible, iff $R_{ml_j}, j = 1, 2, \ldots, n$ are invertible, since a upper diagonal matrix is invertible, iff all its diagonal elements are invertible.

We may rewrite R_{m,L_1} as R_{ml_1}, where $L_1 = \{l_1\}$.

Theorem 3.3.3. *Let* $\mathbb{T} = (T_1, \ldots, T_k)$ *be a commuting k-tuple of operators on $\mathcal{H}$. If $1 \le p, q, j_1, \ldots, j_n \le k$ and $z_m \in \rho(\Lambda_{j_m}) \bigcap \rho(T_{j_m})$, then*

$$[R_{p,J}(z), R_{q,J}(z)] = 0, \qquad (3.3.21)$$

where $z = (z_1, \ldots, z_n)$, $J = \{j_1, \ldots, j_n\}$.

Proof. In the case of $n = 1$, $R_{p,J}(z) = R_{p,j_1}(z_1)$. Thus (3.3.13) implies (3.3.21) for $n = 1$. Hence we only have to prove (3.3.21) for $n \ge 2$.

For the simplicity of notation, denote $A_l = (T_{j_l} - z_l)^{-1}$, $B_p = (T_p^* - \bar{w}_p)^{-1}$, $B_q = (T_q^* - \bar{w}_q)^{-1}$, $\hat{R}_{ti} = (R_{tj_i}(z_i) - \bar{w}_t)^{-1}$ for $w_t \in \rho(T_t)$, $t = p, q$. Denote $\lambda_i = (\Lambda_{j_i} - z_i)^{-1}$ and

$$U_{ml}(p, q) \stackrel{\text{def}}{=} (R_{qj_m}(z_m) - \bar{w}_q)C_{pj_m}\lambda_m \cdots \lambda_l + C_{qj_m}\lambda_m \cdots \lambda_l(R_{pj_l}(z_l) - \bar{w}_p)$$

$$- \sum_{i=l+1}^{m-1} C_{qj_m}\lambda_m \cdots \lambda_i C_{pj_i}\lambda_i \cdots \lambda_l,$$

for $m - l > 1$, and

$$U_{ml}(p, q) \stackrel{\text{def}}{=} (R_{qj_m}(z_m) - \bar{w}_q)C_{pj_m}\lambda_m\lambda_l + C_{qj_m}\lambda_m\lambda_l(R_{pj_l}(z_l) - \bar{w}_p)$$

for $m = l + 1$. It is easy to see that (3.3.21) is equivalent to

$$U_{ml}(p, q) = U_{ml}(q, p), \qquad (3.3.22)$$

for $l < m$ and arbitrary $j_l, \ldots, j_m$.

Denote

$$A_{lm} = \hat{R}_{ql}C_{qj_l} \prod_{i=m}^{l} \lambda_i \qquad \text{and} \qquad B_{lm} = C_{pj_l}(\prod_{i=m}^{l} \lambda_i)\hat{R}_{pm},$$

for $l \geq m$. Define $E_1 = E_1(p,q) = I$ and $E_m = E_m(p,q)$, for $m > 1$ by the recurrence formula

$$E_m = \sum_{i=1}^{m-1} A_{mi} E_i. \qquad (3.3.23)$$

Define $F_1 = I$ and F_m, $m > 1$ by the recurrence formula

$$\begin{aligned}
F_m &= (\lambda_m + \lambda_m \hat{R}_{qm} C_{qj_m} \lambda_m) F_{m-1} \\
&= (R_{j_m q}(w_q)^* - z_m)^{-1} F_{m-1}.
\end{aligned} \qquad (3.3.24)$$

We have to prove that

$$\sum_{i=1}^{m-1} \prod_{s=i}^{m-1} \lambda_s E_i = F_{m-1}\lambda_1, \qquad m = 2, 3, \ldots, \qquad (3.3.25)$$

and

$$E_m = \hat{R}_{qm} C_{qj_m} \lambda_m F_{m-1}\lambda_1, \qquad m = 2, 3, \ldots. \qquad (3.3.26)$$

It is easy to see that (3.3.25) and (3.3.26) hold good for $m = 2$. Suppose that (3.3.25) and (3.3.26) hold good for $m = 2, 3, \ldots, l$, we have to prove that (3.3.25) and (3.3.26) hold good for $m = l + 1$. From (3.3.24) and (3.3.25) it is easy to see that

$$\begin{aligned}
\sum_{i=1}^{l} (\prod_{s=i}^{l} \lambda_s) E_i &= \lambda_l F_{l-1}\lambda_1 + \lambda_l E_l \\
&= \lambda_l F_{l-1}\lambda_1 + \lambda_l \hat{R}_{ql} C_{qj_l} \lambda_l F_{l-1}\lambda_1 \\
&= (\lambda_l + \lambda_l \hat{R}_{ql} C_{qj_l} \lambda_l) F_{l-1}\lambda_1 \\
&= F_l \lambda_1
\end{aligned}$$

which proves (3.3.25) for $m = l + 1$. Then from (3.3.23), we have

$$\begin{aligned}
E_{l+1} &= \hat{R}_{q(l+1)} C_{qj_{l+1}} \lambda_{l+1} \lambda_l \sum_{i=1}^{l-1} (\prod_{s=i}^{l-1} \lambda_s) E_i + A_{(l+1)l} E_l \\
&= \hat{R}_{q(l+1)} C_{qj_{l+1}} \lambda_{l+1} \lambda_l F_{l-1}\lambda_1 + \hat{R}_{q(l+1)} C_{qj_{l+1}} \lambda_{l+1} \lambda_l \hat{R}_{ql} C_{qj_l} \lambda_l F_{l-1}\lambda_1 \\
&= \hat{R}_{q(l+1)} C_{qj_{l+1}} \lambda_{l+1} (\lambda_l + \lambda_l \hat{R}_{ql} C_{qj_l} \lambda_l) F_{l-1}\lambda_1 \\
&= \hat{R}_{q(l+1)} C_{qj_{l+1}} \lambda_{l+1} F_l \lambda_1,
\end{aligned}$$

by (3.3.24), which proves (3.3.26) for $m = l+1$. Hence (3.3.25) and (3.3.26) hold good.

Define

$$N_{ml} = N_{ml}(p,q) \stackrel{\text{def}}{=} \hat{R}_{qm} U_{ml}(p,q) \hat{R}_{pl}. \qquad (3.3.27)$$

Then

$$N_{m(m-1)} = A_{m(m-1)} + B_{m(m-1)} \tag{3.3.28}$$

and

$$N_{ml} = A_{ml} + B_{ml} - \sum_{j=l+1}^{m-1} A_{mj} B_{jl} \qquad \text{for } m > l+1. \tag{3.3.29}$$

Define $N_1 = I$ and N_m for $m > 1$ by the recurrence formula

$$N_m = \sum_{i=1}^{m-1} N_{mi} N_i. \tag{3.3.30}$$

We have to prove that

$$N_m = E_m + \sum_{i=1}^{m-1} B_{mi} N_i, \qquad m = 2, 3, \ldots. \tag{3.3.31}$$

by mathematical induction. It is obvious that (3.3.31) holds good for $m = 2$. Suppose (3.3.31) holds good for $m = 2, 3, \ldots, l-1$. Then from (3.3.29) and (3.3.30), we have

$$N_l - \sum_{i=1}^{l-1} B_{li} N_i = \sum_{i=1}^{l-1} (N_{li} - B_{li}) N_i$$

$$= A_{l(l-1)} N_{l-1} + \sum_{i=1}^{l-2} \left(A_{li} - \sum_{s=i+1}^{l-1} A_{ls} B_{si} \right) N_i$$

$$= A_{l1} + \sum_{i=2}^{l-1} A_{li} \left(N_i - \sum_{s=1}^{i-1} B_{is} N_s \right),$$

which equals to $A_{l1} + \sum_{i=2}^{l-1} A_{li} E_i$ by the hypothesis of the induction. By (3.3.23), it equals to E_l, which proves (3.3.31) for all $m \geq 2$. From (3.3.15), (3.3.24) and (3.3.26), we have

$$\hat{R}_{qm} C_{q j_m} \lambda_m P_K B_q \prod_{i=1}^{m-1} A_i \big|_K = E_m(p,q) \hat{R}_{q1}, \qquad \text{for } m \geq 2. \tag{3.3.32}$$

Define

$$M_m = M_m(p,q) = P_K B_p B_q \prod_{j=1}^{m} A_j \big|_K.$$

From (3.3.14), we have $M_1 = \lambda_1 \hat{R}_{p1} \hat{R}_{q1}$. Then by the commutation relation, (3.3.11) and (3.3.14), we have

$$M_m = P_K B_p B_q A_m C_{qj_m} P_K A_m B_q \prod_{i=1}^{m-1} A_i|_K + P_K B_p A_m B_q \prod_{i=1}^{m-1} A_i|_K$$

$$= \lambda_m \hat{R}_{pm} \hat{R}_{qm} C_{qj_m} \lambda_m P_K B_q \prod_{i=1}^{m-1} A_i|_K$$

$$+ P_K B_p A_m|_K C_{pj_m} P_K A_m B_p B_q \prod_{i=1}^{m-1} A_i|_K + P_K A_m B_p B_q \prod_{i=1}^{m-1} A_i|_K,$$

for $m > 1$. By (3.3.11) and (3.3.12), we have

$$M_m = \lambda_m \hat{R}_{pm} E_m \hat{R}_{q1} + (\lambda_m \hat{R}_{pm} C_{pj_m} \lambda_m + \lambda_m) M_{m-1}. \tag{3.3.33}$$

We have to prove that

$$M_m = \lambda_m \hat{R}_{pm} N_m \hat{R}_{q1} + \lambda_m M_{m-1} \tag{3.3.34}$$

by mathematical induction. It is easy to see that (3.3.34) holds good for $m = 2$, since

$$E_2 \hat{R}_{q1} + C_{pj_2} \lambda_2 M_1 = (E_2 + B_{21}) \hat{R}_{q1} = N_2 \hat{R}_{q1}.$$

Suppose (3.3.34) holds good for $m = 2, 3, \ldots, l-1$. Then from (3.3.33)

$$M_l = \lambda_l \hat{R}_{pl} E_l \hat{R}_{q1} + \lambda_l \hat{R}_{pl} C_{pj_l} \lambda_l (\lambda_{l-1} \hat{R}_{p(l-1)} N_{l-1} \hat{R}_{q1} + \lambda_{l-1} M_{l-2})$$

$$+ \lambda_l M_{l-1}$$

$$= \lambda_l \hat{R}_{pl} (E_l + B_{l(l-1)} N_{l-1}) \hat{R}_{q1} + \lambda_l \hat{R}_{pl} C_{pj_l} \lambda_l \lambda_{l-1} M_{l-2} + \lambda_l M_{l-1}.$$

Continuing this process, we may prove that

$$M_l = \lambda_l \hat{R}_{pl} (E_l + \sum_{j=1}^{l-1} B_{lj} N_j) \hat{R}_{q1} + \lambda_l M_{l-1}. \tag{3.3.35}$$

From (3.3.31) and (3.3.35), we may prove that (3.3.34) holds good for all $m \geq 2$.

From the fact that $[B_p, B_q] = 0$, we have $M_m(p,q) = M_m(q,p)$. Therefore (3.3.34) implies that

$$\hat{R}_{pm} N_m(p,q) \hat{R}_{q1} = \hat{R}_{qm} N_m(q,p) \hat{R}_{p1}. \tag{3.3.36}$$

From (3.3.27) and (3.3.30), we have

$$\hat{R}_{pm} N_m \hat{R}_{q1} = \hat{R}_{pm} \hat{R}_{qm} U_{m1} \hat{R}_{p1} \hat{R}_{q1} + \sum_{l=2}^{m-1} \hat{R}_{pm} \hat{R}_{qm} U_{m1} \hat{R}_{pl} N_l \hat{R}_{q1}. \tag{3.3.37}$$

From $N_{21} = N_2$, (3.3.36) and (3.3.37), we have (3.3.22) for $m = 2$ and $l = 1$. But $j_1, \ldots, j_n$ are arbitrary numbers in $\{1, 2, \ldots, k\}$, therefore (3.3.22) holds good for $m = l + 1$. Assume that (3.3.22) holds good for $m = l + 1, \ldots, l + i, i \geq 1$. Then from (3.3.36) and (3.3.37) in which $l = 1, m = i + 2$, we may prove that (3.3.22) holds good for $m = i + 2$, $l = 1$. Thus (3.3.22) holds good for $m = l + (i + 1)$, which proves (3.3.22) for any $m > l$ and hence the theorem.   □

In order to understand the proof of this theorem better, the author suggests the reader to read this proof in the case $n = 2, 3$ firstly.

Let $\mathbb{T} = \{T_1, \ldots, T_k\}$ be a commuting k-tuple on a Hilbert space $\mathcal{H}$. For the ordered set $L = \{l_1, \ldots, l_n\}$ satisfying $1 \leq l_i \leq k$, let $\rho_L = \rho(T_{L_1}) \times \ldots \times \rho(T_{l_n})$. For the ordered sets $P_m = \{p_1, \ldots, p_m\}$ and $Q_n = \{q_1, \ldots, q_n\}$, $1 \leq p_i, q_j \leq k$, let S_{P_m, Q_n} be the $L(K)$-valued function

$$S_{P_m, Q_n}(z; w) \stackrel{\text{def}}{=} P_K \prod_{i=1}^{m} (T_{p_i}^* - \bar{w}_i)^{-1} \prod_{j=1}^{n} (T_{q_j} - z_j)^{-1}|_K, \qquad (3.3.38)$$

where $z = (z_1, \ldots, z_n) \in \rho_{Q_n}$ and $w = (w_1, \ldots, w_m) \in \rho_{P_m}$.

Define the matrix

$$\mathfrak{S}_{P_m, Q_n} \stackrel{\text{def}}{=} \begin{pmatrix} S_{P_m, Q_1} & S_{P_m, Q_2} & \cdots & S_{P_m, Q_n} \\ S_{P_{m-1}, Q_1} & S_{P_{m-1}, Q_2} & \cdots & S_{P_{m-1}, Q_n} \\ \cdots & \cdots & \cdots & \cdots \\ S_{P_2, Q_1} & S_{P_2, Q_2} & \cdots & S_{P_2, Q_n} \\ S_{P_1, Q_1} & S_{P_1, Q_2} & \cdots & S_{P_1, Q_n} \end{pmatrix} \qquad (3.3.39)$$

where $P_i = \{p_1, \ldots, p_i\}$ and $Q_j = \{q_1, \ldots, q_j\}$ are ordered subsets of P_m and Q_n respectively.

For $w \in \rho(\Lambda_m)$, $z = (z_1, \ldots, z_n)$, if $z_j \in \rho(\Lambda_{q_j})$ and $R_{q_j m}(w)^* - z_j$ is invertible, let

$$X_{m, Q_j} \stackrel{\text{def}}{=} X_{m, Q_j}(z; w) \stackrel{\text{def}}{=} (\prod_{l=1}^{j} (R_{q_l m}(w)^* - z_l)^{-1})(\Lambda_m^* - \bar{w})^{-1}, \qquad (3.3.40)$$

where $Q_j = \{q_1, \ldots, q_j\}$ is an ordered subset of Q_n.

Define $X_{P_i, Q_1}, \ldots, X_{P_i, Q_n}$ by the following formula

$$(X_{P_i, Q_1} \cdots X_{P_i, Q_n}) = (X_{P_1, Q_1} \cdots X_{P_1, Q_n}) \prod_{j=2}^{i} (R_{p_j, Q_n} - \bar{w}_j)^{-1} \qquad (3.3.41)$$

where $P_i = \{p_1, \ldots, p_i\}$ is an ordered subset of $P_m = \{p_1, \ldots, p_m\}$, $1 \leq i \leq m$, the left side of (3.3.41) and the first factor of the right-hand side of (3.3.41) are one row matrices.

Let $\mathfrak{X}_{P_m,Q_n}(z,w)$ be the following matrix

$$\mathfrak{X}_{P_m,Q_n} \overset{\text{def}}{=} \begin{pmatrix} X_{P_m,Q_1} & X_{P_m,Q_2} & \cdots & X_{P_m,Q_n} \\ X_{P_{m-1},Q_1} & X_{P_{m-1},Q_2} & \cdots & X_{P_{m-1},Q_n} \\ \cdots & \cdots & \cdots & \cdots \\ X_{P_2,Q_1} & X_{P_2,Q_2} & \cdots & X_{P_2,Q_n} \\ X_{P_1,Q_1} & X_{P_1,Q_2} & \cdots & X_{P_1,Q_n} \end{pmatrix}. \tag{3.3.42}$$

The following theorem is a generalization of Lemma 1.3.2 to the case of the tuple of commuting operators.

Theorem 3.3.4. *Let* $\mathbb{T} = \{T_1, \ldots, T_k\}$ *be a commuting k-tuple on a Hilbert space* $\mathcal{H}$. *If* $z = (z_1, \ldots, z_n)$ *and* $w = (w_1, \ldots, w_m)$ *satisfy the condition that* $z_i \in \rho(T_{q_i})$, $i = 1, 2, \ldots, n$ *and* $w_j \in \rho(T_{p_j})$, $j = 1, 2, \ldots, m$. *Then*

$$\mathfrak{S}_{P_m,Q_n}(z,w) = \mathfrak{X}_{P_m,Q_n}(z,w). \tag{3.3.43}$$

Proof. In Lemma 3.3.2, the formula (3.3.14) and (3.3.15) are equivalent to (3.3.43) in the case of $n = 1$ and $m = 1$ respectively. Let us prove (3.3.43) by mathematical induction. Suppose (3.3.43) holds good for $m = l - 1 \geq 1$. Let us calculate S_{P_l,Q_n} for any n.

Denote $A_i = (T_{p_i}^* - \bar{w}_i)^{-1}$ and $B_i = (T_{q_i} - z_i)^{-1}$. Then by the commutator formula

$$[A_i, B_j] = B_j A_i|_K C_{p_i q_j} P_K A_i B_j$$

and $A_j|_K = (\Lambda_{p_j}^* - \bar{w}_j)^{-1}$ we have

$$S_{P_l,Q_n} = P_K A_1 \cdots A_l B_1 \cdots B_n|_K$$

$$= \sum_{j=1}^{n} P_K A_1 \cdots A_{l-1} B_1 \cdots B_j|_K (\Lambda_{p_l}^* - \bar{w}_l)^{-1} C_{p_l q_j} P_K A_l B_j \cdots B_n|_K$$

$$+ S_{P_{l-1},Q_n} (\Lambda_{p_l}^* - \bar{w}_l)^{-1}.$$

By (3.3.15), we have

$$S_{P_l,Q_n} = \sum_{j=1}^{n} S_{P_{l-1},Q_j} f_{jn}$$

where $f_{ij} = (\Lambda_{p_l}^* - \bar{w}_l)^{-1} C_{p_l q_i} \prod_{s=i}^{j} (R_{q_s p_l}(w_l)^* - z_s)^{-1} (\Lambda_{p_l}^* - \bar{w}_l)^{-1}$, for $i < j$ and $f_{ii} = (R_{p_l q_i}(z_i)^* - \bar{w}_l)^{-1}$, since

$$(\Lambda_{p_l}^* - \bar{w}_l)^{-1} (C_{p_l q_n} P_K A_l B_n|_K + 1)$$

$$= (\Lambda_{p_l}^* - \bar{w}_l)^{-1} \big((-(R_{p_l q_n}(z_n) - \bar{w}_l) + (\Lambda_{p_l}^* - \bar{w}_l))(R_{p_l q_n}(z_n) - \bar{w}_l)^{-1} + 1 \big)$$

$$= (R_{p_l q_n}(z_n) - \bar{w}_l)^{-1}.$$

Therefore

$$\left(S_{P_l,Q_1}\, S_{P_l,Q_2}\, \cdots\, S_{P_l,Q_n}\right) = \left(S_{P_{l-1},Q_1}\, S_{P_{l-1},Q_2}\, \cdots\, S_{P_{l-1},Q_n}\right) F_n$$

where $F_n = (f_{ij})_{i,j=1,2,\cdots,n}$ and $f_{ij} = 0$ for $i > j$. By the hypothesis of mathematical induction

$$\left(S_{P_{l-1},Q_1}\, \cdots\, S_{P_{l-1},Q_n}\right) = \left(X_{P_{l-1},Q_1}\, \cdots\, X_{P_{l-1},Q_n}\right).$$

Therefore by (3.3.40) to show that $S_{P_l,Q_j} = X_{P_l,Q_j}$, we only have to prove that

$$F_n = (R_{p_l,Q_n} - \bar{w}_l)^{-1}. \qquad (3.3.44)$$

To prove (3.3.44), we only have to show that for any pair (i,j), $1 \le i,j \le n$

$$a_{ii} f_{ii} = I \qquad (3.3.45)$$

and

$$\sum_{s=i}^{j} a_{is} f_{sj} = 0, \qquad i < j, \qquad (3.3.46)$$

where $(a_{ij})_{i,j=1,2,\ldots,n} = R_{p_l,Q_n} - \bar{w}_l$. It is obvious that (3.3.45) holds good. To prove (3.3.46), notice that

$$\sum_{s=i}^{j} a_{is} f_{sj} = I_1 + I_2 + I_3,$$

for $j - i > 0$, where

$$I_1 = a_{ii} f_{ij}$$

$$= (R_{p_l q_i}(z_i) - \bar{w}_l)(\Lambda_{p_l}^* - \bar{w}_l)^{-1} C_{p_l q_i} \prod_{s=i}^{j} (R_{q_s p_l}(w_l)^* - z_s)^{-1} (\Lambda_{p_l}^* - \bar{w}_l)^{-1},$$

$$I_2 = \sum_{s=i+1}^{j-1} a_{is} f_{sj} = \sum_{s=i+1}^{j-1} A_{is} A_{sj},$$

where

$$A_{is} = C_{p_l q_i} \prod_{t=i}^{s} (\Lambda_{q_t} - z_t)^{-1} (\Lambda_{p_l}^* - \bar{w}_l)^{-1},$$

for $1 \le i \le s \le n$, if $j > i+1$, and $I_2 = 0$, if $j = i+1$. Besides,

$$I_3 = -C_{p_l q_i} \prod_{t=i}^{j} (\Lambda_{q_t} - z_t)^{-1} (R_{p_l q_j}(z_j) - \bar{w}_l)^{-1}.$$

However for $j > i + 1$,

$$I_2 = - \sum_{s=i+1}^{j-1} C_{p_l q_i} \prod_{t=i}^{s} (\Lambda_{q_t} - z_t)^{-1} \big(- (R_{q_s p_l}(w_l)^* - z_s) + (\Lambda_{q_s} - z_s) \big)$$

$$\cdot \prod_{t=s}^{j} (R_{q_t p_l}(w_l)^* - z_t)^{-1} (\Lambda_{p_l}^* - \bar{w}_l)^{-1}$$

$$= \sum_{s=i+1}^{j-1} C_{p_l q_i} \prod_{t=i}^{s} (\Lambda_{q_t} - z_t)^{-1} \prod_{t=s+1}^{j} (R_{q_t p_l}(w_l)^* - z_t)^{-1} (\Lambda_{p_l}^* - \bar{w}_l)^{-1}$$

$$- \sum_{s=i+1}^{j-1} C_{p_l q_i} \prod_{t=i}^{s-1} (\Lambda_{q_t} - z_t)^{-1} \prod_{t=s}^{j} (R_{q_t p_l}(w_l)^* - z_t)^{-1} (\Lambda_{p_l}^* - \bar{w}_l)^{-1}.$$

$$(3.3.47)$$

Most of the terms in the two summations of the right-hand side of (3.3.47) canceled each other. Thus $I_2 = J_1 + J_2$, where

$$J_1 = C_{p_l q_i} \prod_{t=i}^{j-1} (\Lambda_{q_t} - z_t)^{-1} (R_{q_j p_l}(w_l)^* - z_j)^{-1} (\Lambda_{p_l}^* - \bar{w}_l)^{-1}$$

$$= C_{p_l q_i} \prod_{t=i}^{j} (\Lambda_{q_t} - z_t)^{-1} (R_{p_l q_j}(z_j) - \bar{w}_l)^{-1}$$

$$= -I_3,$$

since

$$(R_{q_j p_l}(w_l)^* - z_j)^{-1} (\Lambda_{p_l}^* - \bar{w}_l)^{-1} = (\Lambda_{q_j} - z_j)^{-1} (R_{p_l q_j}(z_j) - \bar{w}_l)^{-1}.$$

The term

$$J_2 = -C_{p_l q_i} (\Lambda_{q_i} - z_i)^{-1} \prod_{t=i+1}^{j} (R_{q_t p_l}(w_l)^* - z_t)^{-1} (\Lambda_{p_l}^* - \bar{w}_l)^{-1}$$

$$= \big((R_{p_l q_i}(z_i) - \bar{w}_l) - (\Lambda_{p_l}^* - \bar{w}_l) \big) \prod_{t=i+1}^{j} (R_{q_t p_l}(w_l)^* - z_t)^{-1} (\Lambda_{p_l}^* - \bar{w}_l)^{-1}.$$

$$(3.3.48)$$

But the product of the first four factors in I_1 from the left is

$$(R_{p_l q_i}(z_i) - \bar{w}_l)(\Lambda_{p_l}^* - \bar{w}_l)^{-1} C_{p_l q_i} (R_{q_i p_l}(w_l)^* - z_i)^{-1}$$

$$= (R_{p_l q_i}(z_i) - \bar{w}_l)\big(- (R_{q_i p_l}(w_l)^* - z_i) + (\Lambda_{q_i} - z_i) \big)(R_{q_i p_l}(w_l)^* - z_i)^{-1}$$

$$= -(R_{p_l q_i}(z_i) - \bar{w}_l) + (R_{p_l q_i}(z_i) - \bar{w}_l)(\Lambda_{q_i} - z_i) Q_{p_l q_i}(z_i, w_l)^{-1} (\Lambda_l^* - \bar{w}_l)$$

$$= -(R_{p_l q_i}(z_i) - \bar{w}_l) + (\Lambda_{p_l}^* - \bar{w}_l).$$

$$(3.3.49)$$

Thus comparing to (3.3.49) $I_1 + J_2 = 0$, which proves that $I_1 + I_2 + I_3 = 0$, for $j > i + 1$. If $j = i + 1$, then $I_2 = 0$,

$$
\begin{aligned}
I_1 &= (R_{p_l q_i}(z_i) - \bar{w}_l)(\Lambda_{p_l}^* - \bar{w}_l)^{-1} C_{p_l q_i}(R_{q_i p_l}(w_l)^* - z_i)^{-1} \\
&\quad Q_{p_l q_{i+1}}(z_{i+1}, w_l)^{-1} \\
&= \left(-(R_{p_l q_i}(z_i) - \bar{w}_l) + (\Lambda_{p_l}^* - \bar{w}_l) \right) Q_{p_l q_{i+1}}(z_{i+1}, w_l)^{-1}
\end{aligned}
$$

by (3.3.49) and

$$
I_3 = -C_{p_l q_i}(\Lambda_{q_i} - z_i)^{-1} Q_{p_l q_{i+1}}(z_{i+1}, w_l)^{-1}.
$$

Thus $I_1 + I_3 = 0$ which proves $I_1 + I_2 + I_3 = 0$ as well, and hence it proves (3.3.46). Therefore (3.3.43) is proved. $\qquad\square$

3.4 Analytic model for a subnormal k-tuple of operators

Let $S = (S_1, \ldots, S_k)$ be a subnormal k-tuple of operators on a Hilbert space $\mathcal{H}$ with m.n.e. $\mathbb{N} = (N_1, \ldots, N_k)$ on $\mathcal{K} \supset \mathcal{H}$. Let $\widetilde{\mathcal{H}} = \mathcal{K} \ominus \mathcal{H}$. Let P and $\widetilde{P}$ be the projections from $\mathcal{K}$ to $\mathcal{H}$ and $\widetilde{\mathcal{H}}$ respectively. Let A_j be the operator from $\widetilde{\mathcal{H}}$ to $\mathcal{H}$ defined by

$$
A_j x = P N_j x, \quad x \in \widetilde{\mathcal{H}}, \tag{3.4.1}
$$

$j = 1, 2, \ldots, k$. Let $\widetilde{S}_j$ be the operator from $\widetilde{\mathcal{H}}$ to $\widetilde{\mathcal{H}}$ defined by

$$
\widetilde{S}_j x = N_j^* x, \quad x \in \widetilde{\mathcal{H}}. \tag{3.4.2}
$$

The range of $\widetilde{S}_j$ is in $\widetilde{\mathcal{H}}$, since $\widetilde{\mathcal{H}}$ is an invariant subspace of N_j^*. Then as in §1.3, we have

$$
N_j = \begin{pmatrix} S_j & A_j \\ 0 & \widetilde{S}_j^* \end{pmatrix}. \tag{3.4.3}
$$

The condition (see §3.1)

$$
[N_l, N_m] = 0, \quad l, m = 1, 2, \ldots, k
$$

is equivalent to the following three identities:

$$
[S_l, S_m] = 0, \tag{3.4.4}
$$

$$
[\widetilde{S}_l^*, \widetilde{S}_m^*] = 0 \tag{3.4.5}
$$

and

$$
S_l A_m + A_l \widetilde{S}_m^* = S_m A_l + A_m \widetilde{S}_l^*. \tag{3.4.6}
$$

The Fuglede-Putnam's equality

$$[N_l^*, N_m] = 0, \quad l, m = 1, 2, \ldots, k$$

is equivalent to

$$[S_l^*, S_m] = A_m A_l^*, \tag{3.4.7}$$

$$[\widetilde{S}_l^*, \widetilde{S}_m] = A_m^* A_l, \tag{3.4.8}$$

and

$$A_l \widetilde{S}_m = S_m^* A_l. \tag{3.4.9}$$

As in §3.2, let $M = \mathrm{cl} \vee_{i,j=1}^k [S_i^*, S_j] \mathcal{H}$. By (3.4.7), it is easy to see that

$$M = \mathrm{cl} \vee_{j=1}^k A_j \mathcal{H}. \tag{3.4.10}$$

Then M is an invariant subspace of S_j^*, $j = 1, 2, \ldots, k$. Let $C_{lm} \in L(M)$ defined by

$$C_{lm} x = [S_l^*, S_m] x, \quad x \in M \tag{3.4.11}$$

and $\Lambda_j \in L(M)$ defined by

$$\Lambda_j \overset{\text{def}}{=} (S_j^*|_M)^*. \tag{3.4.12}$$

Example. Let $H^2(\mathbb{T})$, $L^2(\mathbb{T})$, U_+ and N be the spaces and operators defined in the example in §1.3. Let $S_j \overset{\text{def}}{=} U_+^j$, $N_j \overset{\text{def}}{=} N^j$, $j = 1, 2, \ldots, k$. Then $\mathbb{S} = (S_1, \ldots, S_k)$ is a pure subnormal k-tuple of operators on $H^2(\mathbb{T})$ with m.n.e. $\mathbb{N} = (N_1, \ldots, N_k)$. Then $M = \{c_0 + c_1 z + \cdots + c_k z^{k-1} : c_j \in \mathbb{C}, j = 0, 1, 2, \ldots, k-1\}$,

$$\Lambda_j^*(c_0 + c_1 z + \cdots + c_{k-1} z^{k-1}) = c_j + c_{j+1} z + \cdots + c_{k-1} z^{k-j-1},$$

and

$$C_{lm}(c_0 + c_1 z + \cdots + c_k z^{k-1}) = \sum_{j=0}^{\min(l,m)-1} \gamma_{jlm} c_j$$

for some non-negative fixed numbers γ_{jlm}.

Lemma 3.4.1. *Let $\mathbb{S}$ be a subnormal k-tuple of operator with m.n.e. $\mathbb{N}$ on $\mathcal{K}$. Then*

$$\sigma(N_j) \subset \sigma(S_j), \tag{3.4.13}$$

and

$$\sigma(\Lambda_j) \subset \sigma(S_j), \tag{3.4.14}$$

$j = 1, 2, \ldots, k.$

Proof. Suppose $z \in \rho(S_j)$. Let $E_j(\cdot)$ be the spectral measure of the normal operator N_j. Define $L_j = E_j(B(z, \epsilon))$, where $0 < \epsilon < \|(S_j - z)^{-1}\|^{-1}$. Then as in the proof of Theorem 1.1.2, $L_j \perp \mathcal{H}$ and L_j reduces N_j. From

$$[N_j, N_m] = [N_j, N_m^*] = 0, \quad m = 1, \dots, k$$

it concludes that L_j also reduces N_m, $m = 1, 2, \dots, k$. But $\mathbb{N}$ is an m.n.e. of $\mathbb{S}$, therefore $L_j = \{0\}$, i.e. $E_j(B(z, \epsilon)) = 0$ and $z \in \rho(N_j)$ which proves (3.4.13).

The proof of (3.4.14) is similar to the proof of (1.3.8). Let $\lambda \in \rho(S_j^*)$, then $\lambda \in \rho(N_j^*)$ by (3.4.13). For every $f \in \widetilde{\mathcal{H}}$, there is a unique pair of vectors $f_0 \in \mathcal{H}$ and $f_1 \in \widetilde{\mathcal{H}}$ such that

$$f = (\lambda - N_j^*)(f_0 + f_1) = (\lambda - S_j^*)f_0 + (-A_j^* f_0 + (\lambda - \widetilde{S}_j)f_1).$$

Therefore, $(\lambda - S_j^*)f_0 = Pf = 0$, and hence $f_0 = 0$, since $\lambda \in \rho(S_j^*)$. Thus

$$f = (\lambda - \widetilde{S}_j)f_1.$$

Hence,

$$A_l f = A_l(\lambda - \widetilde{S}_j)f_1 = (\lambda - S_j^*)A_l f_1$$

by (3.4.9), for $l = 1, 2, \dots, k$. This shows that $(\lambda - S_j^*)M$ is a dense subset of M, and hence it equals M, since $\lambda - S_j^*$ is invertible. Therefore $(\lambda - \Lambda_j^*)M = M$, i.e. $\lambda \in \rho(\Lambda_j^*)$. Thus $\sigma(\Lambda_j^*) \subset \sigma(S_j^*)$, which proves (3.4.14). $\qquad\square$

Let P_M be the projection to M. Let $\mathrm{sp}(\mathbb{N}) \subset \mathbb{C}^k$ be the spectrum of the normal k-tuple $\mathbb{N}$, and $E(\cdot)$ be the spectral measure of $\mathbb{N}$ on $\mathrm{sp}(\mathbb{N})$. Define a $L(M)$-valued positive measure

$$e(\cdot) = P_M E(\cdot)|_M. \tag{3.4.15}$$

Let $L^2(e)$ be the Hilbert space of all M-valued measurable function f satisfying

$$(f, f) \stackrel{\text{def}}{=} \int_{\mathrm{sp}(\mathbb{N})} (e(du)f(u), f(u))_M < \infty.$$

In $L^2(e)$, two functions f and g are considered as the same vector, if $\|f - g\| = 0$.

Denote

$$Q_{lm}(z, w) = (\bar{w} - \Lambda_l^*)(z - \Lambda_m) - C_{lm}. \tag{3.4.16}$$

Denote $S_j(z) \stackrel{\text{def}}{=} (S_j - z)^{-1}$, for $z \in \rho(S_j)$. Define

$$S_{lm}(z, w) = P_M S_l(w)^* S_m(z)|_M \text{ for } z \in \rho(S_m), w \in \rho(S_l). \tag{3.4.17}$$

Lemma 3.4.2. *Let $\mathbb{S}$ be a subnormal k-tuple of operators. Then $Q_{lm}(z,w)$ is invertible, and*

$$S_{lm}(z,w) = Q_{lm}(z,w)^{-1}, \qquad (3.4.18)$$

for $z \in \rho(S_m)$ and $w \in \rho(S_l)$.

We omit the proof, since it is similar to the proof of Lemma 1.3.2.

Lemma 3.4.3. *Let $\mathbb{S}$ be a subnormal k-tuple of operators on $\mathcal{H}$ with m.n.e. $\mathbb{N}$. Then the measure $e(\cdot)$ has the property that*

$$\int_F Q_{lm}(u_m, u_l)e(du) = \int_F e(du)Q_{lm}(u_m, u_l) = 0 \qquad (3.4.19)$$

for every Borel set $F \subset \mathrm{sp}(\mathbb{N})$, where $u = (u_1, \ldots, u_k)$.

From the proof of (1.3.15), it is easy to see that (3.4.19) holds good for $l = m$. But for $l \neq m$, (3.4.19) is new and important in certain cases.

Proof. We only have to prove that

$$\int Q_{lm}(u_m, u_l)f(u)e(du) = 0 \qquad (3.4.20)$$

for $f(u) = [(\bar{u}_l - \bar{w})(u_m - z)\Pi_{p=1}^r(\bar{u}_{l_p} - \bar{w}_p)\Pi_{q=1}^t(u_{m_q} - z_q)]^{-1}$ for $z \in \rho(S_m)$, $w \in \rho(S_l)$, $z_q \in \rho(S_{m_q})$, $w_p \in \rho(S_{l_p})$, and $l_p, m_q \in \{1, \ldots, k\}$, since then (3.4.20) will be also holds good for any continuous function on $\sigma(S_1) \times \cdots \times \sigma(S_k)$ and

$$\int_F Q_{lm}(u_m, u_l)e(du) = 0 \qquad (3.4.21)$$

holds good for any Borel set $F \subset \sigma(S_1) \times \cdots \times \sigma(S_k)$. By Lemma 3.4.1, (3.4.21) holds good for any Borel set $F \subset \mathrm{sp}(\mathbb{N})$. The rest of (3.4.19) can be proved similarly.

For the simplicity of notation, denote $S_l(w) = (S_l - w)^{-1}$,

$$\hat{S}_p = S_{l_p} - w_p, B_r = \prod_{p=1}^r \hat{S}_p^{*-1} \text{ and } G = \prod_{q=1}^t S_{m_q}(z_q).$$

It is easy to see that the integral in the left hand side of (3.4.20) equals to

$$P_M B_r G|_M + [(\bar{w} - \Lambda_l^*)P_M S_l(w)^* B_r G + (z - \Lambda_m)P_M B_r S_m(z)G$$
$$+ \quad Q_{lm}(z,w)P_M S_l(w)^* B_r S_m(z)G]|_M, \qquad (3.4.22)$$

since

$$Q_{lm}(u_m, u_l)$$
$$= (\bar{u}_l - \bar{w})(u_m - z) + (\bar{w} - \Lambda_l^*)(u_m - z) + (\bar{u}_l - \bar{w})(z - \Lambda_m) + Q_{lm}(z,w).$$

Similar to the commutation relation in the proof of Lemma 1.3.2 (fourth line after (1.3.12)), we have

$$[S_l(w)^*, S_m(z)] = S_m(z)S_l(w)^* C_{lm} P_M S_l(w)^* S_m(z). \tag{3.4.23}$$

Therefore for every operator $T \in L(\mathcal{H})$, we have

$$P_M S_l(w)^* S_m(z)T|_M - (\Lambda_m - z)^{-1} P_m S_l(w)^* T|_M$$
$$= (\Lambda_m - z)^{-1}(\Lambda_l^* - \bar{w})^{-1} C_{lm} P_M S_l(w)^* S_m(z)T|_M,$$

since $P_M S_m(z) = (S_m(z)^* P_M)^* = ((\Lambda_m^* - \bar{z})^{-1} P_M)^* = (\Lambda_m - z)^{-1} P_M.$
Hence

$$Q_{lm}(z,w) P_M S_l(w)^* S_m(z)T|_M = (\Lambda_l^* - \bar{w}) P_M S_l(w)^* T|_M. \tag{3.4.24}$$

Let

$$L_r(T) \stackrel{\text{def}}{=} [Q_{lm}(z,w) P_M S_l(w)^* - P_M(S_m - z)][B_r, S_m(z)]T|_M.$$

Then from (3.4.24) it follows that (3.4.22) equals to $L_r(G)$. Now, we only have to prove that $L_r(T) = 0$ by mathematical induction with respective to the integer r.

It is easy to see that

$$L_1(T) =$$
$$[Q_{lm}(z,w) P_M S_l(w)^* - P_M(S_m - z)]S_m(z)\hat{S}_1^{*-1} C_{l_1 m} P_M \hat{S}_1^{*-1} S_m(z)T|_M$$

by (3.4.23). From (3.4.18), we have

$$P_M S_l(w)^* S_m(z)\hat{S}_1^{*-1}|_M = P_M S_l(w)^* S_m(z)(\Lambda_{l_1}^* - \bar{w}_1)^{-1}|_M$$
$$= Q_{lm}(z,w)^{-1} P_M \hat{S}_1^{*-1}|_M.$$

Thus $L_1(T) = 0$. Notice that by (3.4.23) again, we have

$$[B_r, S_m(z)]T = [B_{r-1}, S_m(z)]\widetilde{S}_r^{*-1} T + B_{r-1} S_m(z)\hat{S}_r^{*-1} C_{l,m} P_M \hat{S}_r^{*-1} S_m(z)T.$$

Hence

$$[B_r, S_m(z)]T = [B_{r-1}, S_m(z)]T_r' + S_m(z)B_{r-1}T_r'', \tag{3.4.25}$$

where $T_r' = \hat{S}_r^{*-1}(T + T_r'')$ and

$$T_r'' = C_{l,m} P_M \hat{S}_r^{*-1} S_m(z)T.$$

The contribution of the first term in the right-hand side of (3.4.25) to the $L_r(T)$ is

$$L_{r-1}(\widetilde{S}_r^{*-1} T) = 0.$$

The contribution of the second term is

$$[Q_{lm}(z,w) P_M S_l(w)^* - P_M(S_m - z)]S_m(z) \prod_{p=1}^{r-1} (\Lambda_{l_p}^* - \bar{w})^{-1} C_{l,m} P_M \hat{S}_r^{*-1} S_m(z)T$$

which equals zero, since

$$[Q_{lm}(z,w) P_M S_l(w)^* - P_M(S_m - z)]S_m(z)|_M = 0$$

by (3.4.18), which proves the lemma. $\square$

Lemma 3.4.4. *Under the condition of Lemma 3.4.3, the measure $e(\cdot)$ has the property that*

$$\int (u_j - \Lambda_j)\Pi_{q=1}^l (u_{m_q} - z_q)^{-1} e(du) = 0 \tag{3.4.26}$$

for $j = 1, 2, \ldots, k$, $z_q \in \rho(S_{m_q})$, and $1 \le m_q \le k$.

Proof. For $f \in \mathcal{H}$ and $\beta \in M$, we have

$$(f, (N_j^* - \Lambda_j^*)\beta) = (f, N_j^*\beta) - (f, S_j^*\beta) = ((N_j - S_j)f, \beta) = 0. \tag{3.4.27}$$

Let $f \stackrel{\text{def}}{=} \Pi_{q=1}^l (N_{m_q} - z_q)^{-1}\alpha$ for $\alpha \in M$. Then $f \in \mathcal{H}$. From (3.4.27), it follows

$$\begin{aligned}
0 &= \left(\int \Pi_{q=1}^l (u_{m_q} - z_q)^{-1} E(du)\alpha, \int (\bar{u}_j E(du)\beta - \Lambda_j^*\beta) \right) \\
&= \left(\int (u_j - \Lambda_j P_M)\Pi_{q=1}^l (u_{m_q} - z_q)^{-1} E(du)\alpha, \beta \right) \\
&= \left(\int (u_j - \Lambda_j)\Pi_{q=1}^l (u_{m_q} - z_q)^{-1} e(du)\alpha, \beta \right)
\end{aligned}$$

which proves (3.4.26). $\qquad\square$

Let $K(\mathbb{S}) \stackrel{\text{def}}{=} \times_{j=1}^k \sigma(S_j)$ and $R^2(K(\mathbb{S}), e)$ be the closure of

$$\vee \left\{ \prod_{j=1}^k (\lambda_j - u_j)^{-1}\alpha : \alpha \in M, \lambda_j \in \rho(S_j) \right\}$$

in $L^2(e)$, where $u = (u_1, \ldots, u_k) \in K(\mathbb{S})$. The following theorem gives an analytic model for pure subnormal tuples of operators.

Theorem 3.4.5. *Let $\mathbb{S} = (S_1, \ldots, S_k)$ be a pure subnormal tuple of operators on a separable Hilbert space $\mathcal{H}$, with an m.n.e. $\mathbb{N}$ on a Hilbert space $\mathcal{K} \supset \mathcal{H}$. Let $M = \mathrm{cl} \vee_{j=1}^k [S_j^*, S_j]\mathcal{H}$. Then there is an $L(M)$-valued positive measure $e(\cdot)$ on $\mathrm{sp}(\mathbb{N})$, defined by (3.4.15) satisfying (3.4.19) and (3.4.26) where $\Lambda_l, C_{lm} \in L(M)$ are defined by (3.4.11) and (3.4.12) such that the operator U defined by*

$$Uf(\mathbb{N})\alpha = f(\cdot)\alpha \tag{3.4.28}$$

for every bounded Baire function f and $\alpha \in M$ extends a unitary operator U from $\mathcal{H}$ onto $R^2(K(\mathbb{S}), e)$. This operator U satisfies condition that

$$(US_j U^{-1}f)(u) = u_j f(u), \tag{3.4.29}$$

$$(US_j^* U^{-1}f)(u) = \bar{u}_j f(u) + (\Lambda_j^* - \bar{u}_j)f(\Lambda), \tag{3.4.30}$$

where $f(\Lambda) = \int e(du)f(u)$, for $f \in R^2(K(\mathbb{S}), e)$.

Proof. It is easy to see that U is a unitary operator from $\mathcal{K}$ onto $L^2(e)$. Let

$$\mathcal{H}_0 \overset{\text{def}}{=} \text{cl} \vee \{ \prod_{j=1}^{k} (\lambda_j - S_j)^{-1} \alpha : \lambda_j \in \rho(S_j), \alpha \in M \}.$$

It is obvious that $\mathcal{H}_0$ is invariant with respect to S_j, $j = 1, 2, \ldots, k$. It is easy to verify that

$$S_l^* \prod_{j=1}^{k} S_j(\lambda_j)\alpha =$$

$$- \sum_{j=1}^{k} \prod_{m=1}^{j} S_m(\lambda_m) C_{lj} \prod_{p=j}^{k} (\Lambda_p - \lambda_p)^{-1}\alpha + \prod_{j=1}^{k} S_j(\lambda_j)\Lambda_l^*\alpha, \quad (3.4.31)$$

where $S_j(z) \overset{\text{def}}{=} (S_j - z)^{-1}$, since

$$[S_l^*, S_j(z)] = -S_j(z) C_{lj} P_M S_j(z)$$

and $P_M \prod_j S_j(\lambda_j)\alpha = \prod_j (\Lambda_j - \lambda_j)^{-1}\alpha$. Therefore $\mathcal{H}_0$ is also invariant with respect to S_l^*, $l = 1, 2, \ldots, k$. Hence $\mathcal{H}_0 = \mathcal{H}$, since $M \subset \mathcal{H}_0$ and $\mathbb{S}$ is pure. Thus

$$U\mathcal{H} = U\mathcal{H}_0 = R^2(K(\mathbb{S}), e).$$

The identity (3.4.29) is a consequence of (3.4.28). To show (3.4.30), notice that

$$(US_l^*U^{-1}) \prod_{i=1}^{k} (u_j - \lambda_j)^{-1}\alpha = - \sum_{j=1}^{k} \prod_{m=1}^{j} (u_m - \lambda_m)^{-1} C_{lj} \prod_{p=j}^{k} (\Lambda_p - \lambda_p)^{-1}\alpha$$

$$+ \prod_{m=1}^{k} (u_m - \lambda_m)^{-1}\Lambda_l^*\alpha, \quad (3.4.32)$$

by (3.4.29) and (3.4.31). However by Lemma 3.4.3, the following equality of vectors in $L^2(e)$ holds:

$$C_{lj}f(u) = (\bar{u}_l - \Lambda_l^*)(u_j - \Lambda_j)f(u)$$

for $f \in \vee\{\Pi_j(u_j - \lambda_j)^{-1}\alpha, \alpha \in M, \lambda_j \in \rho(S_j)\}$. Therefore the right-hand side of (3.4.32) may be rewritten as

$$(\bar{u}_l - \Lambda_l^*) \sum_{j=1}^{k} \left(\prod_{m=1}^{j} (u_m - \lambda_m)^{-1} \prod_{p=j+1}^{k} (\Lambda_p - \lambda_p)^{-1}\alpha \right.$$

$$\left. - \prod_{m=1}^{j-1} (u_m - \lambda_m)^{-1} \prod_{p=j}^{k} (\Lambda_p - \lambda_p)^{-1}\alpha \right) + \prod_{m=1}^{k} (u_m - \lambda_m)^{-1}\Lambda_l^*\alpha,$$

where the product $\prod_{m=1}^{0} (u_m - \lambda_m)^{-1} = \prod_{p=k+1}^{k} (\Lambda_p - \lambda_p)^{-1} = 1$. Thus the formula (3.4.32) may be rewritten as

$$(US_l^*U^{-1}) \prod_{j=1}^{k} (u_j - \lambda_j)^{-1} \alpha = \bar{u}_l \prod_{j=1}^{k} (u_j - \lambda_j)^{-1} \alpha - (\bar{u}_l - \Lambda_l^*) \prod_{j=1}^{k} (\Lambda_j - \lambda_j)^{-1} \alpha.$$

$$(3.4.33)$$

On the other hand, we have

$$\int (e(du)) \prod_{j=1}^{k} (u_j - \lambda_j)^{-1} \alpha, \beta) = (\prod_{j=1}^{k} S_j(\lambda_j) \alpha, \beta) = (\alpha, \prod_{j=1}^{k} (\Lambda_j^* - \bar{\lambda}_j)^{-1} \beta),$$

for $\alpha, \beta \in M$. Hence

$$\int e(du) \prod_{j=1}^{k} (u_j - \lambda_j)^{-1} = \prod_{j=1}^{k} (\Lambda_j - \lambda_j)^{-1}.$$

Thus (3.4.33) implies (3.4.30), which proves the theorem. □

Just like the Corollary 1.3.5, we have

Corollary 3.4.6. *Under the condition of Theorem 3.4.5,*

$$(UP_M U^{-1} f)(u) = f(\Lambda), \qquad (3.4.34)$$

$$(U[S_l^*, S_m]U^{-1} f)(u) = C_{lm} f(\Lambda), \qquad (3.4.35)$$

$$(UA_l^* U^{-1} f)(u) = (\bar{u}_l - \Lambda_l^*) f(\Lambda), \qquad (3.4.36)$$

for $f \in R^2(K(\mathbb{S}), e)$, and

$$UA_j U^{-1} f = \int (v_j - \Lambda_j) e(dv) f(v), \qquad (3.4.37)$$

$$(U\widetilde{S}_j U^{-1} f)(u) = \bar{u}_j f(u), \qquad (3.4.38)$$

$$(U\widetilde{S}_j^* U^{-1} f)(u) = u_j f(u) - \int (v_j - \Lambda_j) e(dv) f(v), \qquad (3.4.39)$$

and

$$([U\widetilde{S}_l^* U^{-1}, U\widetilde{S}_m U^{-1}] f)(u) = (\bar{u}_l - \Lambda_l^*) \int (v_m - \Lambda_m) e(dv) f(v), \quad (3.4.40)$$

for $f \in U\widetilde{\mathcal{H}} = L^2(e) \ominus R^2(K(\mathbb{S}), e)$, where A_j and $\widetilde{S}_j$ are defined by (3.4.3). Besides,

$$U\widetilde{\mathcal{H}} = \mathrm{cl} \vee \{\Pi_{j=1}^{k} (\bar{u}_j - \bar{\lambda}_j)^{-1} (\bar{u}_l - \Lambda_l^*) \alpha : \alpha \in M, \lambda_j \in \rho(S_j), l = 1, 2, \ldots, k\}.$$

$$(3.4.41)$$

Proof. The proof of (3.4.37)-(3.4.40) is similar to the proof of the corresponding part in Corollary 1.3.5. Let us prove (3.4.41) now.

Suppose $\lambda_j, \mu_j \in \rho(S_j)$, $\lambda_j \neq \mu_j$ and $\alpha, \beta \in M$. Then by (3.4.26) we have

$$(\prod_{j=1}^{k}((\cdot)_j - \mu_j)^{-1}\beta, (\overline{(\cdot)}_l - \Lambda_l^*)\prod_{j=1}^{k}(\overline{(\cdot)}_j - \bar{\lambda}_j)^{-1}\alpha)$$

$$= (\int (u_l - \Lambda_l)\prod_{j=1}^{k}((u_j - \mu_j)^{-1}(u_j - \lambda_j)^{-1})e(du)\beta, \alpha) = 0.$$

Thus $(\overline{(\cdot)}_l - \Lambda_l^*)\prod_{j=1}^{k}(\overline{(\cdot)}_j - \bar{\lambda}_j)^{-1}\alpha \perp R^2(K(\mathbb{S}), e)$. Now we only have to prove that if $f_0 \in U\widetilde{\mathcal{H}}$, and

$$f_0 \perp (\overline{(\cdot)}_l - \Lambda_l^*)\prod_{j=1}^{k}(\overline{(\cdot)}_j - \bar{\lambda}_j)^{-1}\alpha, \qquad (3.4.42)$$

for $\alpha \in M$, $\lambda_j \in \rho(S_j)$ and $j = 1, 2, \ldots, k$ then $f_0 = 0$. From (3.4.42), we have

$$\int \frac{(u_l - \Lambda_l)e(du)f_0(u)}{\prod_{j=1}^{k}(u_j - \lambda_j)} = 0, \text{ for } \lambda_j \in \rho(S_j) \text{ and } l = 1, 2, \ldots, k. \quad (3.4.43)$$

Let $F = \bigvee\{\prod_{p=1}^{k}\overline{(\cdot)}_p^{m_p}\prod_l(\cdot)_l^{n_l}f_0 : m_p, n_l = 0, 1, 2, \ldots\}$. We have to prove that $F \subset U\widetilde{\mathcal{H}}$. Then $\mathrm{cl}\, F$ reduces $U\mathbb{N}U^{-1}$ and f_0 must be zero, since $\mathbb{N}$ is an m.n.e.

It is obvious that

$$\{\prod_{p=1}^{k}\overline{(\cdot)}_p^{m_p}f_0 : m_1, \ldots, m_k = 0, 1, 2, \ldots\} \subset U\widetilde{\mathcal{H}},$$

since $U\widetilde{\mathcal{H}}$ is invariant with respect to $UN_p^*U^{-1}$, $p = 1, 2, \ldots, k$. For fixed $m_l, \ldots, m_k$ let us prove that

$$\int \frac{u_1 - \Lambda_1}{u_1 - \lambda_1}e(du)\bar{u}_1^{m_1}\bar{u}_2^{m_2}\cdots\bar{u}_k^{m_k}f_0(u) = 0, \quad \lambda_1 \in \rho(S_1), \qquad (3.4.44)$$

and

$$\int \frac{e(du)\bar{u}_1^{m_1}\bar{u}_2^{m_2}\cdots\bar{u}_k^{m_k}}{u_1 - \lambda_1}f_0(u) = 0, \quad \lambda_1 \in \rho(S_1) \qquad (3.4.45)$$

for $m_1, \ldots, m_k = 0, 1, 2, \ldots$ by mathematical induction with respect to $m_1, m_2, \ldots, m_k$ in succession.

First, from $\overline{(\cdot)}_1^{m_1}\overline{(\cdot)}_2^{m_2}\cdots\overline{(\cdot)}_k^{m_k}f_0 \in U\widetilde{\mathcal{H}}$ it follows that

$$\int e(du)\overline{u}_1^{m_1}\overline{u}_2^{m_2}\cdots\overline{u}_k^{m_k}f_0(u) = 0.$$

From

$$\int \frac{u_1 - \Lambda_1}{u_1 - \lambda_1}e(du)\overline{u}_1^{m_1}\overline{u}_2^{m_2}\cdots\overline{u}_k^{m_k}f_0(u) =$$

$$\int e(du)\overline{u}_1^{m_1}\overline{u}_2^{m_2}\cdots\overline{u}_k^{m_k}f_0(u) + (\lambda_1 - \Lambda_1)\int \frac{e(du)\overline{u}_1^{m_1}\overline{u}_2^{m_2}\cdots\overline{u}_k^{m_k}}{u_1 - \lambda_1}f_0(u)$$

and $\lambda \in \rho(\Lambda_1)$ it follows that (3.4.44) and (3.4.45) are equivalent.

For $m_1 = \cdots = m_k = 0$, (3.4.43) implies (3.4.44) and hence also (3.4.45). Suppose (3.4.44) and (3.4.45) hold good for $m_1,\ldots,m_k$. Let us prove that they also hold good for $m_1,\ldots,m_{l-1}, m_l + 1, m_{l+1},\ldots,m_k$. By (3.4.19), we have

$$\int \frac{(u_1 - \Lambda_1)e(du)\overline{u}_1^{m_1}\cdots\overline{u}_{l-1}^{m_{l-1}}\overline{u}_l^{m_l+1}\overline{u}_{l+1}^{m_{l+1}}\cdots\overline{u}_k^{m_k}f_0(u)}{u_1 - \lambda_1}$$

$$= \int \frac{Q_{l1}(u_1,u_l)e(du)\overline{u}_1^{m_1}\cdots\overline{u}_k^{m_k}f_0(u)}{u_1 - \lambda_1} + C_{l1}\int \frac{e(du)\overline{u}_1^{m_1}\cdots\overline{u}_k^{m_k}}{u_1 - \lambda_1}f_0(u)$$

$$+ \Lambda_l^*\int \frac{(u_1 - \Lambda_1)e(du)\overline{u}_1^{m_1}\cdots\overline{u}_k^{m_k}}{u_1 - \lambda_1}f_0(u) = 0$$

which proves (3.4.44) and (3.4.45) for $m_1,\ldots,m_l + 1,\ldots,m_k$ and hence all non-negative integers $m_1,\ldots,m_k$.

By the same method, we may prove that

$$\int \frac{u_2 - \Lambda_2}{u_2 - \lambda_2}e(du)\frac{\overline{u}_1^{m_1}\cdots\overline{u}_k^{m_k}}{u_1 - \lambda_1}f_0(u) = 0, \quad \lambda_j \in \rho(S_j),\ j = 1,2,$$

and

$$\int \frac{e(du)\overline{u}_1^{m_1}\cdots\overline{u}_k^{m_k}f_0(u)}{(u_2 - \lambda_2)(u_1 - \lambda_1)} = 0, \quad \lambda_j \in \rho(S_j),\ j = 1,2.$$

Continuing this process, eventually we have

$$\int \frac{(u_l - \Lambda_l)e(du)\overline{u}_1^{m_1}\cdots\overline{u}_k^{m_k}f_0(u)}{\Pi_{j=1}^k(u_j - \lambda_j)} = 0, \quad \lambda_j \in \rho(S_j).$$

Following the method of proving (1.3.43) by (1.3.44), we may prove that

$$UA_lU^{-1}\Pi_{j=1}^k(\overline{(\cdot)}_j^{m_j}(\cdot)_j^{n_j})f_0 = 0$$

by (3.4.37) and hence $\Pi_{j=1}^k(\overline{(\cdot)}_j^{m_j}(\cdot)_j^{n_j})f_0 \in U\widetilde{\mathcal{H}}$, which proves (3.4.41). $\quad\square$

Besides, it is easy to verify that

$$C_{lm} = \int (u_m - \Lambda_m)e(du)(\overline{u}_l - \Lambda_l^*) \tag{3.4.46}$$

and

$$\widetilde{C}_{lm} \overset{\text{def}}{=} C_{lm} - [\Lambda_l^*, \Lambda_m] = \int (\overline{u}_l - \Lambda_l^*)e(du)(u_m - \Lambda_m). \tag{3.4.47}$$

The transpose of matrices (C_{lm}) and $(\widetilde{C}_{lm})$ are positive semidefinite, i.e.

$$\sum_{l,m}(C_{lm}x_l, x_m) \geq 0 \text{ and } \sum(\widetilde{C}_{lm}x_m, x_l) \geq 0 \tag{3.4.48}$$

for $x_1, \ldots, x_k \in M$. Besides, we have

$$M = \text{cl} \bigvee_{j=1}^{k} C_{jj}M. \tag{3.4.49}$$

Corollary 3.4.7. *Under the condition of Theorem 3.4.5,*

$$[C_{lm}, \Lambda_n] = [C_{ln}, \Lambda_m] \tag{3.4.50}$$

and

$$\widetilde{C}_{pl}C_{mn} = \widetilde{C}_{pn}C_{ml}. \tag{3.4.51}$$

Proof. It is easy to see that

$$\int \overline{u}_l u_n e(du) = \int [(\overline{u}_l - \Lambda_l^*)(u_n - \Lambda_n) + \Lambda_l^*(u_n - \Lambda_n) + \overline{u}_l\Lambda_n]e(du)$$
$$= C_{ln} + \Lambda_n\Lambda_l^*.$$

Therefore

$$C_{lm}\Lambda_n = \int (\overline{u}_l - \Lambda_l^*)(u_m - \Lambda_m)u_n e(du)$$

$$= \int [\overline{u}_l(u_m - \Lambda_m) - \Lambda_l^*(u_m - \Lambda_m)]u_n e(du)$$

$$= \int \overline{u}_l u_m u_n e(du) - \Lambda_m \int \overline{u}_l u_n e(du)$$

$$= \int \overline{u}_l u_m u_n e(du) - \Lambda_m\Lambda_n\Lambda_l^* - \Lambda_m C_{ln}. \tag{3.4.52}$$

Exchanging m and n, we have

$$C_{ln}\Lambda_m = \int \overline{u}_l u_m u_n e(du) - \Lambda_n\Lambda_m\Lambda_l^* - \Lambda_n C_{lm}. \tag{3.4.53}$$

Subtracting (3.4.53) from (3.4.52), we get (3.4.50) by $[\Lambda_n, \Lambda_m] = 0$.

To prove (3.4.51), notice that for $\alpha \in M$, we have

$$(S_l A_n + A_l \widetilde{S}_n^*)(\bar{u}_m - \Lambda_m^*)\alpha = u_l C_{mn}\alpha + \int (v_l - \Lambda_l)e(dv)(\bar{v}_m - \Lambda_m^*)v_n\alpha$$
$$= u_l C_{mn}\alpha + C_{ml}\Lambda_n\alpha \qquad (3.4.54)$$

by (3.4.37), (3.4.39) and (3.4.46). From (3.4.6) and (3.4.54), we have

$$u_l C_{mn}\alpha + C_{ml}\Lambda_n\alpha = u_l C_{ml}\alpha + C_{nl}\Lambda_l\alpha. \qquad (3.4.55)$$

From (3.4.47), we have

$$\int (\bar{u}_p - \Lambda_p^*)e(du)(u_l C_{mn}\alpha + C_{ml}\Lambda_n\alpha) = \int (\bar{u}_p - \Lambda_p^*)e(du)(u_l - \Lambda_l)C_{mn}\alpha$$
$$= \widetilde{C}_{pl}C_{mn}\alpha. \qquad (3.4.56)$$

From (3.4.55) and (3.4.56) it follows (3.4.51). $\qquad\qquad \square$

The following formula is useful: If $f \in U\mathcal{H}$ then

$$\int (u_j - \Lambda_j)e(du)f(u) = 0, \quad j = 1, 2, \ldots, k. \qquad (3.4.57)$$

This formula is a direct consequence of (3.4.26).

The following theorem is the converse of Theorem 3.4.5.

Theorem 3.4.8. *Let M be a Hilbert space and $e(\cdot)$ be a $L(M)$-valued non-negative measure on a set $\sigma = \times_{j=1}^k \sigma_j$, where σ_j is a compact set of $\mathbb{C}$, satisfying $e(\sigma) = I$, (3.4.26) for $j = 1, 2, \ldots, k$ z_q in the unbounded component D_{m_q} of $\mathbb{C} \setminus \sigma_{m_q}$ and*

$$\int_B e(du)Q_{lm}(u, u) = \int_B Q_{lm}(u, u)e(du) = 0, \quad l, m = 1, 2, \ldots, k,$$

for every Borel set B in σ, where

$$Q_{lm}(u, v) = (\bar{v} - \Lambda_l^*)(u - \Lambda_m) - C_{lm},$$

and $C_{lm}, \Lambda_l \in L(M)$. Let $\mathcal{H}_1 \stackrel{def}{=} \bigvee\{\Pi_{j=1}^k(\lambda_j - \Lambda_j)^{-1}\alpha, \alpha \in M, \lambda_j \in D_j\}$, and $\mathcal{H}$ be the closure of $\mathcal{H}_1$ in $L^2(e)$. Define

$$(S_j f)(u) = u_j f(u), \quad j = 1, 2, \ldots, k, \ f \in \mathcal{H}.$$

Then $\mathbb{S} = (S_1, \ldots, S_k)$ is a subnormal k-tuple of operators on $\mathcal{H}$ with normal extension

$$(N_j f)(u) = u_j f(u), \quad j = 1, 2, \ldots, k,$$

on $L^2(e)$. Besides

$$(S_j^* f)(u) = \bar{u}_j f(u) - (\bar{u}_j - \Lambda_j^*)f(\Lambda), \quad f \in \mathcal{H}$$

where $f(\Lambda) \stackrel{def}{=} \int e(du)f(u)$.

We omit the proof of Theorem 3.4.8, since it is similar to the proof of Theorem 1.3.6.

Let

$$C_{\mathbb{T}} \stackrel{\text{def}}{=} \begin{pmatrix} C_{11} & C_{12} & \cdots & C_{k1} \\ C_{12} & C_{22} & \cdots & C_{k2} \\ \cdots & \cdots & \cdots & \cdots \\ C_{1k} & C_{2k} & \cdots & C_{kk} \end{pmatrix}, \tag{3.4.58}$$

where $C_{ij} = [S_i^*, S_j]|_M$ is an operator in $L(M)$ as in (3.4.11).

Lemma 3.4.9. *If* $\mathbb{S} = (S_1, \ldots, S_k)$ *is a subnormal tuple of operators on a Hilbert space* $\mathcal{H}$. *Then the matrix* $C_{\mathbb{T}}$ *is positively semidefinite, i.e.*

$$\sum_{i,j=1}^{k} (C_{ij} x_i, x_j) \geq 0, \quad x_1, \ldots, x_k \in M. \tag{3.4.59}$$

Proof. From (3.4.7), we have $\sum_{i,j=1}^{k} (C_{ij} x_i, x_j) = \| \sum_{i=1}^{k} A_i^* x_i \|^2 \geq 0$. $\square$

3.5 Mosaic

In §1.4, we studied the mosaic of a single subnormal operator. Now, we generalize this concept to the k-tuple case.

Let $\mathcal{S} = (S_1, \ldots, S_k)$ be a pure subnormal k-tuple of operators on a Hilbert space $\mathcal{H}$ with m.n.e. $\mathbb{N} = (N_1, \ldots, N_k)$ on $\mathcal{K} \supset \mathcal{H}$. Let $l = \{l_1, \ldots, l_n\}$ be a ordered set of natural numbers satisfying $1 \leq l_i \leq k$. Let $l_{i,j}$ be any ordered subset $\{l_i, l_{i+1}, \ldots, l_{j-1}, l_j\}$ of l. Let $z = (z_1, \ldots, z_n)$, satisfying $z_j \in \rho(N_{l_j})$. Let

$$\hat{\mu}_{l_{i,j}}(z; f) \stackrel{\text{def}}{=} P_M (N_{l_i} - S_{l_i} P) \prod_{p=i}^{j} (N_{l_p} - z_p)^{-1} f, \tag{3.5.1}$$

for $f \in \mathcal{K}$ and let

$$\hat{\mu}_{l_{i,j}}(z) \stackrel{\text{def}}{=} P_M (N_{l_i} - S_{l_i} P) \prod_{p=i}^{j} (N_{l_p} - z_p)^{-1}|_M, \tag{3.5.2}$$

where M is the defect space of $\mathcal{S}$, P_M is the projection from $\mathcal{K}$ to M, P is the projection from $\mathcal{K}$ to $\mathcal{H}$. In the analytic model, (3.5.1) and (3.5.2) may be rewritten as the following:

$$\hat{\mu}_{l_{i,j}}(z; f) = \int_{\text{sp}(\mathbb{N})} (u_{l_i} - \Lambda_{l_i}) e(du) \prod_{p=i}^{j} (u_{l_p} - z_p)^{-1} f(u), \tag{3.5.3}$$

and

$$\hat{\mu}_{l_{i,j}}(z) = \int_{\text{sp}(\mathbb{N})} (u_{l_i} - \Lambda_{l_i}) e(du) \prod_{p=i}^{j} (u_{l_p} - z_p)^{-1}. \qquad (3.5.4)$$

In the case $i = j$, we simply write $\mu_{l_i}(z, f)$ and $\mu_{l_i}(z)$, instead of $\hat{\mu}_{l_{i,i}}(z; f)$ and $\hat{\mu}_{l_{i,i}}(z)$ respectively. These essentially are the functions $\mu(z; f)$ and $\mu(z)$ for the single subnormal operator S_{l_i}, but M may contain the defect space of S_{l_i} as a proper subspace.

Let $\theta_{ij} = 0$ for $i > j$ and $\theta_{ij} = 1$ for $i \leq j$. Define the upper triangle matrices

$$\mu_l(z; f) \stackrel{\text{def}}{=} (\theta_{ij} \hat{\mu}_{l_{i,j}}(z, f))_{i,j=1,2,\dots,n}, \qquad (3.5.5)$$

and

$$\mu_l(z) \stackrel{\text{def}}{=} (\theta_{ij} \hat{\mu}_{l_{i,j}}(z))_{i,j=1,2,\dots,n}. \qquad (3.5.6)$$

$\mu_l(z)$ is said to be a *mosaic* of $\mathcal{S}$.

Let us generalize (iv) of Lemma 1.4.1 to the following:

Theorem 3.5.1. *Let $\mathcal{S} = (S_1, \dots, S_k)$ be a pure subnormal tuple of operators on a Hilbert space $\mathcal{H}$ with m.n.e. $\mathbb{N} = (N_1, \dots, N_k)$ on $\mathcal{K} \supset \mathcal{H}$. Then for any ordering set $l = \{l_1, \dots, l_n\}$ satisfying $1 \leq l_j \leq k$, and $z = (z_1, \dots, z_n), z \in \rho(N_{l_j})$,*

$$\mu_l(z; f) = \mu_l(z)\mu_l(z; f), \ \text{for } f \in \mathcal{H}, \qquad (3.5.7)$$

$$\mu_l(z) = \mu_l(z)^2, \qquad (3.5.8)$$

and

$$\mu_l(z) = 0, \ \text{for } z_j \in \rho(S_{l_j}), \ j = 1, 2, \dots, n. \qquad (3.5.9)$$

Proof. For the simplicity of notation, let $\hat{N}_j = N_{l_j} - z_j, \hat{S}_j = S_{l_j} - z_j, \mathring{S}_j = \widetilde{S}_{l_j}^* - z_j, \hat{\Lambda}_j = \Lambda_{l_j} - z_j,$ and $\hat{u}_j = u_{l_j} - z_j$. For $f \in \mathcal{H}$ and fixed $q, 1 \leq q \leq n$, denote

$$f_0 = P \prod_{p=1}^{q} \hat{N}_p^{-1} f \ \text{ and } \ f_1 = \widetilde{P} \prod_{p=1}^{q} \hat{N}_p^{-1} f. \qquad (3.5.10)$$

Define

$$m_f = \prod_{p=1}^{q} \hat{N}_p f_1. \qquad (3.5.11)$$

Then from (3.5.10) and (3.5.11), we have

$$m_f = f - \prod_{p=1}^{q} \hat{N}_p f_0 \in \mathcal{H}. \qquad (3.5.12)$$

Therefore (3.5.11) may be written as

$$m_f = P \prod_{p=1}^{q} \begin{pmatrix} \hat{S}_p & A_{l_p} \\ 0 & \mathring{S}_p \end{pmatrix} \begin{pmatrix} 0 \\ f_1 \end{pmatrix} = \sum_{p=1}^{q} \prod_{j=p+1}^{q} \hat{S}_j A_{l_p} \Pi_{j=1}^{p-1} \mathring{S}_j f_1. \qquad (3.5.13)$$

In (3.5.13), the factors $\prod_{j=p+1}^{q} \hat{S}_j$ and $\prod_{j=1}^{p-1} \mathring{S}_j$ become I for $p = q$ and $p = 1$ respectively. By (3.4.37), (3.4.38) and (3.4.39), in the analytic model, (3.5.13) may be written as

$$m_f(u) = \sum_{p=1}^{q} \prod_{j=p+1}^{q} \hat{u}_j \int (\hat{v}_p - \hat{\Lambda}_p) \prod_{j=1}^{p-1} \hat{v}_j e(dv) f_1(v). \qquad (3.5.14)$$

On the other hand, from (3.5.10), we have $f = \prod_{p=1}^{q} \hat{N}_p (f_0 + f_1)$. Therefore

$$f_1(u) = \prod_{p=1}^{q} \hat{u}_p^{-1} f(u) - f_0(u). \qquad (3.5.15)$$

By (3.4.57), we have

$$\int (\hat{v}_p - \hat{\Lambda}_p) \prod_{j=1}^{p-1} \hat{v}_j e(dv) f_0(v) = 0, \qquad (3.5.16)$$

since $f_0 \in \mathcal{H}$. Hence, from (3.5.14), (3.5.15) and (3.5.16), we have

$$m_f(u) = \sum_{i=1}^{q} \prod_{j=i+1}^{q} \hat{u}_j \hat{\mu}_{l_{i,q}}(z; f). \qquad (3.5.17)$$

From (3.5.12), we have $f_0(u) = \prod_{j=1}^{q} \hat{u}_j^{-1}(f(u) - m_f(u))$. By (3.4.57) again, similar to (3.5.16), we have

$$\int \prod_{j=1}^{p-1} \hat{u}_j (\hat{u}_p - \hat{\Lambda}_p) e(du) \prod_{i=1}^{q} \hat{u}_i^{-1}(f(u) - m_f(u))$$

$$= \int \prod_{j=1}^{p-1} \hat{u}_j (\hat{u}_p - \hat{\Lambda}_p) e(du) f_0(u) = 0. \qquad (3.5.18)$$

From (3.5.17) and (3.5.18), we have

$$\mu_{l_{p,q}}(z; f) - \sum_{i=p}^{q} \mu_{l_{p,i}}(z) \mu_{l_{i,q}}(z; f) = 0,$$

for $1 \leq p \leq q \leq n$, which is equivalent to (3.5.7). In (3.5.7), letting $f \in M$, we have (3.5.8). The formula (3.5.9) is the direct consequence of (3.4.26). $\qquad \square$

For $l = \{l_1, \ldots, l_n\}, 1 \leq l_j \leq k$ and $z = (z_1, \ldots, z_n), z \in \rho(N_{l_j})$, define

$$\mu_l^\dagger(z) = I - (\theta_{ij}\mu_{l_{j'i'}}(z)^*)_{i,j=1,2,\ldots,n} \tag{3.5.19}$$

where $i' = n - i + 1$ and $j' = n - j + 1$.

Corollary 3.5.2. *Under the condition of Lemma 3.5.1,*

$$\mu_l^\dagger(z)^2 = \mu_l^\dagger(z). \tag{3.5.20}$$

Proof. We only have the prove that for any $p, q, 1 \leq p \leq q \leq n$

$$\sum_{j=p}^q \hat{\mu}_{l_{j'},p'}^* \hat{\mu}_{l_{q'},j'}^* = \hat{\mu}_{l_{q'},p'}^*. \tag{3.5.21}$$

However, (3.5.21) is equivalent to

$$\sum_{j=p}^q \hat{\mu}_{l_{q'},j'} \hat{\mu}_{l_{j'},p'} = \hat{\mu}_{l_{q'},p'}. \tag{3.5.22}$$

But the left-hand side of (3.5.22) is equal to $\sum_{j=q'}^{p'} \hat{\mu}_{l_{q'},j} \hat{\mu}_{l_{j,p'}}$. Therefore (3.5.22) is a direct consequence of (3.5.8) which proves the corollary. $\square$

For any ordered set $G = \{g_1, \ldots, g_s\}$ and $L = \{l_1, \ldots, l_t\}$, satisfying $1 \leq g_i, l_j \leq k$, define

$$S_{G,L}(z, w) = P_M \prod_{i=1}^s (N_{g_i}^* - \overline{w}_i)^{-1} \prod_{j=1}^t (N_{l_j} - z_j)^{-1}|_M$$

$$= \int e(du) \prod_{i=1}^s (\overline{u}_{g_i} - \overline{w}_i)^{-1} \prod_{j=1}^t (u_{l_j} - z_j)^{-1}, \tag{3.5.23}$$

where $z = (z_1, \ldots, z_t)$, $w = (w_1, \ldots, w_s)$, $z_j \in \rho(N_{l_j})$ and $w_i \in \rho(N_{g_i})$. This $S_{G,L}$ defined in (3.5.23) is a special case of (3.3.38) while $(T_1, \ldots, T_k) = (N_1, \ldots, N_k)$. Let $G_{i,j} = \{g_i, g_{i+1}, \ldots, g_j\}$ and $L_{i,j} = \{l_i, l_{i+1}, \ldots, l_j\}$. Define a $s \times t$ matrix

$$\mathfrak{S}_{G,L} = \left(S_{G_{1,i'},L_{1,j}} \right)_{i=1,\ldots,s, j=1,\ldots,t} \tag{3.5.24}$$

where $i' = s - i + 1$. Define a $(s + t) \times (s + t)$ matrix which is formed by block matrices:

$$\mathfrak{J}_{G,L} \overset{\text{def}}{=} \begin{pmatrix} \mu_G^\dagger & \mathfrak{S}_{G,L} \\ 0 & \mu_L \end{pmatrix} \tag{3.5.25}$$

where 0 in the lower left corner of the matrix of (3.5.25) is a zero $l \times m$ matrix with all zero entries.

Theorem 3.5.3. *Let $\mathcal{S} = (S_1, \ldots, S_k)$ be a pure subnormal k-tuple of operators on a Hilbert space $\mathcal{H}$ with m.n.e. $\mathbb{N} = (N_1, \ldots, N_k)$ on $\mathcal{K} \supset \mathcal{H}$. Let $G = \{g_1, \ldots, g_s\}$ and $L = \{l_1, \ldots, l_t\}$ be ordered set of integers, $1 \le g_i, l_j \le k$. Let $z_i \in \rho(N_{l_i})$, $w_j \in \rho(N_{g_j})$. Then*

$$\mathfrak{J}_{G,L}^2 = \mathfrak{J}_{G,L}. \tag{3.5.26}$$

Proof. We only have to prove that for any pair $\{p,q\}, 1 \le p \le s, 1 \le q \le t$,

$$\sum_{i=1}^{p} \hat{\mu}_{G_{i,p}}^* S_{G_{1,i},L_{1,q}} = \sum_{i=1}^{q} S_{G_{1,p},L_{1,i}} \hat{\mu}_{L_{i,q}}. \tag{3.5.27}$$

Let us use the same method in the proof of Theorem 3.5.1 to prove (3.5.27). For simplicity of notation, let

$$\hat{S}_j = S_{l_j} - z_j, \hat{\Lambda}_j = \Lambda_{l_j} - z_j, \hat{N}_j = N_{l_j} - z_j, \hat{u}_j = u_{l_j} - z_j,$$

$$\check{S}_j = S_{g_j} - w_j, \check{\Lambda}_j = \Lambda_{g_j} - w_j, \check{N}_j = N_{g_j} - w_j, \mathring{S}_j = \widetilde{S}_{g_j} - \overline{w}_j, \check{u}_j = u_{g_j} - w_j.$$

For $f \in M$, let

$$\Psi = \widetilde{P} \prod_{j=1}^{q} \hat{N}_j^{-1} f, \tag{3.5.28}$$

$$f_0 = P \prod_{i=1}^{p} \check{N}_i^{*-1} \prod_{j=1}^{q} \hat{N}_j^{-1} f, \tag{3.5.29}$$

and

$$f_1 = \widetilde{P} \prod_{i=1}^{p} \check{N}_i^{*-1} \prod_{j=1}^{q} \hat{N}_j^{-1} f. \tag{3.5.30}$$

Then

$$\Psi = \widetilde{P} \prod_{i=1}^{p} \check{N}_i^* (f_0 + f_1) = \widetilde{P} \prod_{i=1}^{p} \begin{pmatrix} \check{S}_i^* & 0 \\ A_{g_i}^* & \mathring{S}_i \end{pmatrix} \begin{pmatrix} f_0 \\ f_1 \end{pmatrix}$$

$$= \prod_{i=1}^{q} \mathring{S}_i f_1 + \Phi, \tag{3.5.31}$$

where

$$\Phi = \widetilde{P} \prod_{i=1}^{p} \begin{pmatrix} \check{S}_i^* & 0 \\ A_{g_i}^* & \mathring{S}_i \end{pmatrix} \begin{pmatrix} f_0 \\ 0 \end{pmatrix} = \sum_{j=1}^{p} (\prod_{i=1}^{j-1} \mathring{S}_i) A_{g_j}^* \prod_{i=j+1}^{p} \check{S}_i^* f_0. \tag{3.5.32}$$

By (3.4.30), (3.4.36) and (3.4.38) we have

$$\Phi(u) = \sum_{j=1}^{p} \prod_{i=1}^{j-1} \bar{u}_i (\bar{u}_j - \Lambda_{g_j}^*) \int \prod_{i=j+1}^{p} \bar{v}_i e(dv) f_0(v), \qquad (3.5.33)$$

since

$$(\int e(dv) (\prod_{i=1}^{j-1} \check{S}_i^* f_0)(v), \alpha) = \int (e(dv) f_0(v), (\prod_{i=j+1}^{p} \check{S}_i \alpha)(v))$$

$$= (\int e(dv) \prod_{i=j+1}^{p} \bar{v}_i f_0(v), \alpha),$$

for $\alpha \in M$.

However, $\prod_{i=j+1}^{p} \check{v}_i \alpha \in \mathcal{H}$, we have

$$(\int \prod_{i=j+1}^{p} \bar{v}_i e(dv) f_0(v), \alpha) = \int (e(du) P(\frac{1}{\prod_{i=1}^{p} \bar{u}_i \prod_{i=1}^{q} \hat{u}_i} f), \prod_{i=j+1}^{p} \check{u}_i \alpha)$$

$$= \int (e(du) (\prod_{i=1}^{p} \bar{u}_i \prod_{i=1}^{q} \hat{u}_i)^{-1} f, \prod_{i=j+1}^{p} \check{u}_i \alpha)$$

$$= (\int e(du) (\prod_{i=1}^{j} \bar{u}_i \prod_{i=1}^{q} \hat{u}_i)^{-1} f, \alpha)$$

$$= (S_{G_{1,j}, L_{1,q}} f, \alpha),$$

for $\alpha \in M$. Therefore

$$\Phi(u) = \sum_{j=1}^{p} \prod_{i=1}^{j} \bar{u}_i^{-1} (\bar{u}_j - \Lambda_{m_j}^*) S_{G_{1,j}, L_{1,q}} f. \qquad (3.5.34)$$

Let $\Psi_1 \stackrel{\text{def}}{=} P \prod_{j=1}^{q} \hat{N}_j^{-1} f$, then $\prod_{j=1}^{q} \hat{N}_j^{-1} f = \Psi_1 + \Psi$. Therefore

$$\prod_{j=1}^{q} \hat{N}_j \Psi = f - \prod_{j=1}^{q} \hat{N}_j \Psi_1 \in \mathcal{H}.$$

That is

$$\widetilde{P} \prod_{j=1}^{q} \begin{pmatrix} \hat{S}_j & A_j \\ 0 & \widetilde{S}_{l_j}^* - z_j \end{pmatrix} \begin{pmatrix} 0 \\ \Psi \end{pmatrix} = \begin{pmatrix} 0 \\ \prod_{j=1}^{q} (\widetilde{S}_{l_j}^* - z_j) \Psi \end{pmatrix} = 0.$$

That means

$$\prod_{j=1}^{q} (\widetilde{S}_{l_j}^* - z_j) \Psi = 0. \qquad (3.5.35)$$

By (3.4.39) and (3.5.28), (3.5.35) is

$$\prod_{j=1}^{q} \hat{u}_j \Psi - \sum_{j=1}^{q} \prod_{i=j+1}^{q} \hat{u}_i \int (\hat{v}_j - \hat{\Lambda}_j) \prod_{r=1}^{j-1} \hat{v}_r e(dv) \Psi$$

$$= \prod_{j=1}^{q} \hat{u}_j \Psi - \sum_{j=1}^{q} \prod_{i=j+1}^{q} \hat{u}_i \mu_{L_{j,q}} f = 0, \qquad (3.5.36)$$

since

$$\int ((\hat{v}_j - \hat{\Lambda}_j) \prod_{r=1}^{j-1} \hat{v}_r e(dv) \Psi, \alpha) = \int (e(dv) \widetilde{P}(\prod_{j=1}^{q} \hat{u}_j^{-1} f), (\overline{\hat{v}}_j - \check{\Lambda}_j^*) \prod_{r=1}^{j-1} \overline{\hat{v}}_r \alpha)$$

$$= \int (e(dv) \prod_{j=1}^{q} \hat{v}_j^{-1} f, (\overline{\hat{v}}_j - \check{\Lambda}_j^*) \prod_{r=1}^{j-1} \overline{\hat{v}}_r \alpha)$$

$$= (\mu_{L_{j,q}} f, \alpha)$$

for $\alpha \in M$, since $(\overline{\hat{v}}_j - \check{\Lambda}_j^*) \prod_{r=1}^{j-1} \overline{\hat{v}}_r \alpha \in \widetilde{\mathcal{H}}$.

From (3.4.38), (3.5.31), (3.5.34) and (3.5.36), we have

$$\prod_{j=1}^{q} \hat{u}_j \Pi_{i=1}^{p} \overline{\hat{u}}_i f_1(u) =$$

$$\sum_{j=1}^{q} \prod_{i=j+1}^{q} \hat{u}_i \mu_{L_{j,q}} f - \Pi_{j=1}^{q} \hat{u}_j \sum_{i=1}^{p} \prod_{r=1}^{i-1} \overline{\hat{u}}_r (\overline{u}_i - \Lambda_{g_i}^*) S_{G_{1,i}, L_{1,q}} f. \qquad (3.5.37)$$

Notice that $\int e(du) f_1(u) = 0$, since $f_1(\cdot) \in \widetilde{\mathcal{H}}$. Dividing both sides of (3.5.37) by $\Pi_{j=1}^{q} \hat{u}_j \Pi_{i=1}^{p} \overline{\hat{u}}_i$ and then integrating by $\int e(du) \cdots$, we obtain (3.5.27), which proves the theorem. □

Theorem 3.5.4. *If* $\mathcal{S} = (S_1, \ldots, S_k)$ *is a pure subnormal k-tuple of operators on a Hilbert space* $\mathcal{H}$ *with m.n.e.* $\mathbb{N} = (N_1, \ldots, N_k)$ *on* $\mathcal{K} \supset \mathcal{H}$. *Then for any ordered set* $L = \{l_1, \ldots, l_n\}, 1 \le l_j \le k$, *and any number* $p, 1 \le p \le k$,

$$[R_{p,L}(z_1, \ldots, z_n), \mu_L(z_1, \ldots, z_n)] = 0 \qquad (3.5.38)$$

where $z_j \in \rho(N_{l_j}), j = 1, \ldots, n$.

In (3.5.38), the matrix $R_{p,L}(z_1, \ldots, z_n)$ is defined in §3.3.

Proof. We use the same notations, $\hat{u}_j$, $\hat{\Lambda}_j$, etc., as in the proof of the previous theorems. Since n is arbitrary, in order to prove (3.5.38), we only have to show that

$$C_{pl_1} \sum_{q=1}^{n} \prod_{j=1}^{q} \hat{\Lambda}_j^{-1} \int \frac{\hat{v}_q - \hat{\Lambda}_q}{\prod_{j=q}^{n} \hat{v}_j} e(dv) - \Lambda_p^* \int \frac{\hat{v}_1 - \hat{\Lambda}_1}{\prod_{j=1}^{n} \hat{v}_j} e(dv)$$

$$= -\int \frac{\overline{v}_p(\hat{v}_1 - \hat{\Lambda}_1)e(dv)}{\prod_{j=1}^{n} \hat{v}_j} + C_{pl_1} \prod_{j=1}^{n} \hat{\Lambda}_j^{-1} \tag{3.5.39}$$

and

$$\sum_{q=1}^{n} \int \frac{(\hat{v}_1 - \hat{\Lambda}_1)e(dv)}{\prod_{j=1}^{q} \hat{v}_j} C_{pl_q} \prod_{j=q}^{n} \hat{\Lambda}_j^{-1} - \int \frac{(\hat{v}_1 - \hat{\Lambda}_1)e(dv)}{\prod_{j=1}^{n} \hat{v}_j} \Lambda_p^*$$

$$= -\int \frac{\overline{v}_p(\hat{v}_1 - \hat{\Lambda}_1)e(dv)}{\prod_{j=1}^{n} \hat{v}_j} + C_{pl_1} \prod_{j=1}^{n} \hat{\Lambda}_j^{-1}. \tag{3.5.40}$$

It is obvious that the first term in the left-hand side of (3.5.39) equals to

$$C_{pl_1}\left(\sum_{q=1}^{n} \prod_{j=1}^{q} \hat{\Lambda}_j^{-1} \int \frac{e(dv)}{\prod_{j=q+1}^{n} \hat{v}_j} - \sum_{q=1}^{n} \prod_{j=1}^{q-1} \hat{\Lambda}_j^{-1} \int \frac{e(dv)}{\prod_{j=q}^{n} \hat{v}_j}\right)$$

$$= C_{pl_1}\left(\prod_{j=1}^{n} \hat{\Lambda}_j^{-1} - \int \frac{e(dv)}{\prod_{j=1}^{n} \hat{v}_j}\right). \tag{3.5.41}$$

On the other hand, by Lemma 3.4.3, we have

$$\int \frac{C_{pl_1} + \Lambda_p^*(\hat{v}_1 - \hat{\Lambda}_1)e(dv)}{\prod_{j=1}^{n} \hat{v}_j} = \int \frac{\overline{v}_p(\hat{v}_1 - \hat{\Lambda}_1)e(dv)}{\prod_{j=1}^{n} \hat{v}_j}. \tag{3.5.42}$$

From (3.5.41) and (3.5.42), it follows (3.5.39).

By means of Lemma 3.4.3, the first term of the left-hand side of (3.5.40) equals to

$$\sum_{q=1}^{n} \int (\hat{v}_1 - \hat{\Lambda}_1)e(dv)(\overline{v}_p - \Lambda_p^*)(\hat{v}_q - \hat{\Lambda}_q) \prod_{j=q}^{n} \hat{\Lambda}_j^{-1} \prod_{j=1}^{q} \hat{v}_j^{-1}$$

$$= \sum_{q=1}^{n} \int (\hat{v}_1 - \hat{\Lambda}_1)e(dv)(\overline{v}_p - \Lambda_p^*)\left(\prod_{j=q}^{n} \hat{\Lambda}_j^{-1} \prod_{j=1}^{q-1} \hat{v}_j^{-1} - \prod_{j=q+1}^{n} \hat{\Lambda}_j^{-1}\Pi_{j=1}^{q}\hat{v}_j^{-1}\right)$$

$$= \int (\hat{v}_1 - \hat{\Lambda}_1)e(dv)(\overline{v}_p - \Lambda_p^*)\left(\prod_{j=1}^{n} \hat{\Lambda}_j^{-1} - \prod_{j=1}^{n} \hat{v}_j^{-1}\right).$$

Hence, it follows (3.5.40), which proves the theorem. $\qquad\square$

3.6 Operator identities for the products of resolvents and mosaic

In the case of $z_i \in \rho(S_{q_i})$ and $w_j \in \rho(S_{p_j})$, it is easy to see that

$$S_{P_m,Q_n}(z,w) = P_M \prod_{i=1}^{m}(S_{p_j}^* - \bar{w}_j)^{-1} \prod_{i=1}^{n}(S_{q_i} - z_i)^{-1}|_M.$$

By Theorem 3.3.4, we have (3.3.43) in which $\mathbb{T}$ is replaced by $\mathbb{S}$. Now we consider the more general case of $z_i \in \rho(N_{q_i})$ and $w_j \in \rho(N_{p_j})$ as follows:

Theorem 3.6.1. *Let* $\mathbb{S} = (S_1,\ldots,S_k)$ *be a pure subnormal k-tuple of operators on a separable Hilbert space* $\mathcal{H}$ *with minimal normal extension* $\mathbb{N} = (N_1,\ldots,N_k)$ *on* $\mathcal{K} \supset \mathcal{H}$. *For integers* $p_i,q_j, i = 1,2,\ldots,m, j = 1,2,\ldots,n$ *satisfying* $1 \le p_i, q_j \le k$. *If* $z_i \in \rho(\Lambda_i) \cap \rho(N_{q_i})$, $i = 1,2,\ldots,n$, $w_j \in \rho(\Lambda_j) \cap \rho(N_{p_j})$, $j = 1,2,\ldots,m$ *satisfy condition that* $Q_{p_j,q_i}(z_i,w_j)$ *are invertible,* $j = 1,2,\ldots,m, i = 1,2,\ldots,n$, *then*

$$\mathfrak{S}_{P_m,Q_n} = \mu_{P_m}^\dagger \mathfrak{X}_{P_m,Q_n} - \mathfrak{X}_{P_m,Q_n}\mu_{Q_n}, \tag{3.6.1}$$

where $\mathfrak{S}_{P_m,Q_n}$ *stands for* $\mathfrak{S}_{\{p_1,\cdots,p_m\};\{q_1,\ldots,q_n\}}(z_1,\ldots,z_n;w_1,\ldots,w_m)$.

This theorem is a generalization of Theorem 1.4.11 to the case of subnormal tuple of operators.

Proof. We will prove (3.6.1) by mathematical induction with respect to m and n. First consider the case $m = n = 1$. If $m = n = 1$, then (3.6.1) is equivalent to the following

Lemma 3.6.2. *If* $z \in \rho(\Lambda_q) \cap \rho(N_q)$, $w \in \rho(\Lambda_p) \cap \rho(N_p)$ *and* $Q_{pq}(z,w)$ *is invertible, then*

$$S_{p;q}(z,w) = (I - \mu_p(w)^*)Q_{pq}(z,w)^{-1} - Q_{pq}(z,w)^{-1}\mu_q(z).$$

Here $S_{p:q}$ means $S_{\{p\},\{q\}}$.

Proof. The proof is similar to that for the Lemma 1.4.11. But in order to make this book readable, we give the details.

By (3.4.19), we have

$$Q_{pq}(z,w)e(du) = ((\bar{w} - \bar{u}_p)(z - \Lambda_q) - (\bar{u}_p - \Lambda_p^*)(u_q - z))e(du).$$

Thus

$$Q_{pq}(z, w)S_{p;q}(z, w) = \int \frac{Q_{pq}(z, w)e(du)}{(\bar{u}_p - \bar{w})(u_q - z)}$$

$$= -\int \frac{(z - u_q + u_q - \Lambda_q)}{u_q - z}e(du) - \int \frac{\bar{u}_p - \Lambda_p^*}{\bar{u}_p - \bar{w}}e(du)$$

$$= I - \mu_q(z) - \int \frac{\bar{u}_p - \Lambda_p^*}{\bar{u}_p - \bar{w}}e(du). \tag{3.6.2}$$

Let us prove that

$$\int \frac{\bar{u}_p - \Lambda_p^*}{\bar{u}_p - \bar{w}}e(du)Q_{pq}(z, w) = Q_{pq}(z, w)\mu_p(w)^*. \tag{3.6.3}$$

Firstly, we have

$$\int \frac{\bar{u}_p - \Lambda_p^*}{\bar{u}_p - \bar{w}}e(du)(\bar{w} - \Lambda_p^*) = (\bar{w} - \Lambda_p^*) + (\bar{w} - \Lambda_p^*)\int \frac{e(du)}{\bar{u}_p - \bar{w}}(\bar{w} - \Lambda_p^*)$$

$$= (\bar{w} - \Lambda_p^*) + (\bar{w} - \Lambda_p^*)\int \frac{e(du)}{\bar{u}_p - \bar{w}}(\bar{u}_p - \Lambda_p^* - (\bar{u}_p - \bar{w}))$$

$$= (\bar{w} - \Lambda_p^*)\mu_p(w)^*. \tag{3.6.4}$$

Next, we have to prove that

$$\int \frac{\bar{u}_p - \Lambda_p^*}{\bar{u}_p - \bar{w}}e(du)((\bar{w} - \Lambda_p^*)\Lambda_q + C_{pq}) = ((\bar{w} - \Lambda_p^*)\Lambda_q + C_{pq})\mu_p(w)^*. \tag{3.6.5}$$

By (3.4.19), the left hand side of (3.6.5) equals to

$$\int \frac{\bar{u}_p - \Lambda_p^*}{\bar{u}_p - \bar{w}}e(du)((\bar{w} - \Lambda_p^*)\Lambda_q + (\bar{u}_p - \Lambda_p^*)(u_q - \Lambda_q))$$

$$= \int \frac{\bar{u}_p - \Lambda_p^*}{\bar{u}_p - \bar{w}}e(du)((\bar{w} - \bar{u}_p)\Lambda_q + (\bar{u}_p - \Lambda_p^*)u_q) \tag{3.6.6}$$

$$= \int (\bar{u}_p - \Lambda_p^*)u_q e(du)\frac{\bar{u}_p - \Lambda_p^*}{\bar{u}_p - \bar{w}},$$

since $\int(\bar{u}_p - \Lambda_p^*)(du) = 0$. By (3.4.19) again, we have

$$(\bar{u}_p - \Lambda_p^*)u_q e(du) = (C_{pq} + (\bar{u}_p - \Lambda_p^*)\Lambda_q)e(du).$$

Thus the right-hand side of (3.6.6) equals to

$$C_{pq}\mu_p(w)^* + \int (\bar{u}_p - \Lambda_p^*)\Lambda_q e(du)(\bar{u}_p - \Lambda_p^*)(\bar{u}_p - \bar{w})^{-1} = C_{pq}\mu_p(w)^*$$

$$+ (\bar{w} - \Lambda_p^*)\Lambda_q\mu_p(w)^* + \int (\bar{u}_p - \bar{w})\Lambda_q e(du)(\bar{u}_p - \Lambda_p^*)(\bar{u}_p - \bar{w})^{-1}. \tag{3.6.7}$$

However, the third term in the right-hand side of (3.6) is zero, which proves (3.6.5). From (3.6.4) and (3.6.5), it follows (3.6.3). From (3.6.2) and (3.6.3), it follows the lemma.

$$\square$$

In the case $m = 1$, Theorem 3.6.1 is equivalent to the following:

Lemma 3.6.3. *If $w \in \rho(\Lambda_p) \cap \rho(N_p)$, $z_j \in \rho(\Lambda_{q_j}) \cap \rho(N_{q_j})$, $1 \le p, q_j \le k$ and $Q_{pq_j}(z_j, w)$, $j = 1, 2, \ldots, n$ are invertible, then*

$$S_{\{p\}, Q_n} = (I - \mu_p^*) X_{p, Q_n} - \sum_{j=1}^{n} X_{p, Q_j} \hat{\mu}_{Q_{j,n}}. \tag{3.6.8}$$

In (3.6.8), X_{p, Q_n} is defined by (3.3.40), $Q_j = \{q_1, \ldots, q_j\}$, $Q_{j,l} = \{q_j, \ldots, q_l\}$, $j \le l$ and $Q_{q,n}$ is defined in (3.5.2).

Proof. Let us prove it by the mathematical induction with respect to the number of q's. For the case that there is only one q's, say, q_1, (3.6.8) is equivalent to Lemma 3.6.2. Assume that (3.6.8) holds good for $q_2, \ldots, q_n$ (there are $n - 1$ q's), i.e.

$$S_{\{p\}, Q_{2,n}} = (I - \mu_p^*) X_{p, Q_{2,n}} - \sum_{j=2}^{n} X_{p, Q_{2,j}} \hat{\mu}_{Q_{j,n}}. \tag{3.6.9}$$

We have to prove that (3.6.8) holds good for $Q_n = \{q_1, q_2, \ldots, q_n\}$.

By (3.4.19) again, we have

$$Q_{pq_j}(z_1, w) e(du) = ((\Lambda_p^* - \bar{w})(u_{q_1} - z_1) - (\bar{u}_p - \bar{w})(u_{q_1} - \Lambda_{q_1})) e(du).$$

Therefore

$$Q_{pq_1} S_{\{p\}, Q_n} = \int \frac{Q_{pq_1}(z, w) e(du)}{(\bar{u}_p - \bar{w}) \prod_{j=1}^{n} (u_{q_j} - z_j)} = (\Lambda_p^* - \bar{w}) S_{\{p\}, Q_{2,n}} - \hat{\mu}_{Q_{1,n}}. \tag{3.6.10}$$

From

$$Q_{pq}(z, w)^{-1}(\Lambda_p^* - \bar{w}) = (R_{qp}(w)^* - z)^{-1}, \tag{3.6.11}$$

$$[(R_{qp}(w)^* - z)^{-1}, I - \mu_p(w)^*] = 0,$$

(see (3.5.38)), (3.6.9) and (3.6.10), it follows that

$$\begin{aligned}
S_{\{p\}, Q_n} &= (R_{q_1 p}(w)^* - z_1)^{-1} S_{\{p\}, Q_{2,n}} - Q_{pq_1}^{-1} \hat{\mu}_{Q_{1,n}} \\
&= (I - \mu_p(w)^*)(R_{q_1 p}(w)^* - z_1)^{-1} X_{\{p\}, Q_{2,n}} \\
&\quad - \sum_{j=2}^{n} (R_{q_1 p}(w)^* - z_1)^{-1} X_{\{p\}, Q_{2,j}} \hat{\mu}_{\{p\}, Q_{j,n}} - Q_{pq_1}^{-1} \hat{\mu}_{Q_{1,n}}.
\end{aligned} \tag{3.6.12}$$

By the Hermitian of $Q_{pq}(z, w)$, (3.3.6) and (3.4.40), we have $X_{\{p\}, \{q_1\}} = Q_{pq_1}^{-1}$ and

$$X_{\{p\}, Q_{1,j}} = (R_{q_1 p}(w)^* - z_1)^{-1} X_{\{p\}, Q_{2,j}}.$$

Thus (3.6.12) implies (3.6.8), which proves the lemma. $\square$

In the case of $n = 1$, Theorem 3.6.1 is equivalent to the following:

Lemma 3.6.4. *If $w_j \in \rho(\Lambda_{p_j}) \cap \rho(N_{p_j})$, $z \in \rho(\Lambda_q) \cap \rho(N_q)$, $1 \le p_j, q \le k$ and $Q_{p_j q}(z, w_j)$, $j = 1, 2, \ldots, m$ are invertible, then*

$$S_{P_m, \{q\}}(z; w_1, \ldots, w_m) = -\sum_{j=1}^{m} \mu^*_{P_{j,m}} X_{P_j, \{q\}} + (I - \mu_q) X_{P_m, \{q\}} \quad (3.6.13)$$

where $P_{j,m} = \{p_j, \ldots, p_m\}$.

Proof. The proof of this lemma is similar to the proof of Lemma 3.6.3. For the case $m = 1$, (3.6.13) is just the Lemma 3.6.2. Assume that (3.6.13) holds for $P_{2,m} = \{p_2, \ldots, p_m\}$, i.e.

$$S_{P_{2,m}, \{q\}}(z; w_2, \ldots, w_m) = -\sum_{j=2}^{m} \mu^*_{P_{j,m}} X_{P_{2,j}, \{q\}} + (I - \mu_q) X_{P_{2,m}, \{q\}}.$$
$$(3.6.14)$$

From $e(du) Q_{p_1 q}(z, w_1) = e(du)((\bar{u}_{p_1} - \bar{w}_1)(\Lambda_q - z) - (\bar{u}_{p_1} - \Lambda^*_{p_1})(u_q - z))$, it follows that

$$S_{P_m, \{q\}} Q_{p_1 q} = \int \frac{e(du)((\bar{u}_{p_1} - \bar{w}_1)(\Lambda_q - z) - (\bar{u}_{p_1} - \Lambda^*_{p_1})(u_q - z))}{(u_q - z) \prod_{j=1}^{m} (\bar{u}_{p_j} - \bar{w}_j)}$$

$$= S_{P_{2,m}, \{q\}} (\Lambda_q - z) - \hat{\mu}^*_{P_m}.$$

From $(\Lambda_q - z) Q_{p_1 q}(z, w_1)^{-1} = (R_{p_1 q}(z) - \bar{w}_1)^{-1}$, (3.5.38) and (3.6.14), it follows that

$$S_{P_m, \{q\}} = S_{P_{2,m}, \{q\}} (R_{p_1 q}(z) - \bar{w}_1)^{-1} - \hat{\mu}^*_{P_m} Q_{p_1 q}(z, w_1)^{-1}$$

$$= -\sum_{j=2}^{m} \mu^*_{P_{j,m}} X_{P_{2,j}, \{q\}} (R_{p_1 q}(z) - \bar{w}_1)^{-1}$$

$$- (I - \mu_q) X_{P_{2,m}, \{q\}} (R_{p_1 q}(z) - \bar{w}_1)^{-1} - \hat{\mu}^*_{P_m} Q_{p_1 q}(z, w_1)^{-1}.$$
$$(3.6.15)$$

But from (3.3.40) and (3.3.41) it is easy to see that

$$X_{P_{2,j}, \{q\}} (R_{p_1 q}(z) - \bar{w}_1)^{-1} = (\Lambda_q - z)^{-1} \prod_{i=1}^{j} (R_{p_i q}(z) - \bar{w}_i)^{-1} = X_{P_j, \{q\}}.$$

Therefore (3.6.15) is equivalent to (3.6.14), which proves the lemma. $\qquad \square$

(3.6.13) also can be proved by taking the adjoint of the both sides of (3.6.8).

Now, let us continue to prove Theorem 3.6.1. It is easy to see that (3.6.1) is equivalent to the following

$$S_{P_m,Q_n} = (I - \mu_{p_m}^*)X_{P_m,Q_n} - \sum_{j=1}^{m-1} \hat{\mu}_{P_{j,m}}^* X_{P_j,Q_n} - \sum_{j=1}^{n} X_{P_m,Q_j}\hat{\mu}_{Q_{j,n}},$$

$$(3.6.16)$$

for any natural numbers m and n.

Lemma 3.6.3 shows that (3.6.16) holds for $m = 1$, and any n. Assume that (3.6.16) holds for $m = v - 1 \geq 1$. Let us prove that (3.6.16) holds for $m = v$ and any n, by mathematical induction with respect to n. Lemma 3.6.4 shows that (3.6.16) is true of $n = 1$. We assume that (3.6.16) is true for $m = v - 1$ and that n is replaced by $n - 1$. Notice that

$$e(du)Q_{p_1q_n}(z_n,w_1) = e(du)\big((\bar{u}_{p_1} - \bar{w}_1)$$
$$(\Lambda_{q_n} - z_n) - (\bar{u}_{p_1} - \bar{w}_1)(u_{q_n} - z_n) + (\Lambda_{p_1}^* - \bar{w}_1)(u_{q_n} - z_n)\big).$$

If $n > 1$, then

$$S_{P_v,Q_n}Q_{p_1q_n}$$
$$= \int (e(du)\Big((\bar{u}_{p_1} - \bar{w}_1)(\Lambda_{q_n} - z_n) - (\bar{u}_{p_1} - \bar{w}_1)(u_{q_n} - z_n)$$
$$+ (\Lambda_{p_1}^* - \bar{w}_1)(u_{q_n} - z_n))\Big)\Big(\prod_{j=1}^{v}(\bar{u}_{p_j} - \bar{w}_j)\prod_{j=1}^{n}(u_{q_j} - z_j) \Big)^{-1}$$
$$= S_{\hat{P}_v,Q_n}(\Lambda_{q_n} - z_n) - S_{\hat{P}_v,Q_{n-1}} + S_{P_v,Q_{n-1}}(\Lambda_{p_1}^* - \bar{w}_1),$$

where $\hat{P}_v = \{p_2,\ldots,p_v\}$. Since there are only $v - 1$ natural numbers in $\hat{P}_v$, we may apply (3.6.16) to $S_{\hat{P}_v,Q_n}$ and $S_{\hat{P}_v,Q_{n-1}}$. Besides, by the hypothesis of mathematical induction with respect to n, we may also use the formula (3.6.16) for $S_{P_v,Q_{n-1}}$. Thus

$$S_{P_v,Q_n} = (I_1 + I_2 + I_3)Q_{p_1q_n}^{-1}, \qquad (3.6.17)$$

where

$$I_1 = \big((I - \mu_{p_v}^*)X_{\hat{P}_v,Q_n} - \sum_{j=2}^{v-1}\hat{\mu}_{P_{j,v}}^* X_{\hat{P}_j,Q_n} - \sum_{j=1}^{n} X_{\hat{P}_v,Q_j}\hat{\mu}_{Q_{j,n}}\big)(\Lambda_{q_n} - z_n),$$

$$I_2 = -\big((I - \mu_{p_v}^*)X_{\hat{P}_v,Q_{n-1}} - \sum_{j=2}^{v-1}\hat{\mu}_{P_{j,v}}^* X_{\hat{P}_j,Q_{n-1}} - \sum_{j=1}^{n-1} X_{\hat{P}_v,Q_j}\hat{\mu}_{Q_{j,(n-1)}}\big),$$

and

$$I_3 = \left((I - \mu_{p_v}^*) X_{P_v, Q_{n-1}} - \sum_{j=1}^{v-1} \hat{\mu}_{P_{j,v}}^* X_{P_j, Q_{n-1}} \right.$$

$$\left. - \sum_{j=1}^{n-1} X_{P_v, Q_j} \hat{\mu}_{Q_{j,(n-1)}} \right) (\Lambda_{p_1}^* - \bar{w}_1).$$

Let us rearrange the terms in the summation of (3.6.17). Then

$$S_{P_v, Q_n} = J_1 + J_2, \tag{3.6.18}$$

where

$$J_1 = (I - \mu_{p_v}^*) \widetilde{X}_{P_v, Q_n} - \sum_{j=2}^{v-1} \hat{\mu}_{P_{j,v}}^* \widetilde{X}_{P_j, Q_n} - \hat{\mu}_{P_v}^* \widetilde{X}_{\{p_1\}, Q_n}, \tag{3.6.19}$$

where

$$\widetilde{X}_{P_j, Q_n} = \left(X_{\hat{P}_j, Q_n}(\Lambda_{q_n} - z_n) - X_{\hat{P}_j, Q_{n-1}} + X_{P_j, Q_{n-1}}(\Lambda_{p_1}^* - \bar{w}_1) \right) Q_{p_1 q_n}^{-1}, \tag{3.6.20}$$

for $n > 1$ and $\widetilde{X}_{P_j, Q_1} = X_{\hat{P}_j, Q_1}(R_{p_1 q_1}(z_n) - \overline{w_1})^{-1}$. Besides,

$$J_2 = -\sum_{j=1}^{n} X_{\hat{P}_v, Q_j} \hat{\mu}_{Q_{j,n}} (R_{p_1 q_n}(z_n) - \bar{w}_1)^{-1} + \sum_{j=1}^{n-1} X_{\hat{P}_v, Q_j} \hat{\mu}_{Q_{j,(n-1)}} Q_{p_1 q_n}^{-1}$$

$$- \sum_{j=1}^{n-1} X_{P_v, Q_j} \hat{\mu}_{Q_{j,(n-1)}} (\Lambda_{p_1}^* - \bar{w}_1) Q_{p_1 q_n}^{-1}, \tag{3.6.21}$$

since $(\Lambda_{q_n} - z_n) Q_{p_1 q_n}^{-1} = (R_{p_1 q_n}(z_1) - \bar{w}_1)^{-1}$.

Now, let us prove that

$$\widetilde{X}_{P_j, Q_n} = X_{P_j, Q_n}. \tag{3.6.22}$$

According to (3.3.41), we only have to prove that

$$\left(\widetilde{X}_{P_j, Q_1} \cdots \widetilde{X}_{P_j, Q_n} \right) (R_{p_1, Q_n} - \overline{w_1}) = \left(X_{\hat{P}_j, Q_1} \cdots X_{\hat{P}_j, Q_n} \right), \tag{3.6.23}$$

by mathematical induction with respect to n, where $j \geq 2$, and the left hand side and the first factor of the right hand side of (3.6.23) are one row matrices. From (3.3.41), (3.6.23) holds for $n = 1$ and $\widetilde{X}_{P_j, Q_1}$ can be replaced by $\widetilde{X}_{P_j, Q_1}$. Assume that (3.6.23) holds while n is replaced by $n-1$,when $n \geq 2$. According to the definition of X's of (3.4.41), $\tilde{X}_{P_j,Q_i} = X_{P_j,Q_i}$ for $i = 1, 2, \ldots, n-1$. In order to prove that (3.6.23) holds good for n, we only have to prove that $L_n = 0$, where

$$L_n \overset{def}{=} -\sum_{i=1}^{n-1} X_{P_j,Q_i} C_{p_1 q_i} \prod_{l=i}^{n} (\Lambda_{q_l} - z_l)^{-1} + \tilde{X}_{P_j,Q_n} (R_{p_1 q_n}(z_n) - \bar{w}_1) - X_{\hat{P}_j,Q_n}.$$

By the hypothesis of mathematical induction: $L_{n-1} = 0$ and $\tilde{X}_{P_j,Q_{n-1}} = X_{P_j,Q_{n-1}}$,

$$-\sum_{i=1}^{n-1} X_{P_j,Q_i} C_{p_1 q_i} \prod_{l=i}^{n} (\Lambda_{q_l} - z_l)^{-1}$$

$$= -\left(\sum_{i=1}^{n-1} X_{P_j,Q_i} C_{p_1 q_i} \prod_{l=i}^{n} (\Lambda_{q_l} - z_l)^{-1}\right)(\Lambda_{q_n} - z_n)^{-1}$$

$$= \left[X_{\hat{P}_j,Q_{n-1}} - X_{P_j,Q_{n-1}}(R_{p_1 q_{n-1}}(z_{n-1}) - \bar{w}_1)\right.$$

$$\left. - X_{P_j,Q_{n-1}} C_{p_1 q_{n-1}}(\Lambda_{q_{n-1}} - z_{n-1})^{-1}\right] \cdot (\Lambda_{q_n} - z_n)^{-1}$$

$$= \left(X_{\hat{P}_j,Q_{n-1}} - X_{P_j,Q_{n-1}}(\Lambda_{p_1}^* - \bar{w}_1)\right)(\Lambda_{q_n} - z_n)^{-1},$$

since $R_{p_1 q_{n-1}}(z_{n-1}) - \bar{w}_1 = -C_{p_1 q_{n-1}}(\Lambda_{q_{n-1}} - z_{n-1})^{-1} + \Lambda_1^* - \bar{w}_1$. Thus

$$L_n = \{\tilde{X}_{P_j,Q_n} - X_{\hat{P}_j,Q_n}(R_{p_1 q_n}(z) - \bar{w}_j)^{-1} + X_{\hat{P}_j,Q_{n-1}} Q_{p_1 q_n}^{-1}$$

$$- X_{P_j,Q_{n-1}}(\Lambda_{p_1}^* - \bar{w}_1)Q_{p_1 q_n}^{-1}\}(R_{p_1 q_n}(z_n) - \bar{w}_1)$$

which equals to zero by (3.6.20). Therefore (3.6.23) is proved and so does (3.6.22).
From (3.6.19) and (3.6.22), we have

$$J_1 = (I - \mu_{p_v}^*)X_{P_v,Q_n} - \sum_{j=2}^{v-1} \hat{\mu}_{P_j,v}^* X_{P_j,Q_n} - \hat{\mu}_{P_v}^* X_{\{p_1\},Q_n}. \qquad (3.6.24)$$

Next, let us study J_2. From $L_n = 0$ and (3.6.22), we have

$$X_{\hat{P}_v,Q_j} = -\sum_{s=1}^{j-1} X_{P_v,Q_s} C_{p_1 q_s} \prod_{i=s}^{j} (\Lambda_{q_i} - z_i)^{-1} + X_{P_v,Q_j}(R_{p_1 q_j}(z_j) - \bar{w}_1).$$

Thus

$$J_2 = -\sum_{j=1}^{n}[-\sum_{s=1}^{j-1}X_{P_v,Q_s}C_{p_1q_s}\prod_{i=s}^{j}(\Lambda_{q_i} - z_i)^{-1}$$

$$+ X_{P_v,Q_j}(R_{p_1q_j}(z_j) - \bar{w}_1)]\hat{\mu}_{Q_{j,n}}(R_{p_1q_n}(z_n) - \bar{w}_1)^{-1}$$

$$+\sum_{j=1}^{n-1}\big(-\sum_{s=1}^{j-1}X_{P_v,Q_s}C_{p_1q_s}\prod_{i=s}^{j}(\Lambda_{q_i} - z_i)^{-1}$$

$$+ X_{P_v,Q_j}(R_{p_1q_j}(z_j) - \bar{w}_1)\big)\hat{\mu}_{Q_{j,(n-1)}}Q_{p_1q_n}^{-1}$$

$$-\sum_{j=1}^{n-1}X_{\hat{P}_v,Q_j}\hat{\mu}_{Q_{j,(n-1)}}(\Lambda_{p_1}^* - \bar{w}_1)Q_{p_1q_n}^{-1}. \tag{3.6.25}$$

For $n - 1 \geq s \geq 1$, let us group all the terms with coefficient X_{P_v,Q_s} in the right side of (3.6.25). That is

$$X_{P_v,Q_s}(K_1 + K_2 + K_3), \tag{3.6.26}$$

where

$$K_1 = (\sum_{j=s+1}^{n}C_{p_1q_s}\prod_{i=s}^{j}(\Lambda_{q_i} - z_i)^{-1}\hat{\mu}_{Q_{j,n}}$$

$$- (R_{p_1q_s}(z_s)\bar{w}_1)\hat{\mu}_{Q_{s,n}})(R_{p_1q_n}(z_n) - \bar{w}_1)^{-1},$$

$$K_2 = \big(-\sum_{j=s+1}^{n-1}C_{p_1q_s}\prod_{i=s}^{j}(\Lambda_{q_i} - z_i)^{-1}\hat{\mu}_{Q_{j,(n-1)}}$$

$$+ (R_{p_1q_s}(z_s) - \bar{w}_1)\hat{\mu}_{Q_{s,(n-1)}}\big)Q_{p_1q_n}^{-1},$$

and

$$K_3 = -\hat{\mu}_{Q_{s,(n-1)}}(\Lambda_{p_1}^* - \bar{w}_1)Q_{p_1q_n}^{-1}.$$

By (3.5.38), we have

$$K_1 = (\sum_{j=s}^{n-1}\hat{\mu}_{Q_{sj}}C_{p_1q_j}\prod_{i=j}^{n}(\Lambda_{q_i} - z_i)^{-1}$$

$$- \hat{\mu}_{Q_{sn}}(R_{p_1q_n}(z_n) - \bar{w}_1))(R_{p_1q_n}(z_n) - \bar{w}_1)^{-1}$$

and

$$K_2 = \big(-\sum_{j=s}^{n-2}\hat{\mu}_{Q_{sj}}C_{p_1q_j}\prod_{i=j}^{n-1}(\Lambda_{q_i} - z_i)^{-1}$$

$$+ \hat{\mu}_{Q_{s(n-1)}}(R_{p_1q_{n-1}}(z_{n-1}) - \bar{w}_1))Q_{p_1q_n}^{-1}.$$

Notice that $Q_{p_1 q_n}^{-1} = (\Lambda_{q_n} - z_n)^{-1}(R_{p_1 q_n}(z_n) - \bar{w}_1)^{-1}$ and

$$R_{p_1 q_{n-1}}(z_{n-1}) - \bar{w}_1 = -C_{p_1 q_{n-1}}(\Lambda_{q_{n-1}} - z_{n-1})^{-1} + (\Lambda_{p_1}^* - \bar{w}_1),$$

we have

$$
\begin{aligned}
K_1 + K_2 + K_3 + \hat{\mu}_{Q_{sn}} &= \hat{\mu}_{Q_{s(n-1)}}\big(C_{p_1 q_{n-1}}(\Lambda_{q_{n-1}} - z_{n-1})^{-1} \\
&\quad - \Lambda_{p_1}^* + \bar{w}_1 + R_{p_1 q_{n-1}}(z_{n-1}) - \bar{w}_1\big)Q_{p_1 q_n}^{-1} = 0.
\end{aligned}
\tag{3.6.27}
$$

From (3.6.25), (3.6.26) and (3.6.27), it follows

$$J_2 = -\sum_{s=1}^{n} X_{P_v, Q_s} \hat{\mu}_{Q_{sn}}. \tag{3.6.28}$$

From (3.6.18), (3.6.24) and (3.6.28), it follows (3.6.16), which proves the theorem. $\qquad\square$

3.7 Subnormal tuple of operators with compact self-commutators

Let $\mathcal{S} = (S_1, \ldots, S_k)$ be a subnormal tuple of operators on a Hilbert space $\mathcal{H}$. If $[S_i^*, S_j] \in L_0(\mathcal{H}), i, j = 1, 2, \ldots, k$, then $\mathcal{S}$ is said to be with compact self-commutators.

Lemma 3.7.1. *Let $\mathcal{S} = (S_1, \ldots, S_k)$ be a subnormal tuple of operators on a Hilbert space $\mathcal{H}$. If $[S_i^*, S_i] \in L_0(\mathcal{H}), i = 1, 2, \ldots, k$, then $[S_i^*, S_j] \in L_0(\mathcal{H}), i, j = 1, 2, \ldots, k$.*

Proof. From $[S_i^*, S_i] \in L_0(\mathcal{H}), i = 1, 2, \ldots, k$, it is easy to see there is a sequence of projections $\{P_n\}$ satisfying the condition that $\mathrm{rank}(P_n) < \infty, n = 1, 2, \ldots$ and

$$\lim_{n \to \infty} \|(I - P_n)[S_i^*, S_i](I - P_n)\| = 0, \quad i = 1, 2, \ldots, k. \tag{3.7.1}$$

From (3.4.7), we may prove that

$$
\begin{aligned}
&\|(I - P_n)[S_i^*, S_j](I - P_n)\|^2 \\
&\le \|(I - P_n)[S_i^*, S_i](I - P_n)\|\|(I - P_n)[S_j^*, S_j](I - P_n)\|,
\end{aligned}
$$

and that by (3.7.1), we have $\lim_{n \to \infty} \|(I - P_n)[S_i^*, S_j](I - P_n)\| = 0$. Hence $[S_i^*, S_j] \in L_0(\mathcal{H})$. $\qquad\square$

By means of the method in the proof of the Theorem 1.6.1, we may prove the following:

Theorem 3.7.2. *Let $S = (S_1, \ldots, S_k)$ be a pure subnormal k-tuple of operators on $\mathcal{H}$, and $l = \{l_1, \ldots, l_n\}, 1 \leq l_j \leq k$. If $[S_{l_j}^*, S_{l_j}] \in L_0(\mathcal{H}), j = 1, 2, \ldots, n$, then*

$$\text{rank}(\mu_l(z_1, \ldots, z_n)) < \infty,$$

for $z_j \in \rho(N_{lj}), j = 1, 2, \ldots, n$.

We omit the proof. Let us generalize Lemma 1.6.4, Lemma 1.6.5 and Theorem 1.6.6 to the subnormal k-tuple case.

Lemma 3.7.3. *Let $S = (S_1, \ldots, S_k)$ be a subnormal k-tuple of operators on a Hilbert space $\mathcal{H}$. If $[S_i^*, S_i] \in L^1(\mathcal{H}), i = 1, 2, \ldots, k$, then*

$$[\prod_{i=1}^{k}(\overline{w}_i - S_i^*)^{-1}, \prod_{j=1}^{k}(z_j - S_j)^{-1}] \in L^1(\mathcal{H}) \tag{3.7.2}$$

and

$$\text{tr}([\prod_{i=1}^{k}(\overline{w} - S_i^*)^{-1}, \prod_{j=1}^{k}(z_j - S_j)^{-1}]\prod_{l=1}^{k}(\overline{v}_l - S_l^*)^{-1})$$

$$= \sum_{i=1}^{k} \text{tr}\left(\int \frac{e(du)(\overline{u}_i - \Lambda_i^*)}{(\overline{w}_i - \overline{u}_i)\prod_{j=1}^{k}[(\overline{w}_j - \overline{u}_j)(\overline{v}_j - \overline{u}_j)(z_j - u_j)]}\right) \tag{3.7.3}$$

for $z_i, v_i, w_i \in \rho(S_i), i = 1, 2, \ldots, k$.

Proof. Similar to Lemma 3.7.1, we may conclude that $[S_i^*, S_j] \in L^1(\mathcal{H})$ for $i, j = 1, \ldots, k$. For simplicity, let $A_i = \overline{w}_i - S_i^*, B_i = z_i - S_i$ and $C_i = \overline{v}_i - S_i^*$, then

$$[A_i, B_j] = C_{ij}P_M.$$

It is easy to calculate that

$$[A_i^{-1}, \prod_{j=1}^{k} B_j^{-1}] = \sum_{j=1}^{k} A_i^{-1}(\prod_{p=j}^{k} B_p^{-1})C_{ij}P_M(\prod_{q=1}^{j} B_q^{-1})A_i^{-1}.$$

Therefore

$$[\prod_{j=1}^{k} A_j^{-1}, \prod_{j=1}^{k} B_j^{-1}] = \sum_{i,j=1}^{k} \prod_{u=1}^{i} A_u^{-1} \prod_{p=j}^{k} B_p^{-1} C_{ij}P_M\Pi_{q=1}^{j}B_q^{-1}\prod_{t=i}^{k} A_t^{-1},$$

and it belongs to $L^1(\mathcal{H})$. Thus from $[A_i, C_l] = 0$, we have

$$\text{tr}([\prod_{i=1}^{k} A_i^{-1}, \prod_{j=1}^{k} B_j^{-1}]\prod_{l=1}^{k} C_l^{-1})$$

$$= \sum_{i,j=1}^{k} \text{tr}(A_i^{-1}\prod_{l=1}^{k} C_l^{-1}\prod_{u=1}^{k} A_u^{-1}\Pi_{p=j}^{k}B_p^{-1}C_{ij}P_M \prod_{q=1}^{j} B_q^{-1}). \tag{3.7.4}$$

It is obvious that $P_M B_j^{-1} = (z_j - \Lambda_j)^{-1} P_M$ and for $a \in M$

$$P_M A_i^{-1} \prod_{l=1}^{k} C_l^{-1} \prod_{u=1}^{k} A_u^{-1} \prod_{p=j}^{k} B_p^{-1} a$$

$$= \int \frac{e(du)a}{(\overline{w}_i - \overline{u}_i) \prod_{l=1}^{k}[(\overline{w}_l - \overline{u}_l)(\overline{v}_l - \overline{u}_l)] \prod_{p=j}^{k}(z_p - u_p)}.$$

Therefore (3.7.4) equals to

$$\sum_{i,j=1}^{k} \mathrm{tr}\Big(\int \frac{e(du)C_{ij} \prod_{q=1}^{j}(z_q - \Lambda_q)^{-1}}{(\overline{w}_i - \overline{u}_i) \prod_{l=1}^{k}[(\overline{w}_l - \overline{u}_l)(\overline{v}_l - \overline{u}_l)] \prod_{p=j}^{k}(z_p - u_p)} \Big). \qquad (3.7.5)$$

By (3.4.19), $e(du)C_{ij} = e(du)(\overline{u}_i - \Lambda_i^*)(u_j - \Lambda_j)$, (3.7.5) is reduced to

$$\sum_{i=1}^{k} \mathrm{tr}\Big(\int \frac{e(du)(\overline{u}_i - \Lambda_i^*)}{(\overline{w}_i - \overline{u}_i) \prod_{l=1}^{k}[(\overline{w}_l - \overline{u}_l)(\overline{v}_l - \overline{u}_l)]} \cdot$$

$$\sum_{j=1}^{k} \Big(\frac{\prod_{q=1}^{j-1}(z_q - \Lambda_q)^{-1}}{\Pi_{p=j}^{k}(z_j - u_j)} - \frac{\prod_{q=1}^{j}(z_q - \Lambda_q)^{-1}}{\Pi_{p=j+1}^{k}(z_p - u_p)} \Big) \Big)$$

$$= \sum_{i=1}^{k} \mathrm{tr}\Big(\int \frac{e(du)(\overline{u}_i - \Lambda_i^*)}{(\overline{w}_i - \overline{u}_i)\Pi_{j=1}^{k}[(\overline{w}_j - \overline{u}_j)(\overline{v}_j - \overline{u}_j)(z_j - u_j)]}$$

$$- \int \frac{e(du)(\overline{u}_i - \Lambda_i^*)\Pi_{q=1}^{k}(z_q - \Lambda_q)^{-1}}{(\overline{w}_i - \overline{u}_i)\Pi_{j=1}^{k}[(\overline{w}_j - \overline{u}_j)(\overline{v}_j - \overline{u}_j)]} \Big) \qquad (3.7.6)$$

which equals to the right hand side of (3.7.3), since the second term of the
most right hand side of (3.7.6) equals to zero by (3.4.26). $\qquad \square$

Lemma 3.7.4. *If $[S_i^*, S_i]^{1/2} \in L^1(\mathcal{H}), i = 1, \ldots, k,$ then for every bounded
Baire function $f(\cdot)$ on $\mathrm{sp}(\mathbb{N})$, where $\mathbb{N}$ is the m.n.e. of $\mathcal{S}$,*

$$\int f(u)e(du)(\overline{u}_i - \Lambda_i^*) \in L^1(M), \quad i = 1, 2, \ldots, k, \qquad (3.7.7)$$

and there is a complex measure $v_i(\cdot)$ with finite total variation such that

$$\mathrm{tr}\Big(\int f(u)e(du)(\overline{u}_i - \Lambda_i^*) \Big) = \int f(u)v_i(du). \qquad (3.7.8)$$

The proof of this lemma is similar to that of Lemma 1.6.5, so we omit
it.

Let $R(\mathcal{S})$ be the algebra of functions of variables $u = (u_1, \ldots, u_k)$ generated by $(\overline{u}_i - \overline{z}_i)^{-1}$ and $(u_i - z_i)^{-1}, z \in \rho(S_i), i = 1, \ldots, k.$ For

$$f(u) = f(\overline{u}_1, \ldots, \overline{u}_k, u_1, \ldots, u_k) \in R(\mathcal{S}),$$

let

$$f(\mathcal{S}^*, \mathcal{S}) = f(S_1^*, \ldots, S_k^*; S_1, \ldots, S_k).$$

It is obvious that, in general, this operator $f(\mathcal{S}^*, \mathcal{S})$ depends on the ordering of the product. It is easy to see that if $[S_i^*, S_i] \in L^1(\mathcal{H}), i = 1, 2, \ldots, k$, then $[f(\mathcal{S}^*, \mathcal{S}), h(\mathcal{S}^*, \mathcal{S})] \in L^1(\mathcal{H})$ for any $f, h \in R(\mathcal{S})$ and

$$\text{tr}[f(\mathcal{S}^*, \mathcal{S}), h(\mathcal{S}^*, \mathcal{S})]$$

is independent of the ordering of the product in the functional calculus $f(\mathcal{S}^*, \mathcal{S})$ and $h(\mathcal{S}^*, \mathcal{S})$.

Theorem 3.7.5. *Let $\mathcal{S} = (S_1, \ldots, S_k)$ be a subnormal k-tuple of operators on $\mathcal{H}$ with m.n.e. $\mathbb{N}$, satisfying $[S_i^*, S_i]^{1/2} \in L^1(\mathcal{H}), i = 1, 2, \ldots, k$. Then there are complex measures with finite total variations*

$$v_j(du) = i \, \text{tr}((u_j - \Lambda_j)e(du)), \quad j = 1, 2, \ldots, k,$$

on $\text{sp}(\mathbb{N})$ *such that for $f, h \in R(\mathcal{S})$, and any ordering of the product of $f(\mathcal{S}^*, \mathcal{S})$ and $h(\mathcal{S}^*, \mathcal{S})$,*

$$i \, \text{tr}([f(\mathcal{S}^*, \mathcal{S}), h(\mathcal{S}^*, \mathcal{S})]) = \int_{\text{sp}(\mathbb{N})} f(\overline{u}, u) d_v h(\overline{u}, u), \qquad (3.7.9)$$

where

$$d_v h(\overline{u}, u) = \sum_{j=1}^{k} \left(\frac{\partial h(\overline{u}, u)}{\partial u_j} v_j(du) + \frac{\partial h(\overline{u}, u)}{\partial \overline{u}_j} \overline{v_j(du)} \right). \qquad (3.7.10)$$

Proof. From Lemma 3.7.3 and Lemma 3.7.4, it is easy to see that for $f(\overline{u}), h(u), q(\overline{u}) \in R(\mathcal{S})$, we have

$$i \, \text{tr}([f(\mathcal{S}^*), h(\mathcal{S})]q(\mathcal{S}^*)) = \sum_{j=1}^{k} \int_{\text{sp}(\mathbb{N})} q(\overline{u}) h(u) \frac{\partial f(\overline{u})}{\partial \overline{u}_j} \overline{v_j(du)}. \qquad (3.7.11)$$

Taking the conjugate of both sides of (3.7.11), it is easy to verify for $r(u) \in R(\mathcal{S})$, we have

$$i \, \text{tr}([f(\mathcal{S}^*), h(\mathcal{S})]r(\mathcal{S})) = \sum_{j=1}^{n} \int_{\text{sp}(\mathbb{N})} r(u) f(\overline{u}) \frac{\partial h(u)}{\partial u_j} v_j(du). \qquad (3.7.12)$$

The rest of the proof is similar to the last part of the proof of Theorem 1.6.6, we omit it. □

Chapter 4

Subnormal Tuple of Operators with Finite Dimension Defect Space

4.1 Spectrum of the minimal normal extension

In this chapter, we study subnormal k-tuple of operators $\mathcal{S} = (S_1, \ldots, S_k)$ on a Hilbert space $\mathcal{H}$ with m.n.e. $\mathbb{N}$ on $\mathcal{K} \supset \mathcal{H}$ and with finite rank self commutators, i.e.

$$\operatorname{rank}[S_i^*, S_j] < \infty, \ i, j = 1, 2, \ldots, k, \tag{4.1.1}$$

i.e., the defect space M is of finite dimension m. By (3.4.7), (4.1.1) is equivalent to

$$\operatorname{rank} A_i^* A_i = \operatorname{rank}[S_i^*, S_i] < \infty, \ i = 1, 2, \ldots, k,$$

where A_j is defined by (3.4.3). In this case, $C_{ij}, \Lambda_i, i, j = 1, 2, \ldots, k$ are $m \times m$ matrices. Define

$$P_{ij}(z, w) = \det Q_{ij}(z, w). \tag{4.1.2}$$

The polynomial $P_{ij}(z, w)$ is with leading term $(z\overline{w})^m$, where $m = \dim M$, and is hermitian: $\overline{P_{ij}(z, w)} = P_{ji}(w, z)$.

Lemma 4.1.1. *If* (4.1.1) *holds, then*

$$\operatorname{sp}(\mathbb{N}) \subset P(\mathcal{S}), \tag{4.1.3}$$

where

$$P(\mathcal{S}) \overset{def}{=} \{(u_1, \ldots, u_k) \in \mathbb{C}^k : P_{ij}(u_j, u_i) = 0, i, j = 1, 2, \ldots, k\}.$$

Proof. Let $E(\cdot)$ be the spectral measure of the m.n.e. $\mathbb{N} = \{N_1, \ldots, N_k\}$. Define $e_j(\cdot)$ by

$$e_j(F_j) = \int_{F_j} e_j(du) = P_M E_j(F_j)|_M, \tag{4.1.4}$$

where $E_j(\cdot)$ is the spectral measure of N_j, or

$$E_j(F_j) = \int_{\{u=(u_1,\ldots,u_k)\in \mathrm{sp}(N):u_j\in F_j\}} E(du).$$

We may rewrite (4.1.4) as

$$e_j(F_j) = \int_{\{u\in \mathrm{sp}(N):u_j\in F_j\}} e(du).$$

From §2.2, we may conclude that the measure $e_j(\cdot)$ does not have singular part, i.e. it contains only absolutely continuous part and atom part. Therefore the measure $e(\cdot)$ only contains absolutely continuous part and atom part.

Suppose $v = (v_1, \ldots, v_k) \notin P(\mathcal{S})$. Then there are $i, j, 1 \leq i, j \leq k$ and a number $\epsilon > 0$ such that if $u = (u_1, \ldots, u_k)$ and

$$u \in V_{ij}(\epsilon) = \{(u_1, \ldots, u_k) \in \mathrm{sp}(\mathbb{N}) : |u_i - v_i| < \epsilon, |u_j - v_j| < \epsilon\},$$

then $P_{ij}(u_j, u_i) \neq 0$. Thus $Q_{ij}(u_j, u_i)^{-1}$ exists. By (3.4.19), we have

$$\int_{V_{ij}(\epsilon)} e(du) = \int_{V_{ij}(\epsilon)} Q_{ij}(u_j, u_i)^{-1} Q_{ij}(u_j, u_i) e(du) = 0.$$

Thus $V_{ij}(\epsilon) \cap \mathrm{sp}(\mathbb{N}) = \emptyset$.

Therefore the spectrum of $\mathbb{N}$ is in $P(\mathcal{S})$. Thus the support of $e(\cdot)$ is in $P(\mathcal{S})$. From the fact that $\mathcal{K}$ is unitarily equivalent to $L^2(e)$, we may conclude that (4.1.3) holds good. $\qquad\square$

Denote $\mathrm{sp}_{\mathrm{ess}}(\mathbb{N})$ be the absolutely continuous part of the $\mathrm{sp}(\mathbb{N})$ and $\mathrm{sp}_{\mathrm{p}}(\mathbb{N})$ be the point spectrum. Then $\mathrm{sp}(\mathbb{N}) = \mathrm{sp}_{\mathrm{ess}}(\mathbb{N}) \cup \mathrm{sp}_{\mathrm{p}}(\mathbb{N})$, if (4.1.1) is satisfied.

Proposition 4.1.2. *If (4.1.1) holds then* $\mathrm{sp}_{\mathrm{ess}}(\mathbb{N})$ *consists of a finite set of real algebraic arcs and* $\mathrm{sp}_{\mathrm{p}}(\mathbb{N})$ *is a finite set.*

Proof. From §2.2, we know that $\sigma(N_j)$ consists of a finite collection of real algebraic arcs and a possible finite point spectrum. It is obvious that

$$\sigma(N_1) = \{u_1 : (u_1, \ldots, u_k) \in \mathrm{sp}(\mathbb{N})\},$$

or $\sigma(N_1)$ is the projection of $\mathrm{sp}(\mathbb{N})$ to the first coordinate plane. Suppose γ is a real algebraic arc in $\sigma(N_1)$. For any point $u_1 \in \gamma$, by (4.1.3), the point $u = (u_1, \ldots, u_k) \in \mathrm{sp}(\mathbb{N})$ must satisfy

$$P_{1l}(u_l, u_1) = 0, \ l = 2, 3, \ldots, k. \tag{4.1.5}$$

Therefore except finite points in γ there are at most m solutions of u_l of the equation (4.1.5). Thus there are at most m^{k-1} points in sp($\mathbb{N}$) with the same first coordinate u_1. Besides the coordinate u_j of those points must be a real algebraic function. Similarly, for any point $u_1 \in \sigma_p(N_1)$, except the case that there is a real algebraic arc in sp($\mathbb{N}$) on that arc every point has the same first coordinate u_1, there are only finite points in the sp$_p$($\mathbb{N}$) with the same first coordinate u_1, which proves the proposition. $\square$

Proposition 4.1.3. *Let $S = (S_1, \ldots, S_k)$ be a pure subnormal k-tuple of operators on a Hilbert space $\mathcal{H}$. If*

$$\dim \bigvee_{j=1}^{k} [S_j^*, S_j]\mathcal{H} = 1, \tag{4.1.6}$$

then there are complex numbers $\alpha_j, \beta_j, j = 1, 2, \ldots, k$ satisfying $\sum_{j=1}^{k} |\beta_j| > 0$ such that

$$S_j = \alpha_j + \beta_j U_+, j = 1, 2, \ldots, k, \tag{4.1.7}$$

where U_+ is a unilateral shift with multiplicity 1.

Proof. From (4.1.6) and (3.4.49), we have $\dim M = 1$. We may suppose that $M = \mathbb{C}$. Therefore C_{ij} and $\Lambda_i, i = 1, 2, \ldots, k$ are complex numbers and

$$P_{ij}(z, w) = (\overline{w} - \overline{\Lambda}_i)(z - \Lambda_j) - C_{ij}.$$

From (4.1.3)

$$\text{sp}(\mathbb{N}) \subset \{(u_1, \ldots, u_k) : (\overline{u}_i - \overline{\Lambda}_i)(u_j - \Lambda_j) = C_{ij}, i, j = 1, 2, \ldots, k\}. \tag{4.1.8}$$

Let $C_{ii} = r_i^2$, where $r_i \geq 0, i = 1, 2, \ldots, k$, since $C_{ii} = [S_i^*, S_i]|_M \geq 0$. Therefore the i-th coordinate of the points in sp($\mathbb{N}$) is in a circle $\{u_i : |u_i - \Lambda_i| = r_i\}$, if $r_i > 0$ or a single point set $\{\Lambda_i\}$ if $r_i = 0$. For $(u_1, \ldots, u_k) \in$ sp($\mathbb{N}$), let $u_j = \Lambda_j + r_j e^{i\theta_j}, \theta_j \in \mathbb{R}$. Then from (4.1.8)

$$C_{lj} = r_l r_j e^{i(\theta_j - \theta_l)}. \tag{4.1.9}$$

It is obvious that sp($\mathbb{N}$) cannot be a single set. Therefore at least one of r's, say $r_1 > 0$. Let

$$e_1(\gamma) \overset{\text{def}}{=} e(\{(u_1, \ldots, u_k) : u_1 \in \gamma\}).$$

Then the mosaic of S_1 is

$$\mu_1(z) = \int_{\text{sp}(\mathbb{N})} \frac{(u_1 - \Lambda_1)e(du)}{u_1 - z} = \int_{\sigma(N_1)} \frac{u_1 - \Lambda_1}{u_1 - z} e_1(du), \tag{4.1.10}$$

where $\sigma(N_1) \subset \{u_1 : |u_1 - \Lambda_1| = r_1\}$. As a mosaic, the complex function $\mu_1(z)$ is idempotent, i.e., $\mu_1(z)$ is either 0 or 1. The function $\mu_1(z)$ is analytic on $\mathbb{C} \setminus \{u_1 : |u_1 - \Lambda_1| = r_1\}$. The function $\mu_1(z)$ must be constant on $\{u_1 : |u_1 - \Lambda_1| < r_1\}$. From

$$\lim_{z \to \infty} \mu_1(z) = 0,$$

we may conclude that $\mu_1(z) = 0$ for $|z - \Lambda_1| > r_1$. Then $\mu_1(z) = 1$ for $|z - \Lambda_1| < r_1$, since $\mu_1(z)$ cannot be identically zero. From (4.1.10) and the Plemelj's formula or the method in §2.2, we may conclude that

$$(u_1 - \Lambda_1)e_1(du) = r_1 \frac{du_1}{2\pi i}$$

and $\mathrm{sp}(\mathbb{N}_1) = \{u_1 : |u_1 - \Lambda_1| = r_1\}$. Therefore

$$S_1 = r_1 U_+ + \Lambda_1, \tag{4.1.11}$$

where U_+ is a unilateral shift with multiplicity one.

By (4.1.10), if $j > 1$ and $r_j > 0$, $C_{1j} = r_1 r_j e^{iv_j}$, then

$$u_j = \Lambda_j + r_j e^{i(\theta_1 + v_j)}, \text{ for } u_1 = \Lambda_1 + r_1 e^{i\theta_1}.$$

It is

$$u_j = \Lambda_j + r_j e^{iv_j}(u_1 - \Lambda_1)/r_1. \tag{4.1.12}$$

From (4.1.11), (4.1.12) we may prove (4.1.7) where $\beta_j = r_j e^{iv_j}/r_1$ and $\alpha_j = \Lambda_j - \Lambda_1 r_j e^{iv_j}/r_1$. $\qquad\square$

Let us examine the spectrum of the m.n.e. N of the unilateral shift. It satisfies

$$P_{11}(u, u) = \overline{u}u - 1 = 0, \ u \in \sigma(N). \tag{4.1.13}$$

In the same time, the subnormal operator satisfies the following operator identity

$$U_+^* U_+ - I = 0. \tag{4.1.14}$$

The equation (4.1.14) can be considered as a replacement of $u, \overline{u}$ and 1 by U_+, U_+^* and I in (4.1.14) respectively. The following proposition shows that this kind replacement works for any pure subnormal k-tuple of operators with finite rank self-commutator.

Let $p(z, w) = \sum p_{ln} \overline{w}^l z^n$ be a polynomial of z and $\overline{w}$ where $p_{ln} \in \mathbb{C}$. Let A and B be operators. We adopt the following Weyl ordering

$$p(B, A) = \sum p_{ln} A^{*l} B^n.$$

For any two polynomials

$$P(w) = \sum_{j=0}^{k} p_j w^j, \text{ and } Q(w) = \sum_{j=0}^{k} q_j w^j$$

with coefficients p_l and q_j which are polynomials of a variable u for $u \in \sigma$, let us define the determinant $R(P, Q)$ as follows

$$R(P, Q) \overset{\text{def}}{=} \det(r_{ij})_{i,j=1,2,\ldots,2(k+1)}, \tag{4.1.15}$$

where

$$r_{ij} = p_{j-i}, \text{ for } 1 \le i \le k+1 \text{ and } i \le j \le i+k,$$

$$r_{ij} = q_{j-i+k}, \text{ for } k+2 \le i \le 2(k+1) \text{ and } i-k \le j \le i$$

and $r_{ij} = 0$ elsewhere. Then $R(P, Q)$ is said to be the resolvent of P and Q. If the polynomials P and Q have a common zero for every $u \in \sigma$ then $R(P, Q) = 0$ for $u \in \sigma$.

Proposition 4.1.4. *Let $S = (S_1, \ldots, S_k)$ be a pure subnormal k-tuple of operators with finite dimension defect space. Let P_{ij} be the polynomials defined in (4.1.2) and $R(P_{ii}(u_i, \cdot), P_{ij}(u_j, \cdot))$ be the resolvent of the polynomials $P_{ii}(u_i, \cdot)$ and $P_{ij}(u_j, \cdot)$ defined in (4.1.15). Then for $i, j = 1, 2, \ldots, k$,*

$$P_{ij}(S_j, S_i) = 0 \tag{4.1.16}$$

and

$$R(P_{ii}(S_i, \cdot), P_{ij}(S_i, \cdot)) = 0. \tag{4.1.17}$$

Proof. Let $P_{ij}(z, w) = \sum p_{ln} w^l z^n$, where p_{ln} actually also depends on i and j then

$$P_{ij}(u_j, u_i) = \sum p_{ln} \bar{u}_i^l u_j^n = 0 \text{ for } (u_1, \ldots, u_k) \in \text{sp}(\mathbb{N}) \tag{4.1.18}$$

by (4.1.3). Thus for $f, g \in \mathcal{H}$

$$(P_{ij}(S_j, S_i)f, g) = \sum p_{ln}(S_i^{*l} S_j^n f, g) = \sum p_{ln}(S_j^n f, S_i^l g)$$

$$= \int_{\text{sp}(N)} \sum p_{ln}(e(du) u_j^n f, u_i^l g)$$

$$= \int_{\text{sp}(N)} p_{ij}(u_j, u_i)(e(du)f, g) = 0$$

by (4.1.18).

For $(u_1, \ldots, u_k) \in \text{sp}(\mathbb{N})$, the polynomials $P_{ii}(u_i, \overline{w})$ and $P_{ij}(u_j, \overline{w})$ with variable w have a common zero $\overline{u}_i$, therefore

$$R(P_{ii}(u_i, \cdot), P_{ij}(u_j, \cdot)) = 0 \text{ for } (u_1, \ldots, u_k) \in \text{sp}(\mathbb{N}).$$

Let $P_{ii}(u_i, \overline{z}) = \sum_{l=0}^{k} f_l(u_i) z^l$ and $P_{ij}(u_j, z) = \sum g_l(u_j) z^l$. Then $R(P_{ii}(u_i, \cdot), P_{ij}(u_j, \cdot))$ is a polynomial of u_i and u_j. Therefore for any $f \in \mathcal{H}$,

$$R(P_{ii}(u_i, \cdot), P_{ij}(u_j, \cdot)) f(u) = 0, \text{ for } u \in \text{sp}(\mathbb{N}).$$

Thus

$$R(P_{ii}(S_i, \cdot), P_{ij}(S_j, \cdot)) f = 0$$

which proves (4.1.17). $\qquad\qquad\qquad\qquad\qquad\qquad\qquad\qquad\qquad\quad \square$

4.2   Joint point spectrum and joint eigenvectors

Let $\mathbb{T} = (T_1, \ldots, T_k)$ be a k-tuple of operators on $\mathcal{H}$. Let

$$\text{sp}_{jp}(\mathbb{T}) = \{(\lambda_1, \ldots, \lambda_k) \in \mathbb{C}^k : \text{ there is } h \neq 0 \in \mathcal{H} \text{ such that}$$
$$(T_j - \lambda_j)h = 0, j = 1, \ldots, k\}.$$

$\text{sp}_{jp}(\mathbb{T})$ is said to be the joint point spectrum of $\mathbb{T}$. For $F \subset \mathbb{C}^k$, let $F^* = \{(\lambda_1, \ldots, \lambda_k) \in \mathbb{C}^k : (\overline{\lambda}_1, \ldots, \overline{\lambda}_k) \in F\}$, let $\mathbb{T}^* = (T_1^*, \ldots, T_k^*)$ and

$$\text{sp}_{jp}(\mathbb{T}^*)^* = \{(\lambda_1, \ldots, \lambda_k) \in \mathbb{C}^k : (\overline{\lambda}_1, \ldots, \overline{\lambda}_k) \in \text{sp}_{jp}(\mathbb{T}^*)\}.$$

It is obvious that

$$\text{sp}_r(\mathbb{T}) \supset \text{sp}_{jp}(\mathbb{T}^*)^*.$$

Proposition 4.2.1. *Let $\mathcal{S} = (S_1, \ldots, S_k)$ be a subnormal k-tuple of operators on $\mathcal{H}$ with finite rank self-commutators and with m.n.e. $\mathbb{N}$ on $\mathcal{K} \supset \mathcal{H}$. If $\lambda = (\lambda_1, \ldots, \lambda_k) \in \text{sp}_r(\mathcal{S})$, and $\lambda_j \in \sigma(S_j) \cap \rho(N_j), j = 1, \ldots, k$. Then*

$$\lambda \in \text{sp}_{jp}(\mathcal{S}^*)^*.$$

Proof. There is a sequence $\{f_n\} \subset \mathcal{H}$ satisfying $\|f_n\| = 1$ such that

$$\lim_{n \to \infty} \sum_{j=1}^{k} (S_j - \lambda_j)(S_j - \lambda_j)^* f_n = 0,$$

since 0 is in the approximate point spectrum of the self-adjoint operator $\sum_{j=1}^{k}(S_j - \lambda_j)(S_j - \lambda_j)^*$. Thus

$$\lim_{n \to \infty} \|(S_j - \lambda_j)^* f_n\| = 0, \ j = 1, 2, \ldots, k \qquad (4.2.1)$$

since $\|(S_j - \lambda_j)^* f_n\|^2 \leq (\sum_j (S_j - \lambda_j)(S_j - \lambda_j)^* f_n, f_n)$. On the other hand from (3.4.30),

$$(S_j^* - \overline{\lambda}_j) f_n(u) = (\overline{u}_j - \overline{\lambda}_j) f_n(u) - (\overline{u}_j - \Lambda_j^*) f_n(\Lambda), \qquad (4.2.2)$$

where

$$f_n(\Lambda) = \int_{\mathrm{sp}(\mathbb{N})} e(du) f_n(u). \qquad (4.2.3)$$

It is easy to show that

$$\|f_n(\Lambda)\| \leq \|f_n\| = 1.$$

As a sequence of bounded vectors $\{f_n(\Lambda)\}$ in the finite dimensional defect space M, we may choose a subsequence $\{m_n\}$, such that

$$f_{m_n}(\Lambda) \to a \in M \text{ as } n \to \infty. \qquad (4.2.4)$$

From (4.2.2),

$$f_{m_n}(u) = \frac{\overline{u}_j - \Lambda_j^*}{\overline{u}_j - \overline{\lambda}_j} f_{m_n}(\Lambda) + \frac{1}{\overline{u}_j - \overline{\lambda}_j} (S_j^* - \overline{\lambda}_j) f_{m_n}, \qquad (4.2.5)$$

and

$$|u_j - \lambda_j| \geq \mathrm{dist}(\lambda_j, \sigma(N_j)) > 0, \text{ for } \lambda_j \in \rho(N_j),$$

it follows that as a sequence of vectors in $\mathcal{H}$, $\{f_{m_n}\}$ converges to

$$g(u) = \frac{\overline{u}_j - \Lambda_j^*}{\overline{u}_j - \overline{\lambda}_j} a, \; j = 1, 2, \dots, k. \qquad (4.2.6)$$

On the other hand, from (4.2.3) and (4.2.5), we have

$$f_{m_n}(\Lambda) = \mu_j(\lambda_j)^* f_{m_n}(\Lambda) + \int \frac{e(du)}{\overline{u}_j - \overline{\lambda}_j}((S_j^* - \overline{\lambda}_j) f_{m_n})(u), \qquad (4.2.7)$$

where

$$\mu_j(z) = \int \frac{(u_j - \Lambda_j)e(du)}{u_j - z}$$

is the mosaic of the single subnormal operator S_j. In (4.2.7), letting $n \to \infty$, by (4.2.1) and (4.2.4), we have

$$a = \mu_j(\lambda_j)^* a. \qquad (4.2.8)$$

From Theorem 1.4.5, (4.2.6) and (4.2.8) we conclude that

$$S_j^* g = \overline{\lambda}_j g, \; j = 1, 2, \dots, k.$$

The vector $g \in \mathcal{H}$ cannot be zero since $\|g\| = \lim_{n \to \infty} \|f_{m_n}\| = 1$. Therefore g is a joint eigenvector corresponding to the joint point spectrum $(\overline{\lambda}_1, \dots, \overline{\lambda}_k)$ of the operator tuple $\mathcal{S}^* = (S_1^*, \dots, S_k^*)$. $\qquad \square$

Lemma 4.2.2. *Let $S = (S_1, \ldots, S_k)$ be a pure subnormal k-tuple of operators on $\mathcal{H}$ with m.n.e. $\mathbb{N} = (N_1, \ldots, N_k)$. Let $z_l \in \rho(N_{v_l}), l = 1, 2, \ldots, n$, $1 \leq v_l \leq k$, $w \in \rho(N_j) \cap \rho(\Lambda_j)$ and $Q_{jv_l}(z_l, w)$ is invertible for $l = 1, 2, \ldots, n$. Then*

$$\int \frac{e(du)(\overline{u}_j - \Lambda_j^*)}{\prod_{l=1}^n (u_{v_l} - z_l)(\overline{u}_j - \overline{w})}$$
$$= \prod_{l=1}^n (R_{v_l j}(w)^* - z_l)^{-1} \mu_j(w)^* + \sum_{q=1}^n \int \frac{e(du)}{\prod_{l=1}^q (u_{v_l} - z_l)}$$
$$(I + (z_q - \Lambda_{v_q})(R_{v_q j}(w)^* - z_q)^{-1}) \prod_{p=q+1}^n (R_{v_p j}(w)^* - z_p)^{-1}. \quad (4.2.9)$$

Proof. For the simplicity of notation, let us assume that $v_l = l$. Let us prove (4.2.9) by mathematical induction with respect to n. First, let us prove it for $n = 1$. From Lemma 3.6.2, we have

$$\int \frac{e(du)}{(u_1 - z_1)(\overline{u}_j - \overline{w})} = (I - \mu_j(w)^*)Q_{j1}(z_1, w)^{-1} - Q_{j1}(z_1, w)^{-1}\mu_1(z).$$

Therefore, by (1.4.32), (3.3.6) and (3.6.11),

$$\int \frac{e(du)(\overline{w} - \Lambda_j^*)}{(u_1 - z_1)(\overline{u}_j - \overline{w})} = (I - \mu_j(w)^*)(z - R_{1j}(w)^*)^{-1}$$
$$- (z_1 - \Lambda_1)^{-1}(\overline{w} - R_{j1}(z))^{-1}\mu_1(z)(\overline{w} - \Lambda_j^*)$$
$$= (I - \mu_j(w)^* - (z_1 - \Lambda_1)^{-1}\mu_1(z)(z_1 - \Lambda_1))(z_1 - R_{1j}(w)^*)^{-1}. \quad (4.2.10)$$

From $(z_1 - \Lambda_1)^{-1}\mu_1(z_1) = (z_1 - \Lambda_1)^{-1} + \int (u_1 - z_1)^{-1}e(du)$ and (4.2.10), we have (4.2.9) for $n = 1$.

Suppose that (4.2.9) holds good for $n = p$. By (3.4.19), we have

$$\int \frac{e(du)Q_{j(p+1)}(z_{p+1}, w)}{\prod_{l=1}^{p+1}(u_l - z_l)(\overline{u}_j - \overline{w})}$$
$$= \int \frac{e(du)(-(\overline{u}_j - \Lambda_j^*)(u_{p+1} - z_{p+1}) - (\overline{u}_j - \overline{w})(z_{p+1} - \Lambda_{p+1}))}{(\overline{u}_j - \overline{w}) \prod_{l=1}^{p+1}(u_l - z_l)}$$
$$= -\int \frac{e(du)(\overline{u}_j - \Lambda_j^*)}{(\overline{u}_j - \overline{w}) \prod_{l=1}^p(u_l - z_l)} - \int \frac{e(du)(z_{p+1} - \Lambda_{p+1})}{\prod_{l=1}^{p+1}(u_l - z_l)}. \quad (4.2.11)$$

From $(\overline{w} - \Lambda_j^*) = Q_{j(p+1)}(z_{p+1}, w)(z_{p+1} - R_{(p+1)j}(w)^*)^{-1}$ and (4.2.11), we have

$$\int \frac{e(du)(\overline{u}_j - \Lambda_j^*)}{(\overline{u}_j - \overline{w}) \prod_{l=1}^{p+1}(u_l - z_l)} = \int \frac{e(du)}{\prod_{l=1}^{p+1}(u_l - z_l)}$$

$$+ \int \frac{e(du)Q_{j(p+1)}(z_{p+1}, w)}{(\overline{u}_j - \overline{w}) \prod_{l=1}^{p+1}(u_l - z_l)}(z_{p+1} - R_{(p+1)j}(w)^*)^{-1}$$

$$= \int \frac{e(du)}{\prod_{l=1}^{p+1}(u_l - z_l)}(I + (z_{p+1} - \Lambda_{p+1})(R_{(p+1)j}(w)^* - z_{p+1})^{-1})$$

$$+ \int \frac{e(du)(\overline{u}_j - \Lambda_j^*)}{(\overline{u}_j - \overline{w}) \prod_{l=1}^{p}(u_l - z_l)}(R_{(p+1)j}(w)^* - z_{p+1})^{-1}$$

which proves (4.2.9) for $n = p + 1$ by the hypothesis of mathematical induction. $\qquad \square$

Proposition 4.2.3. *Let* $\mathcal{S} = (S_1, \ldots, S_k)$ *be a pure subnormal k-tuple of operators on $\mathcal{H}$ with finite rank self-commutators. Let $w_j \in \sigma(S_j) \cap \rho(N_j), j = 1, 2$. Suppose there are $c_v \in \mathbb{C}, v = 1, 2, \ldots, k$ and a vector $a \in M$ satisfying*

$$\mu_j(w_j)^* a = a, \ j = 1, 2, \tag{4.2.12}$$

and

$$R_{vj}(w_j)^* a = c_v a, \ j = 1, 2; v = 1, 2, \ldots, k. \tag{4.2.13}$$

Let

$$f_j(u) \stackrel{def}{=} \frac{\overline{u}_j - \Lambda_j^*}{\overline{u}_j - \overline{w}_j} a, \ u \in \text{sp}(\mathbb{N}), \ j = 1, 2.$$

Then $f_1(\cdot) = f_2(\cdot)$ as vectors in $\mathcal{H}$ and

$$S_j^* f_j = \overline{w}_j f_j, \ j = 1, 2. \tag{4.2.14}$$

Proof. From (4.2.9), (4.2.12) and (4.2.13), we have

$$\int \frac{e(du)f_1(u)}{\prod_{l=1}^{k}(u_l - z_l)} = \int \frac{e(du)f_2(u)}{\prod_{l=1}^{k}(u_l - z_l)}, \ z_l \in \rho(N_l), \ l = 1, 2, \ldots, k. \tag{4.2.15}$$

From (4.1.3) we may conclude that for every continuous function $h(\cdot)$ on sp($\mathbb{N}$), there exists a sequence $\{h_n\}$, where

$$h_n \in \bigvee \{\Pi_{l=1}^{k}(u_l - z_l)^{-1} : z_l \in \rho(N_l), \ det(Q_{jl}(z_l, z_j)) \neq 0 \ j = 1, 2, \ldots, k,$$

$$l = 1, 2, \ldots, k\}$$

such that

$$\lim_{n \to \infty} \max_{u \in \text{sp}(\mathbb{N})} |h(u) - h_n(u)| = 0. \tag{4.2.16}$$

From (4.2.15) and (4.2.16), we have that

$$\int_{\mathrm{sp}(\mathbb{N})} h(u)e(du)f_1(u) = \int_{\mathrm{sp}(\mathbb{N})} h(u)e(du)f_2(u) \tag{4.2.17}$$

holds good for every continuous function $h(u)$ on $\mathrm{sp}(\mathbb{N})$ and hence for every bounded Baire function $g(u)$ on $\mathrm{sp}(\mathbb{N})$, which proves that $f_1(\cdot) = f_2(\cdot)$ as vectors in $\mathcal{H}$. (4.2.14) comes from Theorem 1.4.5. $\qquad\qquad\square$

4.3   Domains in some analytic manifold

In order to make this section readable, let us firstly give the definition of an analytic manifold. Let $\mathcal{A}$ be a connected Housdorff space and Φ be a family of pairs (O, f) satisfying the following conditions: (i) O is an open set in $\mathcal{A}$ and

$$\mathcal{A} = \bigcup_{(O,f)\in\Phi} O,$$

(ii) for every $(O, f) \in \Phi$, f is a homeomorphism from O onto an open set $f(O) \subset \mathbb{C}$, and (iii) if $(O_i, f_i) \in \Phi, i = 1, 2$ and $O_1 \cap O_2 \neq \emptyset$, then the map $f_2 \circ f_1^{-1}$ from $f_1(O_1 \cap O_2)$ onto $f_2(O_1 \cap O_2)$ is a conformal mapping, i.e., analytic and univalent. Then $(\mathcal{A}, \Phi)$ or simply $\mathcal{A}$ is said to be an analytic manifold (with local complex dimension one) or analytic surface, Φ is said to be the coordinate system and each $(O, f) \in \Phi$ is said to be a coordinate chart. Suppose D is an open set in $\mathcal{A}$ and h is a mapping from D to $\mathbb{C}$ (or $\mathbb{C} \cup \{\infty\}$). If for every $(O, f) \in \Phi$ satisfying $O \cap D \neq \emptyset$, the mapping $h \circ f^{-1}$ from $f(D \cap O)$ to $h(D \cap O)$ is analytic (or meromorphic), then g is said to be an analytic (or meromorphic, correspondingly) function on D.

Secondly, let us quote some knowledge from § 2.2 and § 2.3. Let $\mathcal{S} = (S_1, \ldots, S_k)$ be a pure subnormal k-tuple of operators on $\mathcal{H}$ with finite rank self-commutators. Let $\mathbb{N} = \{N_1, \ldots, N_k\}$ be its m.n.e.. For every $S_j, j = 1, 2, \ldots, k$, there are union $\mathcal{D}_j$ of quadratic domains, with boundary curves $\mathcal{L}_j$, and idempotent valued meromorphic function $\nu_j(\cdot)$ on $\mathcal{D}_j$, Schwartz function $S_j(\cdot)$ and projection $\Psi_j(\cdot)$ on $\mathcal{D}_j$ satisfying

$$\mu_j(z) = \sum_{\Psi_j(\xi)=z} \nu_j(\xi), \quad z \in \sigma(S_j) \cap \rho(N_j), \tag{4.3.1}$$

and

$$\nu_j(\xi) = 2\pi i(\Psi_j(\xi) - \Lambda_j)\frac{e_j(d\Psi_j(\xi))}{d\Psi_j(\xi)}, \text{ for almost every } \xi \in \mathcal{L}_j \tag{4.3.2}$$

where $e_j(\cdot)$ is the positive measure $e(\cdot)$ corresponding to the operator S_j.

Let

$$\widetilde{\mathfrak{S}} \stackrel{\text{def}}{=} \{(\xi_1, \ldots, \xi_k) \in \mathcal{D}_1 \times \cdots \times \mathcal{D}_k : P_{ln}(\Psi_n(\xi), \bar{S}_l(\xi)) = 0, l, n = 1, \ldots, k\}$$

and

$$\mathfrak{S} \stackrel{\text{def}}{=} \{(\Psi_1(\xi_1), \ldots, \Psi_k(\xi_k)) \in (\sigma(S_1) \cap \rho(N_1)) \times \cdots \times (\sigma(S_k) \cap \rho(N_k))$$
$$: (\xi_1, \ldots, \xi_k) \in \widetilde{\mathfrak{S}}\}.$$

Let Ψ be the projection

$$\Psi(\xi_1, \ldots, \xi_k) = (\Psi_1(\xi_1), \ldots, \Psi_k(\xi_k)).$$

Let $h_j(z), j = 1, \ldots, k$ be homeomorphisms from a domain $\mathcal{V}$ in $\mathbb{C}$ into $\mathcal{D}_j$ satisfying the conditions that (i) h_j^{-1} is an analytic function on an open set in $\mathcal{D}_j$, (ii) $(h_1(z), \ldots, h_k(z)) \in \widetilde{\mathfrak{S}}$ for $z \in \mathcal{V}$. Define $O = \{(h_1(z), \ldots, h_k(z)) : z \in \mathcal{V}\}$ and the mapping f as

$$f(h_1(z), h_2(z), \ldots, h_k(z)) = z.$$

Let $\mathcal{A}$ be the union of all such sets O in $\widetilde{\mathfrak{S}}$ and Φ be the family of all such pairs (O, f). Then $\mathcal{A}$ is a union of analytic manifolds with part of Φ as the coordinate system of those analytic manifolds. It is obvious that $\Psi \mathcal{A} \subset \mathfrak{S}$ is also a union of analytic manifolds.

Now, for simplicity, let us assume that $\mathcal{S}$ has the property (A): the restriction of the projection from $\mathbb{C}^k$ to $\mathbb{C}$

$$P_j(z_1, \ldots, z_k) \stackrel{\text{def}}{=} z_j,$$

in the continuous part $\mathrm{sp}_{\mathrm{ess}}(\mathbb{N})$ is one to one from $\mathrm{sp}_{\mathrm{ess}}(\mathbb{N})$ to $L_j = \sigma_{\mathrm{ess}}(N_j)$ which is the continuous part of $\sigma(N_j)$, for $j = 1, \ldots, k$.

Let γ be a small algebraic arc in $\mathrm{sp}_{\mathrm{ess}}(\mathbb{N})$. Let $\gamma_j = P_j \gamma \subset L_j$. Without loss of generality we may assume that γ and $\gamma_j, j = 1, 2, \ldots, k$ do not have node point. There exist functions $\gamma_l(z), l = 2, \ldots, k$ such that the algebraic arc γ can be parametrized as $\gamma : (z, \gamma_2(z), \ldots, \gamma_k(z)), z \in \gamma_1$. These functions $\gamma_l(\cdot)$ satisfy

$$P_{l1}(z, \gamma_l(z)) = 0, \; z \in \gamma_1 \tag{4.3.3}$$

by (4.1.3). On the other hand, the polynomial $P_{l1}(z, \overline{w})$ is a polynomial of z and w. From (4.3.3), it is easy to see there is a domain $O_1 \subset D_1 \subset \Psi_1 \mathcal{D}_1$, satisfying $\gamma_1 \subset \partial O_1$ such that there are univalent analytic functions $f_l(\cdot), l = 2, \ldots, k$ on O_1 which are continuous on $\gamma_1 \cup O_1$ satisfying

$$P_{l1}(z, \overline{f_l(z)}) = 0, \; z \in \gamma_1 \cup O_1, \tag{4.3.4}$$

and $\overline{f_1(z)} = \gamma_1(z)$ for $z \in \gamma_1$. Thus there are $\hat{O}_j \subset \mathcal{D}_j, j = 1, 2, \ldots, k$ such that

$$P_{11}(\Psi_1(\xi_1), \overline{S_l(\xi_l)}) = 0, \ \xi_1 \in \hat{O}_1 \text{ and } \xi_l \in \hat{O}_l$$

where $\Psi_l(\hat{O}_1) = O_1$, and the variables ξ_1 and ξ_l satisfy the relation

$$\xi_l = h_l(\xi_1) \tag{4.3.5}$$

where $h_l = S_l^{-1} \circ f_l \circ \Psi_1$. We choose γ, O_1 so small such that h_l is analytic and univalent. Let $O_\gamma = \{(\xi_1, h_2(\xi_1), \ldots, h_k(\xi_1)) : \xi_1 \in \hat{O}_1\}$ and f_γ be the mapping

$$f_\gamma(\xi_1, h_2(\xi_1), \ldots, h_k(\xi_1)) = \Psi_1(\xi_1).$$

Then $(O_\gamma, f_\gamma) \in \Phi$. Let $\widetilde{\mathfrak{S}}_0$ be the union of all such domains O_γ in $\mathcal{A}$. Let $\mathcal{A}_1$ be the union of analytic manifolds in $\mathcal{A}$ which meet $\widetilde{\mathfrak{S}}_0$.

On the other hand, from (4.3.2), we have

$$e(F) = \frac{1}{2\pi i} \int_{\Psi_j(\xi_j) \in P_j F} (\Psi_j(\xi_j) - \Lambda_j)^{-1} \nu_j(\xi_j) d\Psi_j(\xi_j), \tag{4.3.6}$$

where F is a Borel set in γ. Thus for $1 \leq l, j \leq k$

$$(\Psi_l(\xi_l) - \Lambda_l)^{-1} \nu_l(\xi_l) \frac{d\Psi_l(\xi_l)}{d\Psi_j(\xi_j)} = (\Psi_j(\xi_j) - \Lambda_j)^{-1} \nu_j(\xi_j) \tag{4.3.7}$$

where $\xi_l = h_l(h_j^{-1}(\xi_j))$ for $\Psi_j(\xi_j) \in \gamma_j$ by (4.3.5). But the both sides of (4.3.7) are meromorphic functions, therefore (4.3.7) holds good for $\xi_j \in \hat{O}_j$. Thus (4.3.7) holds good also for $(\xi_1, \ldots, \xi_k)$ in a domain in $\widetilde{\mathfrak{S}}_0$ and hence also in $\mathcal{A}$. Let

$$M_\xi \stackrel{\text{def}}{=} \text{range } \nu_l(\xi_l)^*$$

where $\xi = (\xi_1, \ldots, \xi_k) \in \mathcal{A}_0$. Then by (4.3.7), M_ξ is well-defined on $\mathcal{A}$ except a finite set, where ξ_l is a pole of $\nu_l(\cdot)$ for some l. From

$$(I - \mu_l(\Psi_l(\xi))^*) \nu_l(\xi)^* = 0,$$

we have

$$\mu_l(\Psi_l(\xi_l))^* a = a, \ l = 1, 2, \ldots, k,$$

where $a \in M_\xi$ for $(\xi_1, \ldots, \xi_k) \in \mathcal{A}$. From (4.3.2) and (4.3.6), we have

$$(R_{lj}(\Psi_j(\xi_j))^* - \overline{S_l(\xi_l)}) \nu_j(\xi_j)^*$$
$$= 2\pi i \frac{e(d\Psi(\xi_1, \ldots, \xi_k))}{d\Psi_j(\xi_j)} Q_{jl}(\Psi_l(\xi_l), \Psi_j(\xi_j)) = 0, \tag{4.3.8}$$

for $(\Psi_1(\xi_1), \ldots, \Psi_k(\xi_k)) \in \mathrm{sp}_{\mathrm{ess}}(\mathbb{N})$ by (3.4.19). On the other hand,

$$\nu_j(\xi_j)(R_{lj}(\Psi_j(\xi_j)) - S_l(\xi_l))$$

is a meromorphic function on $\mathcal{A}$. Thus (4.3.8) implies that

$$R_{lj}(\Psi_j(\xi_j))^* a = \overline{S_l(\xi_l)}a, \quad l, j = 1, 2, \ldots, k, \quad a \in M_\xi. \tag{4.3.9}$$

Let

$$f_j(u) = \frac{\overline{u}_j - \Lambda_j^*}{\overline{u}_j - \Psi_j(\xi_j)} a, \quad u \in \mathrm{sp}(\mathbb{N}), \quad a \in M_\xi, \quad a \neq 0. \tag{4.3.10}$$

By Proposition 4.2.3, $f_j(u), j = 1, 2, \ldots, k$ represent one vector f in $\mathcal{H}$ and

$$S_j^* f = \overline{\Psi_j(\xi_j)} f. \tag{4.3.11}$$

Theorem 4.3.1. *Let* $S = (S_1, \ldots, S_k)$ *be a pure subnormal k-tuple of operators on $\mathcal{H}$ with finite rank self-commutators. Let $\mathbb{N}$ be its m.n.e.. If S has property (A). Then there is a union $\mathcal{A}_1$ of analytic manifolds whose projection $\mathcal{A}_0$ is with boundary* $\mathrm{sp}_{\mathrm{ess}}(\mathbb{N})$. *For every* $\xi \in \mathcal{A}_0$, *there is a subspace M_ξ such that for every $a \in M_\xi$, there is a joint eigenvector* (4.3.10) *of S_j^* with eigenvalues* $\overline{\Psi_j(\xi_j)}$ *respectively, where* $(\Psi_1(\xi_1), \ldots, \Psi_k(\xi_k)) \in \mathcal{A}_0$.

Proposition 4.3.2. *Let* $S = (S_1, \ldots, S_k)$ *be a pure subnormal k-tuple of operators on $\mathcal{H}$ with finite rank self-commutators. Let $\mathbb{N} = (N_1, \ldots, N_k)$ be its m.n.e.. Suppose S has property (A). For $(z_1, \ldots, z_k) \notin \mathrm{sp}(\mathbb{N})$, if there exists a vector $f \in \mathcal{H}, f \neq 0$ satisfying*

$$S_j^* f = \overline{z}_j f, \quad j = 1, 2, \ldots, k, \tag{4.3.12}$$

then $(z_1, \ldots, z_k) \in \mathcal{A}_0$ *and there are points* $\xi^{(l)} = (\xi_1^{(l)}, \ldots, \xi_k^{(l)}) \in \mathcal{A}_1, l = 1, 2, \ldots, p$ *satisfying*

$$z_j = \Psi_j(\xi_j^{(l)}), \quad l = 1, \ldots, p, \quad j = 1, \ldots, k, \tag{4.3.13}$$

and as a vector in $\mathcal{H}$,

$$f = \frac{\overline{u}_j - \Lambda_j^*}{\overline{u}_j - z_j} a, \quad j = 1, 2, \ldots, k,$$

where $a = \sum_{l=1}^p a^{(l)}$ *and* $a^{(l)} \in M_{\xi^{(l)}}$.

Proof. By Theorem 3.4.5, there is a vector $a = \int e(du) f(u), a \in \mu_j(z_j)^* M$ such that

$$f = \frac{\overline{u}_j - \Lambda_j^*}{\overline{u}_j - z_j} a, \quad j = 1, 2, \ldots, k,$$

if $z \notin \sigma(N_j)$. Then $a = \sum a_j^{(l)}$, where $a_j^{(l)} = \nu_j(\xi_j^{(l)})^*a$ and

$$\mu_j(z_j)^* = \sum_{\Psi_j(\xi_j^{(l)})=z_j} \nu_j(\xi_j^{(l)})^*.$$

By (4.3.8),

$$(R_{nj}(z_j)^* - \overline{S_n(\xi_n^{(l)})})a_j^{(l)} = 0. \tag{4.3.14}$$

From Lemma 4.2.2,

$$\int \frac{e(du)}{u_n - z}f(u) = \int \frac{e(du)(\overline{u_j} - \Lambda_j^*)}{(u_n - z)(\overline{u_j} - \overline{z_j})}a$$

$$= \int \frac{e(du)}{u_n - z}a + \tilde{\mu}_n(z)(R_{nj}(z_j)^* - z)^{-1}a, \tag{4.3.15}$$

where

$$\tilde{\mu}_n(z) \stackrel{\text{def}}{=} \int \frac{e(du)(u_n - \Lambda_n)}{u_n - z}.$$

From (4.3.14) and (4.3.15), we have

$$\int \frac{e(du)}{u_n - z}f(u) - \int \frac{e(du)}{u_n - z}a = \tilde{\mu}_n(z)\sum_l \frac{a_j^{(l)}}{S_n(\xi_n^{(l)}) - z}.$$

Therefore $a_j^{(l)}$ is independent of j and we denote it by $a^{(l)}$. Thus

$$a = \sum a^{(l)}.$$

Denote

$$f_j^{(l)} = \frac{\overline{u_j} - \Lambda_j^*}{\overline{u_j} - \overline{z_j}}a^{(l)}.$$

Then

$$\mu_j(z_j)^*a^{(l)} = a^{(l)}, \; j = 1, 2, \ldots, k, \tag{4.3.16}$$

and

$$R_{nj}(z_j)^*a^{(l)} = \overline{S_n(\xi_n^{(l)})}a^{(l)}, \; n = 1, 2, \ldots, k. \tag{4.3.17}$$

By Proposition 4.2.3, (4.3.16) and (4.3.17), we can conclude that as a vector in $\mathcal{H}$, $f_j^{(l)} = f_{j'}^{(l)}$, for $j \neq j'$. Denote it by $f^{(l)}$. Thus $f = \sum f^{(l)}$ where

$$f^{(l)} = \frac{\overline{u_j} - \Lambda_j^*}{\overline{u_j} - \overline{z_j}}a^{(l)}.$$

Therefore $(z_1, \ldots, z_k) \in \mathcal{A}_0$ and $(\xi_1^{(l)}, \ldots, \xi_k^{(l)}) \in \mathcal{A}_1$. $\qquad\qquad \square$

4.4 Trace formulas

Now, let us generalize Theorem 2.4.3 to the k-tuple case. Let $R(\mathcal{S})$ be the algebra of functions on $K(\mathcal{S})$ generated by rational functions $f(u_j)$ with poles off $\sigma(S_j)$ and its conjugate. For any function $f(\overline{u}, u) = \Pi_j f_j^{(1)}(\overline{u}_j) \Pi f_j^{(2)}(u) \in R(\mathcal{S})$, let $f(\mathcal{S}^*, \mathcal{S}) = \prod_j f_j^{(1)}(S_j^*) \prod f_j^{(2)}(S_j)$.

Theorem 4.4.1. *Let $\mathcal{S} = (S_1, \dots, S_k)$ be a subnormal k-tuple of operators on $\mathcal{H}$ with m.n.e. extension $\mathbb{N}$. Suppose the defect space of $\mathcal{S}$ is of finite dimension. Then the continuous part L of $\mathrm{sp}(\mathbb{N})$ consists of finite set of closed curves which consist of finite algebraic arcs. Let $m(u), u \in L$ be the multiplicity of $\mathbb{N}$ at u. Then $m(\cdot)$ is piecewise constant on L and for any pair of functions $f(\overline{u}, u)$ and $h(\overline{u}, u)$ in $\mathcal{A}(\mathcal{S})$,*

$$\mathrm{tr}[f(\mathcal{S}^*, \mathcal{S}), h(\mathcal{S}^*, \mathcal{S})] = \frac{1}{2\pi} \int_L m f dh. \tag{4.4.1}$$

Proof. Firstly, we assume that $\mathcal{S}$ has the property (A). Similar to the proof of Theorem 2.4.3, we have to study the measure $v_j(du)$ in Theorem 3.7.5. Let us use the notations in the §4.3. Let γ be an algebraic arc in $\mathrm{sp}_{\mathrm{ess}}(\mathbb{N})$, and $\gamma_j = P_j\gamma$, the projection of γ to the j-th coordinate plane. Then

$$\frac{v_j(du)}{du_j} = i \, \mathrm{tr}((u_j - \Lambda_j) \frac{e(du)}{du_j}) = \frac{1}{2\pi} \mathrm{tr} \, v_j(\xi_j),$$

where $u_j = \Psi_j(\xi_j)$. The function $v_j(\cdot)$ is an idempotent valued meromorphic function. Therefore on γ_j, $\mathrm{tr} \, v_j(\xi_j) = m_j$ is an integer which equals to the local multiplicity of N_j, since $e_j(du_j) = P_M E_j(du_j)|_M$ and $E_j(\cdot)$ is the spectral measure of N_j. Therefore

$$v_j(du) = m_j du_j.$$

Suppose $\mathrm{sp}_p(\mathbb{N}) = \{a_l : l = 1, \dots, n\}$ and $v_j(a_l) = \lambda_{jl}$. As in the proof of Theorem 2.4.3, we have

$$2\pi i \, \mathrm{tr}[f(\mathcal{S}^*, \mathcal{S}), h(\mathcal{S}^*, \mathcal{S})] = \int_{\mathrm{sp}_{\mathrm{ess}}(\mathbb{N})} m f dh + J(f, h),$$

where

$$J(f, h) = 2\pi i \sum_{l,j} f(\overline{a}_l, a_l)(\frac{\partial h}{\partial \overline{u}_j}(\overline{a}_l, a_l)\overline{\lambda}_{jl} + \frac{\partial h}{\partial u_j}(\overline{a}_l, a_l)\lambda_{jl}).$$

By the similar calculation in the proof of Theorem 2.4.3, we may show that $J(f, h) = 0$, which proves (4.4.1).

Now, we have to release the restriction: property (A) of $\mathcal{S}$. For any $\mathcal{S}$ with m.n.e. $\mathbb{N}$, there is a non-singular linear transformation π in $\mathbb{C}^k$: for $c = (c_1, \ldots, c_k) \in \mathbb{C}^k$,

$$\pi(c_1, \ldots, c_k) = \left(\sum_j \pi_{j1} c_j, \sum_j \pi_{j2} c_j, \ldots, \sum_j \pi_{jk} c_j\right),$$

such that for the set $\pi(\mathrm{sp}(\mathbb{N}))$ the projection to its coordinate planes is one to one. This π exists since $\mathrm{sp}(\mathbb{N})$ consists of only finite set of algebraic curves and a possible finite set of isolated points. Then we obtain a subnormal k-tuple $\hat{\mathcal{S}} = (\hat{S}_1, \ldots, \hat{S}_k)$ of operators with m.n.e. $\hat{\mathbb{N}} = \{\hat{N}_1, \ldots, \hat{N}_k\}$ where

$$\hat{S}_n = \sum_j \pi_{jn} S_j \text{ and } \hat{N}_n = \sum_j \pi_{jn} N_j.$$

This subnormal k-tuple $\hat{\mathcal{S}}$ of operators has the property (A). Therefore for $\hat{\mathcal{S}}$, the formula (4.4.1) is established. By the linear transformation π^{-1}, we can show that (4.4.1) holds good for the original $\mathcal{S}$. □

For $z = \{z_1, \ldots, z_k\} \in \mathrm{sp}_{jp}(\mathcal{S}^*)^*$, let $E_z = \{f \in \mathcal{H} : S_j^* f = \bar{z}_j f, j = 1, 2, \ldots, k\}$ be the space of joint eigenvectors of $\mathcal{S}^*$, corresponding to the eigenvalue $\bar{z} = (\bar{z}_1, \ldots, \bar{z}_k)$. Let $m(z) \stackrel{\text{def}}{=}$ dimension of E_z, the *multiplicity of the joint eigenvalue* $\bar{z}$.

Theorem 4.4.2. *Let $\mathcal{S} = (S_1, \ldots, S_k)$ be a subnormal k-tuple of operators with finite rank self-commutators. Let $m(z), z = (z_1, \ldots, z_k) \in \mathrm{sp}_{jp}(\mathcal{S}^*)^*$ be the multiplicity function of joint eigenvalue $(\bar{z}_1, \ldots, \bar{z}_k)$ of $\mathcal{S}^*$. Then for any pair of functions f and h in $\mathcal{A}(\mathcal{S})$,*

$$2\pi i \, \mathrm{tr}[f(\mathcal{S}^*, \mathcal{S}), h(\mathcal{S}^*, \mathcal{S})] = \int_{\mathrm{sp}_{jp}(\mathcal{S}^*)^*} m \, df \wedge dh. \qquad (4.4.2)$$

Proof. For $\xi = (\xi_1, \xi_2, \ldots, \xi_k) \in \mathcal{A}_1$, let $\mathcal{E}_\xi$ be the space of eigenfunctions f defined in (4.3.11) corresponding to the vector $a \in M_\xi$ in (4.3.14) and the eigenvalues $(\Psi_1(\xi_1), \Psi_2(\xi_2), \ldots, \Psi_k(\xi_k))$. Let $\hat{m}(\xi) \stackrel{\text{def}}{=} \dim \mathcal{E}_\xi = \dim M_\xi$. Then the boundary value of the function $\hat{m}(\xi)$ on the boundary $\mathrm{sp}_{\mathrm{ess}}(\mathbb{N})$ is just the function $m(\cdot)$ in Theorem 4.4.1. Let us apply well-known Cartan's formula (or just Stokes' Theorem)

$$\int_\Omega d\omega = \int_{\partial\Omega} \omega$$

to the integral on the right hand side of (4.4.1). The union of analytic cycles L is the projection boundary of $\mathcal{A}_1$. Thus we have

$$\int_L m \hat{f} d\hat{h} = \int_{\partial \mathcal{A}_1} \hat{m} d(\hat{f} d\hat{h}) = \int_{\mathcal{A}_1} \hat{m} d\hat{f} \wedge d\hat{h}. \qquad (4.4.3)$$

Actually, in (4.4.3) the functions $\hat{f}$ and $\hat{h}$ are $f \circ \Psi$ and $h \circ \Psi$ respectively. By Theorem 4.3.1 and Theorem 4.3.2, the difference between $\mathcal{A}_0$ and $\mathrm{sp}_p(\mathcal{S}^*)^*$ is only a finite set of points and the multiplicity function

$$m(z_1, \ldots, z_k) = \sum_{\psi_j(\xi_j) = z_j} \hat{m}(\xi_1, \ldots, \xi_k).$$

By means of the projection $\Psi : (\xi_1, \ldots, \xi_k) \to (\Psi_1(\xi_1), \ldots, \Psi_k(\xi_k))$ from $\mathcal{A}_1$ to $\mathcal{A}_0$, we may conclude that (4.4.3) equals to $\int_{\mathcal{A}_0} m \, df \wedge dh$, which proves (4.4.2). $\square$

Chapter 5

Hyponormal Operators with Finite Rank Self-Commutators

5.1 Hyponormal operators with rank one self-commutator

Let H be an operator on a Hilbert space $\mathcal{H}$. Then H is said to be hyponormal if

$$([H^*, H]x, x) \geq 0, \text{ for } x \in \mathcal{H},$$

or equivalently, $\|H^*x\| \leq \|Hx\|$, $x \in \mathcal{H}$. Subnormal operator is hyponormal. In this section, we will give an analytic model of some pure hyponormal operators with rank one self-commutator $[H^*, H]$. As before, we denote

$$M = M_H \overset{\text{def}}{=} \text{closure of } [H^*, H]\mathcal{H}.$$

We will show that there are a lot of hyponormal operators even with rank one self-commutators which are not subnormal. We denote $[H^*, H]|_M \in L(M)$ by D, instead of C. Then

$$[H^*, H] = DP_M, \tag{5.1.1}$$

where P_M is the projection from $\mathcal{H}$ to M. Let $k \in M$ satisfying $\|k\|^2 = \|D\|$. Then $DP_M x = (x, k)k$, for $x \in \mathcal{H}$.

Lemma 5.1.1. *Let H be a hyponormal operator on $\mathcal{H}$ with rank one self-commutator. Then for $z, w \in \rho(H)$,*

$$\det((H^* - \bar{w})(H - z)(H^* - \bar{w})^{-1}(H - z)^{-1}) = 1 + ((H - z)^{-1}k, (H - w)^{-1}k). \tag{5.1.2}$$

Proof. From $[H - z, (H^* - \bar{w})^{-1}] = (H^* - \bar{w})^{-1}DP_M(H^* - \bar{w})^{-1}$ by (5.1.1), we have

$$(H^* - \bar{w})(H - z)(H^* - \bar{w})^{-1}(H - z)^{-1}$$
$$= I + (H^* - \bar{w})[H - z, (H^* - \bar{w})^{-1}](H - z)^{-1}$$
$$= I + DP_M(H^* - \bar{w})^{-1}(H - z)^{-1}. \tag{5.1.3}$$

Let $\{e_n\}$ be a complete orthonormal system of $\mathcal{H}$ with $e_1 = k/\|k\|$. Let
$$a_{ij} \overset{\text{def}}{=} ((H^* - \bar{w})(H - z)(H^* - \bar{w})^{-1}(H - z)^{-1}e_i, e_j).$$
Then by (5.1.3),
$$a_{ij} = \delta_{ij} + ((H^* - \bar{w})^{-1}(H - z)^{-1}e_i, DP_M e_j), \tag{5.1.4}$$
where δ_{ij} is the Kronecker δ. Thus from (5.1.4) we have $a_{ij} = 0$ for $j > 1$, since $P_M e_j = 0$ for $j > 1$. Hence
$$\det(a_{ij}) = 1 + ((H^* - \bar{w})^{-1}(H - z)^{-1}e_1, De_1),$$
which implies (5.1.2), since $De_1 = \|k\|^2 e_1$.   □

Let H be a pure hyponormal operator on $\mathcal{H}$ with rank one self-commutator. Let k be a vector in M normalized by $\|k\|^2 = \|D\|$. Define a complex function of H:
$$S(z, w) \overset{\text{def}}{=} ((H - z)^{-1}k, (H - w)^{-1}k), \quad z, w \in \rho(H).$$
Let us introduce the function
$$T(z, w) \overset{\text{def}}{=} ((H^* - \bar{w})^{-1}k, (H^* - \bar{z})^{-1}k).$$

Lemma 5.1.2. *Under the condition of Lemma 5.1.1,*
$$(1 - T(z, w))(1 + S(z, w)) = 1. \tag{5.1.5}$$

Proof. By the commutation relation
$$[(H^* - \bar{w})^{-1}, (H - z)^{-1}] = (H - z)^{-1}(H^* - \bar{w})^{-1}[H^*, H](H^* - \bar{w})^{-1}(H - z)^{-1}, \tag{5.1.6}$$
we have
$$S(z, w) - T(z, w) = T(z, w)S(z, w),$$
which proves (5.1.5).   □

Lemma 5.1.2 is similar to Lemma 1.3.2.

Lemma 5.1.3. *Under the condition of Lemma 5.1.1, for $z_i, w_i \in \rho(H)$, $i = 1, 2$,*
$$((H - z_1)^{-1}(H^* - \bar{w}_1)^{-1}k, (H - w_2)^{-1}(H^* - \bar{z}_2)^{-1}k) \tag{5.1.7}$$
$$= (T_1 + T_2(1 + S(z_2, w_2))T_3)/(\overline{w_1} - \overline{w_2})(z_1 - z_2)$$

where
$$T_1 = T(z_1, w_1) + T(z_2, w_2) - T(z_2, w_1) - T(z_1, w_2),$$
$$T_2 = T(z_1, w_2) - T(z_2, w_2),$$
and
$$T_3 = T(z_2, w_1) - T(z_2, w_2).$$

Proof. By (5.1.6), the left-hand side of (5.1.7) equals to

$$((H - z_2)^{-1}(H^* - \bar{w}_2)^{-1}(H - z_1)^{-1}(H^* - \bar{w}_1)^{-1}k, k)$$
$$= ((H - z_1)^{-1}(H - z_2)^{-1}(H^* - \bar{w}_2)^{-1}(H^* - \bar{w}_1)^{-1}k, k)$$
$$+ (A[H^*, H]Bk, k), \tag{5.1.8}$$

where

$$A = (H - z_1)^{-1}(H - z_2)^{-1}(H^* - \overline{w}_2)^{-1}$$

and

$$B = (H^* - \overline{w}_2)^{-1}(H - z_2)^{-1}(H^* - \overline{w}_1)^{-1}.$$

By (5.1.6) again,

$$(Bk, k) = ((H - z_2)^{-1}(H^* - \bar{w}_1)^{-1}(H^* - \bar{w}_2)^{-1}k, k)$$
$$+ ((H - z_2)^{-1}(H^* - \bar{w}_2)^{-1}[H^*, H]Bk, k).$$

Besides, $(Ak, k) = (T(z_1, w_2) - T(z_2, w_2))(z_1 - z_2)^{-1}$ and

$$(Bk, k) = (T(z_2, w_1) - T(z_2, w_2))[(\bar{w}_1 - \bar{w}_2)(1 - T(z_2, w_2))]^{-1} \tag{5.1.9}$$

From (5.1.5), (5.1.8), (5.1.9) and

$$((H - z_1)^{-1}(H - z_2)^{-1}(H^* - \bar{w}_2)^{-1}(H^* - \bar{w}_1)^{-1}k, k)$$
$$= \frac{T(z_1, w_1) + T(z_2, w_2) - T(z_1, w_2) - T(z_2, w_1)}{(z_1 - z_2)(\bar{w}_1 - \bar{w}_2)},$$

it follows (5.1.7). □

Proposition 5.1.4. *Let H be a pure hyponormal operator on $\mathcal{H}$ with rank one self-commutator. Then $S(\cdot, \cdot)$ is a complete unitary invariant of H.*

Proof. It is obvious that $S(\cdot, \cdot)$ is a unitary invariant. To prove its completeness, assume that there is a pure hyponormal operator $\widetilde{H}$ on $\widetilde{\mathcal{H}}$ with rank one $\widetilde{D}$ which is the restriction of the self-commutator on its image. Let $\widetilde{D}x = (x, \widetilde{k})\widetilde{k}$. Let

$$\widetilde{S}(z, w) \overset{\text{def}}{=} ((\widetilde{H} - z)^{-1}\widetilde{k}, (\widetilde{H} - w)^{-1}\widetilde{k}), \quad z, w \in \rho(\widetilde{H}).$$

If $\widetilde{S}(\ ,\ ) = S(\cdot, \cdot)$, then $\rho(H) = \rho(\widetilde{H})$ and

$$\widetilde{S}(z, w) = S(z, w), \quad z \in \rho(H).$$

From Lemma 5.1.2 and Lemma 5.1.3. We have

$$((H - z_1)^{-1}(H^* - \bar{w}_1)^{-1}k, (H - w_2)^{-1}(H^* - \bar{z}_2)^{-1}k)$$
$$= ((\widetilde{H} - z_1)^{-1}(\widetilde{H}^* - \bar{w}_1)^{-1}\widetilde{k}, (\widetilde{H} - w_2)^{-1}(\widetilde{H}^* - \bar{z}_2)^{-1}\widetilde{k}), \tag{5.1.10}$$

for $z_i, w_i \in \rho(H), i = 1, 2$. Let us define a linear operator U from $H_0 \overset{\text{def}}{=}$ $\bigvee \{(H-z)^{-1}(H^*-\bar{w})^{-1}k : z, w \in \rho(H)\}$ to $\widetilde{H}_0 \overset{\text{def}}{=} \bigvee \{(\widetilde{H}-z)^{-1}(\widetilde{H}^*-\bar{w})^{-1}\widetilde{k} :$ $z, w \in \rho(\widetilde{H})\}$ by

$$U(H - z)^{-1}(H^* - \bar{w})^{-1}k = (\widetilde{H} - z)^{-1}(\widetilde{H}^* - \bar{w})^{-1}\widetilde{k}.$$

Then U is well-defined and is unitary from the closure $\mathcal{H}_0$ onto the closure of $\widetilde{\mathcal{H}}_0$. But by Propsition 1.2.1, closure of $\mathcal{H}_0 = \mathcal{H}$ and closure of $\widetilde{\mathcal{H}}_0 = \widetilde{\mathcal{H}}$, since H and $\widetilde{H}$ are pure, which proves the Proposition. □

By the method of proving the Theorem 4.3 in p.177 of D. Xia [1], we have the following Pincus' determinant formula:

Proposition 5.1.5. *Let H be a pure hyponormal operator on $\mathcal{H}$ with rank one self-commutator. Then there is a unique measurable function $g_H(\cdot)$ on $\sigma(H)$ satisfying*

$$0 \leq g_H(\cdot) \leq 1$$

such that

$$\det((H^* - \bar{w})(H - z)(H^* - \bar{w})^{-1}(H - z)) = \exp \frac{1}{\pi} \int_{\sigma(H)} \frac{g_H(\zeta)dA(\zeta)}{(\zeta - z)(\bar{\zeta} - \bar{w})},$$
$$(5.1.11)$$

where $A(\cdot)$ is the Lebesgue planar measure.

We omit the proof, since it depends on the theory of hyponormal operators. Sometimes we denote $g_H(\cdot)$ simply by $g(\cdot)$. We call $g_H(\cdot)$ the *Pincus principal function* of H. But in some papers, $-g_H(\cdot)$ is called the Pincus principal function.

From (5.1.2) and (5.1.11) we have

$$S(z, w) = \exp\{\frac{1}{\pi} \int_{\sigma(H)} \frac{g(\zeta)dA(\zeta)}{(\zeta - z)(\bar{\zeta} - \bar{w})}\} - 1, \quad z, w \in \rho(H). \quad (5.1.12)$$

Thus $g_H(\cdot)$ is another complete unitary invariant of H.

5.2 Analytic model of hyponormal operator with rank one self-commutator

Let H be a pure hyponormal operator on $\mathcal{H}$ with rank one self-commutator. Let $g(\cdot)$ be its Pincus principal function. Denote the spectrum of H, $\sigma(H)$ by σ. Let $\mathcal{A}(\sigma)$ be the linear algebra of all analytic function with domain $D_f \supset \sigma$. For functions $f \in \mathcal{A}(\sigma)$, if γ is a smooth positive oriented contour in D_f consisting of piecewise smooth Jordan curves, satisfying the condition that σ is in the interior domain of γ, and $f(z) = \frac{1}{2\pi i} \int_{\gamma_f} \frac{f(\xi)d\xi}{\xi - z}$, $z \in \sigma$, then γ is called suitable for f. Define a sesquilinear from $(\cdot, \cdot)_H$ on $\mathcal{A}(\sigma)$:

$$(f, h)_g \overset{\text{def}}{=} \frac{1}{(2\pi)^2} \int_{\gamma_f} \int_{\gamma_h} f(z) S(z, w) \overline{h(w)} dz d\bar{w}, \qquad (5.2.1)$$

where $S(z, w)$ is the determine function of H and γ_f, γ_h are suitable for f and h respectively. It is easy to see that $(f, h)_g$ defined in (5.2.1) is independent of the choice of γ_f and γ_h. Denote

$$\Gamma(f; \zeta_1) \overset{\text{def}}{=} f(\zeta_1), \quad \zeta_1 \in D_f, \qquad (5.2.2)$$

and

$$\Gamma(f; \zeta_1, \dots, \zeta_n) \overset{\text{def}}{=} \sum_{k=1}^n f(\zeta_k) \prod_{j \neq k} \frac{1}{\zeta_k - \zeta_j}, \quad n \geq 2. \qquad (5.2.3)$$

It is easy to calculate that

$$\Gamma(f; \zeta_1, \dots, \zeta_n) = \frac{1}{(2\pi i)^n} \int_{\gamma_f} \frac{f(z)}{\prod_{j=1}^n (z - \zeta_j)} dz. \qquad (5.2.4)$$

From (5.1.2), we have

$$S(z, w) = \sum_{n=1}^{\infty} \frac{1}{n!} \frac{1}{\pi^n} \int_{\sigma} \cdots \int_{\sigma} \prod_{j=1}^n \frac{g(\zeta_j) dA(\zeta_j)}{(z - \zeta_j)(\bar{w} - \zeta_j)}.$$

Thus

$$(f, h)_g = \sum_{n=1}^{\infty} \frac{1}{n! \pi^n} \int_{\sigma} \cdots \int_{\sigma} \Gamma(f; \zeta_1, \dots, \zeta_n) \overline{\Gamma(h; \zeta_1, \dots, \zeta_n)} \prod_{j=1}^n g(\zeta_j) dA(\zeta_j).$$

Let $\mathcal{H}_g$ be the Hilbert space completion of $\mathcal{A}(\sigma)$ with respect to $(\cdot, \cdot)_g$. Denote

$$\hat{\mathcal{H}} \overset{\text{def}}{=} \text{closure of } \bigvee \{f(H)k : f \in \mathcal{A}(\sigma)\},$$

where $f(H) = \frac{1}{2\pi i} \int_{\gamma_f} (\lambda - H)^{-1} f(\lambda) d\lambda$, γ_f is a suitable for f, $\sigma = \sigma(H)$ and $k \in \text{range}[H^*, H]$ satisfying $\|k\|^2 = \|[H^*, H]\|$. Define a linear operator U from $\bigvee \{f(H)k : f \in \mathcal{A}(\sigma)\}$ to $\mathcal{A}(\sigma)$ by

$$U f(H)k = f(\cdot). \qquad (5.2.5)$$

It is easy to prove the following:

Lemma 5.2.1. *Let H be a pure hyponormal operator with rank one self-commutator on $\mathcal{H}$. Then U extends a unitary operator (denoted by U still) from $\hat{\mathcal{H}}$ onto $\mathcal{H}_g$ satisfying*

$$(U H U^{-1} f)(z) = z f(z), \quad f \in \mathcal{H}_g. \qquad (5.2.6)$$

The operator H is said to be C-cyclic if $\hat{\mathcal{H}} = \mathcal{H}$.

Corollary 5.2.2. *Let H be a pure hyponormal operator with rank one self-commutator on $\mathcal{H}$. If H is C-cyclic then U is unitary from $\mathcal{H}$ onto H_g.*

Define an operator $\hat{H}$ from $\mathcal{A}(\sigma)$ to a space of continuous functions on σ as follows: for $h \in \mathcal{A}(\sigma)$,

$$(\hat{H}h)(z) = \frac{1}{\pi} \int_\sigma \frac{g(\zeta)h(\zeta)dA(\zeta)}{\zeta - z} + \bar{z}h(z). \qquad (5.2.7)$$

Lemma 5.2.3. *Let H be a pure hyponormal operator with rank one self-commutator on $\mathcal{H}$. If $\hat{H}\mathcal{A}(\sigma) \subset \mathcal{H}_g$. Then $\hat{\mathcal{H}} = \mathcal{H}$ and*

$$UH^*U^{-1}f = \hat{H}f. \qquad (5.2.8)$$

Proof. Firstly, we have to prove

$$(UHU^{-1}f, h)_g = (f, \hat{H}h)_g, \quad f, h \in \mathcal{A}(\sigma), \qquad (5.2.9)$$

under the condition $\hat{H}\mathcal{A} \subset \mathcal{H}_g$. For $f, h \in \mathcal{A}(\sigma)$ we have

$(UHU^{-1}f, h)_g$

$$= \sum_{n=1}^\infty \frac{1}{n!\pi^n} \int_\sigma \cdots \int_\sigma \Gamma((\cdot)f(\cdot); \zeta_1, \dots) \overline{\Gamma(h; \zeta_1, \dots)} \prod_{j=1}^n (g(\zeta_j)dA(\zeta_j)). \qquad (5.2.10)$$

It is obvious that

$$(f, \hat{H}h)_g = I_1 + I_2,$$

where

$$I_1 = \sum_{n=1}^\infty \frac{1}{n!\pi^n} \int_\sigma \cdots \int_\sigma \Gamma(f; \zeta_1, \dots) \overline{\Gamma((\overline{(\cdot)}h(\cdot); \zeta_1, \cdots)} \prod_{j=1}^n (g(\zeta_j)dA(\zeta_j)), \qquad (5.2.11)$$

and I_2 equals to

$$\sum_{n=1}^\infty \frac{1}{n!\pi^n} \int_\sigma \cdots \int_\sigma \Gamma(f; \zeta_1, \dots) \overline{\Gamma(\frac{1}{\pi} \int_\sigma \frac{g(\xi)h(\xi)dA(\xi)}{\xi - (\cdot)}; \zeta_1, \dots)}$$
$$\prod_{j=1}^n (g(\zeta_j)dA(\zeta_j)). \qquad (5.2.12)$$

Let us calculate (5.2.10) firstly. It is obvious that

$$\Gamma((\cdot)f(\cdot); \zeta_1) \overline{\Gamma(h; \zeta_1)} = \Gamma(f; \zeta_1) \overline{\Gamma((\overline{(\cdot)}h(\cdot); \zeta_1)},$$

and for $n \geq 2$

$$\Gamma((\cdot)f(\cdot); \zeta_1, \cdots, \zeta_n)\overline{\Gamma(h; \zeta_1, \ldots, \zeta_n)}$$

$$= \sum_{p=1}^{n} \sum_{l=1}^{n} \zeta_p f(\zeta_p) \prod_{i \neq p} (\zeta_p - \zeta_i)^{-1} \overline{h(\zeta_l)} \prod_{j \neq l} (\bar{\zeta}_l - \bar{\zeta}_j)^{-1}$$

$$= \sum_{l=1}^{n} \sum_{p \neq l} (\zeta_p - \zeta_l) f(\zeta_p) \prod_{i \neq p} (\zeta_p - \zeta_i)^{-1} \overline{h(\zeta_l)} \prod_{j \neq l} (\bar{\zeta}_l - \bar{\zeta}_j)^{-1}$$

$$+ \Gamma(f; \zeta_1, \ldots)\overline{\Gamma((\overline{\cdot})h(\cdot); \zeta_1, \ldots)}$$

$$= \sum_{l=1}^{n} \Gamma(f; \zeta_1, \ldots, \zeta_{l-1}, \zeta_{l+1}, \ldots)\overline{h(\zeta_l)} \prod_{j \neq l} (\bar{\zeta}_l - \bar{\zeta}_j)^{-1}$$

$$+ \Gamma(f; \zeta_1, \ldots)\overline{\Gamma((\overline{\cdot})h(\cdot); \zeta_1, \cdots)}.$$

$$(5.2.13)$$

On the other hand, the term $n = 1$ in (5.2.12) is

$$\frac{1}{\pi} \int_{\sigma} \Gamma(f; \zeta_1)\overline{\Gamma(\frac{1}{\pi} \int_{\sigma} \int_{\sigma} \frac{g(\xi)h(\xi)dA(\xi)}{\xi - (\cdot)}; \zeta_1)} g(\zeta_1) dA(\zeta_1)$$

$$= \frac{1}{\pi^2} \int_{\sigma} \int_{\sigma} f(\zeta_1) \frac{\overline{h(\zeta_2)}}{\bar{\zeta}_2 - \bar{\zeta}_1} \prod_{j=1}^{2} (g(\zeta_j) dA(\zeta_j))$$

$$= \frac{1}{2!\pi^2} \int_{\sigma} \int_{\sigma} (\Gamma(f; \zeta_1) \frac{\overline{h(\zeta_1)}}{\bar{\zeta}_2 - \bar{\zeta}_1} + \Gamma(f; \zeta_2) \frac{\overline{h(\zeta_2)}}{\bar{\zeta}_1 - \bar{\zeta}_2}) \prod_{j=1}^{2} g(\zeta_j) dA(\zeta_j).$$

Similarly, for $n \geq 2$, we have

$$\frac{1}{n!\pi^n} \int_{\sigma} \cdots \int_{\sigma} \Gamma(f; \zeta_1, \ldots)\overline{\Gamma(\frac{1}{\pi} \int \frac{g(\xi)h(\xi)dA(\xi)}{\xi - \zeta}; \zeta_1, \ldots)} \prod_{j=1}^{n} g(\zeta_j) dA(\zeta_j)$$

$$= \frac{1}{n!\pi^{n+1}} \int_{\sigma} \cdots \int_{\sigma} \Gamma(f; \zeta_1, \ldots)\overline{\sum_{p=1}^{n} \frac{h(\zeta_{n+1})}{(\zeta_{n+1} - \zeta_p) \prod_{j \neq p} (\zeta_p - \zeta_j)}} \prod_{j=1}^{n+1} (g\zeta_j) dA(\zeta_j)$$

$$= \frac{1}{n!\pi^{n+1}} \int_{\sigma} \cdots \int_{\sigma} \Gamma(f; \zeta_1, \ldots) \frac{\overline{h(\zeta_{n+1})}}{\prod_{p \neq n+1} (\bar{\zeta}_{n+1} - \bar{\zeta}_p)} \prod_{j=1}^{n+1} g(\zeta_j) dA(\zeta_j)$$

$$= \frac{1}{(n+1)!\pi^{n+1}} \int_{\sigma} \cdots \int_{\sigma} \sum_{l=1}^{n+1} \Gamma(f; \ldots, \zeta_{l-1}, \zeta_{l+1}, \ldots) \frac{\overline{h(\zeta_l)}}{\prod_{p \neq l} (\bar{\zeta}_l - \bar{\zeta}_p)})$$

$$\prod_{j=1}^{n+1} g(\zeta_j) dA(\zeta_j).$$

$$(5.2.14)$$

From (5.2.10) – (5.2.14), we have (5.2.9). Thus
$$(UHU^{-1})^* = \hat{H},$$
as operators on $\mathcal{H}_g$.

It is easy to see that
$$([\hat{H}, UHU^{-1}]h)(z) = \frac{1}{\pi} \int_\sigma g(\zeta)h(\zeta)dA(\zeta). \tag{5.2.15}$$

Therefore $1 \in$ range of $[UH^*U^{-1}, UHU^{-1}]$. It is easy to calculate that
$$((UHU^{-1}-z)^{-1}1, (UHU^{-1}-w)^{-1}1)_g = (((\cdot)-z)^{-1}, ((\cdot)-w)^{-1})_g = S(z,w).$$
By Proposition 5.1.4, UHU^{-1} is unitary equivalent to H on $\mathcal{H}$ not only on $\hat{\mathcal{H}}$. Thus $\hat{\mathcal{H}} = \mathcal{H}$, which prove Lemma 5.2.3. $\square$

(5.2.6) and (5.2.7) represent an *analytic model* of the hyponormal operator H and its adjoint H^* under the condition that $\hat{H}\mathcal{A}(\sigma(H)) \subset \mathcal{H}_g$.

Lemma 5.2.4. *If σ is the closure of a finitely connected bounded domains, $g(\zeta) \equiv 1$ for $\zeta \in \sigma$. Let H be the multiplication operator $(Hf)(z) = zf(z), f \in \mathcal{H}_g$. Then for $\hat{h} \in \mathcal{A}(\sigma)$,*
$$(\hat{H}h)(z) = \frac{1}{2\pi i} \int_{\partial\sigma} \frac{\bar{\zeta}h(\zeta)d\zeta}{\zeta - z}, \quad z \in interior\ of\ \sigma. \tag{5.2.16}$$

This is a consequence of Weyl's lemma or Bocher-Martinelli's formula; for any C^2 function $f(\cdot)$
$$f(z) = \frac{1}{2\pi i} \int_{\partial\sigma} \frac{f(\zeta)d\zeta}{\zeta - z} - \frac{1}{\pi} \int_\sigma \frac{\frac{\partial f}{\partial\bar\zeta}dA(\zeta)}{\zeta - z}. \tag{5.2.17}$$
Actually, (5.2.17) can be proved by Greeen's formula. In (5.2.17), let $f(\zeta) = \bar{\zeta}h(\zeta)$, $h \in \mathcal{A}(\sigma)$, we have (5.2.16). $\square$

In this case, it is easy to calculate that from (5.2.6) and (5.2.16), for operator H in Lemma 5.2.1, we have
$$(U[\hat{H}, H]U^{-1}h)(z) = \frac{1}{\pi} \int_\sigma h(\zeta)dA(\zeta). \tag{5.2.18}$$

Later, if $g(\cdot)$ is the characteristic function of σ, we also sometimes denote $(f, h)_g$ by $(f, h)_\sigma$.

Corollary 5.2.5. *Under the condition of Lemma 5.2.4, $1 \in$ range of $U[H^*, H]U^{-1}$ and*
$$(f, 1)_\sigma = \frac{1}{\pi} \int_\sigma f(\zeta)dA(\zeta) = \frac{1}{2\pi i} \int_{\partial\sigma} \bar{\zeta}f(\zeta)d\zeta. \tag{5.2.19}$$

Proof. From (5.2.18), $1 \in U[H^*, H]U^{-1}$. (5.2.19) comes from the definition of $(f, 1)_\sigma$, since $\Gamma(1, \zeta_1, \zeta_2, \ldots, \zeta_n) = 0$, for $n \geq 2$. $\square$

5.3 Hyponormal operator associated with a quadrature domain

A hyponormal operator H is said to be associated with a quadrature domain, if it is pure with rank one self-commutator, satisfying $g_H \equiv 1$ on $\sigma(H)$ and its spectrum $\sigma = \sigma(H)$ is the closure of a quadrature domain in $\mathbb{C}$.

Theorem 5.3.1. *Let H be a hyponormal operator on $\mathcal{H}$ associated with a quadrature D. Then there is a unitary operator U from $\mathcal{H}$ onto $\mathcal{H}_g$ where g is the characteristic function of the closure of D such that*

$$(UHU^{-1}f)(z) = zf(z), \quad f \in \mathcal{A}(\sigma), \tag{5.3.1}$$

and

$$(UH^*U^{-1}f)(z) = \frac{1}{2\pi i}\int_L \frac{\bar{\zeta}f(\zeta)d\zeta}{\zeta - z}, \quad z \in D, f \in \mathcal{A}(\sigma) \tag{5.3.2}$$

where $\sigma = \sigma(H) =$ closure of D and L is the boundary contour of D with positive orientation.

Proof. From Lemma 5.2.1, Lemma 5.2.3 and Lemma 5.2.4, we only have to prove that for $f \in \mathcal{A}(\sigma)$

$$(\hat{H}f)(z) = \frac{1}{2\pi i}\int_L \frac{\bar{\zeta}f(\zeta)d\zeta}{\zeta - z} \in \mathcal{H}_g. \tag{5.3.3}$$

Let $S(\cdot)$ be the Schwartz function of D. Suppose $\{a_1, \ldots, a_n\}$ are the poles of $S(\cdot)$ with orders $m_1, m_2, \ldots, m_n$, respectively. Let the singular part of $S(\cdot)$ at the pole a_j be

$$\sum_{i=1}^{m_j} \frac{c_{ij}}{(z - a_j)^i}. \tag{5.3.4}$$

Then by the calculus of residues, we have

$$\frac{1}{2\pi i}\int_L \frac{\bar{\zeta}f(\zeta)d\zeta}{\zeta - z} = \frac{1}{2\pi i}\int_L \frac{S(\zeta)f(\zeta)}{\zeta - z}d\zeta$$

$$= S(z)f(z) - \sum_{j=1}^{n}\sum_{i=0}^{m_j - 1} \frac{\beta_{ij}}{(a_j - z)^{i+1}} \tag{5.3.5}$$

where $\beta_{ij} = (-1)^i i! \sum_{l=i}^{m_j - 1} \frac{1}{l!}f^{(l-i)}(a_j)c_{(l+1)j}$. It is easy to see that the singular parts of $S(\cdot)f(\cdot)$ at possible pole a's are cancelled by the second term in the right hand side (5.3.5). At the boundary L, the function $S(\cdot)$ is analytic except at a finite set in L. But at those points $S(\cdot)$ is still continuous. By some limiting process we may prove that the right hand side of (5.3.5) is in $\mathcal{H}_g$. $\square$

Define

$$Q_D(z) = \Pi_{j=1}^n (z - a_j)^{m_j},$$

where a_j and m_j are the numbers in (5.3.4). Then $Q_D(\cdot)$ is said to be the denominator of the Schwartz function $S(z)$. The polynomial $Q_D(\cdot)$ can be characterized as the polynomial $P(\cdot)$ with leading coefficient 1 and making $P(\cdot)S(\cdot)$ analytic in D with minimal degree. The degree $n_D \overset{\text{def}}{=} \sum_{j=1}^k m_j$ of $Q_D(\cdot)$ is said to be the order of D.

Let T be an operator on a Hilbert space $\mathcal{H}$. If

$$K = K_H \overset{\text{def}}{=} \text{closure of } \bigvee\{T^{*n}[T^*, T]\mathcal{H} : n = 0, 1, 2, \dots\}$$

is of finite dimensional, then T is said to be *of finite type*.

Lemma 5.3.2. *Let H be a pure hponormal operator on $\mathcal{H}$ associated with a quadrature domain D. Then H is of finite type and*

$$\dim K = \text{order of } D.$$

Proof. From now on we may identify H with UHU^{-1} and identify $\mathcal{H}$ with $\mathcal{H}_g$.

From Corollary 5.2.5, $1 \in$ range of $[H^*, H]$. On the other hand, for $f \in \mathcal{H}_g$,

$$(Q_D(H)f, 1)_\sigma = \frac{1}{2\pi i} \int_L \bar{\zeta} Q_D(\zeta) f(\zeta) d\zeta = \frac{1}{2\pi i} \int_L S(\zeta) Q_D(\zeta) f(\zeta) d\zeta = 0,$$

by Cauchy's theorem. Therefore $(f, Q_D(H)^* 1)_\sigma = 0$, for all $f \in \mathcal{H}_g$. Thus

$$Q_D(H)^* 1 = 0.$$

Let $n_D =$ degree of $Q_D(\cdot)$. Then

$$K = \bigvee\{H^{*m} 1 : m = 0, 1, , 2, \dots, n_D - 1\},$$

which proves $\dim K \leq n_D$. On the other hand if $\dim K < n_D$. Then there exists a polynomial Q with degree less the order of D satisfying $Q(H)^* 1 = 0$. Then

$$\frac{1}{2\pi i} \int_L S(\zeta) Q(\zeta) f(\zeta) d\zeta = (Q(H)f, 1)_\sigma = (f, Q(H)^* 1)_\sigma = 0 \qquad (5.3.6)$$

for every $f \in \mathcal{A}(\sigma)$. But the total order of the poles of $S(\cdot)$ in D is greater than the degree of $Q(\cdot)$, (5.3.6) cannot be true for every $f \in \mathcal{A}(\sigma)$. This leads to a contradiction, which proves this lemma. $\square$

As before, we define

$$C = C_H = [H^*, H]|_K \quad \text{and} \quad \Lambda = \Lambda_H = (H^*|_k)^*. \tag{5.3.7}$$

Define

$$Q(z, w) = Q_H(z, w) = (\bar{w} - \Lambda^*)(z - \Lambda) - C, \tag{5.3.8}$$

and

$$P(z, w) = P_D(z, w) = P_H(z, w) = \det Q(z, w). \tag{5.3.9}$$

Proposition 5.3.3. *Let H be a hyponormal operator on $\mathcal{H}$ associated with a quadrature domain D. Then*

$$Q_D(z) = \det(z - \Lambda), \tag{5.3.10}$$

the determining function of H is

$$S(z, w) = \frac{Q_D(z)\overline{Q_D(w)}}{P_D(z, w)} - 1, \tag{5.3.11}$$

and

$$\partial D \subset \{z : P_D(z, z) = 0\}. \tag{5.3.12}$$

Proof. Let $G(z) = \det(z - \Lambda^*)$. Then $G(\Lambda^*) = 0$. i.e. $G(H^*)1 = 0$. Let $\widetilde{Q}(z) \overset{\text{def}}{=} \overline{Q_D(\bar{z})}$, then $\widetilde{Q}(H^*) = Q_D(H)^*$. For $f \in \mathcal{A}(D \cup L)$, by (5.2.10),

$$(f, \widetilde{Q}(H)1) = (Q_D(H)f, 1) = \frac{1}{2\pi i} \int_L Q_D(\zeta) S(\zeta) f(\zeta) d\zeta = 0.$$

Thus $\widetilde{Q}(H^*)1 = 0$. Both $G(\cdot)$ and $\widetilde{Q}(\cdot)$ have the same leading term z^n, where n is the order of D. If $G(\cdot) \neq \widetilde{Q}(\cdot)$, then from $(G(H^*) - \widetilde{Q}(H^*))1 = 0$, we may conclude that dimension of K is less than or equals to the degree $G(\cdot) - \widetilde{Q}(\cdot)$, which leads a contradiction. Thus $\overline{Q_D(\bar{z})} = G(z)$, i.e. $Q_D(z) = \overline{(\det(\bar{z} - \Lambda^*))}$ which proves (5.3.10).

From (5.3.8) and (5.3.10), we have

$$\frac{P_D(z, w)}{Q_D(z)\overline{Q_D(w)}} = \det(I - (\bar{w} - \Lambda^*)^{-1}C(z - \Lambda)^{-1}). \tag{5.3.13}$$

For fixed $z \in \rho(\Lambda)$, choose an orthonormal basis $\{e_1, \ldots, e_n\}$ of K such that

$$e_1 = \frac{(\bar{z} - \Lambda^*)^{-1}1}{\|(\bar{z} - \Lambda^*)^{-1}1\|}.$$

Then

$$(\bar{w} - \Lambda^*)^{-1}C(z - \Lambda)^{-1}e_j = (e_j, (\bar{z} - \Lambda^*)^{-1}1)(\bar{w} - \Lambda^*)^{-1}1/\|(\bar{z} - \Lambda^*)^{-1}1\| = 0,$$

for $j = 2, \ldots, n$, since $Cx = (x, 1)$. Thus for all $w \in \rho(\Lambda)$ and $|z| > \|\Lambda\|$,

$$
\begin{aligned}
\det(I - (\bar{w} - \Lambda^*)^{-1} C (z - \Lambda)^{-1}) &= 1 - ((\bar{w} - \Lambda^*)^{-1} C (z - \Lambda)^{-1} e_1, e_1) \\
&= 1 - ((z - \Lambda)^{-1} e, 1)((\bar{w} - \Lambda^*)^{-1} 1, e) \\
&= 1 - ((\bar{w} - \Lambda^*)^{-1} 1, (\bar{z} - \Lambda^*)^{-1} 1).
\end{aligned}
$$

Since

$$((z - \Lambda_1)^{-1} e_1, 1) = ((z - \Lambda)^{-1} (\bar{z} - \Lambda^*)^{-1} 1, 1) \| (\bar{z} - \Lambda^*) \|^{-1} = \| (\bar{z} - \Lambda^*)^{-1} 1 \|,$$

hence the above equality also hold for all $z, w \in \rho(H)$, therefore

$$
\begin{aligned}
\det(I - (\bar{w} - \Lambda^*)^{-1} C (z - \Lambda)^{-1}) &= 1 - ((\bar{w} - H^*)^{-1} 1, (\bar{z} - H^*)^{-1} 1) \\
&= 1 - T(z, w), \quad (5.3.14)
\end{aligned}
$$

since $\|1\|^2 = \|[H^*, H]\|$.

From (5.1.5), (5.3.13) and (5.3.14), it follows (5.3.11).

From (5.3.13) and (5.1.12), we have

$$\frac{P_D(z, z)}{|Q_D(z)|^2} = \exp\{-\frac{1}{\pi} \int\int_D \frac{dA(\zeta)}{|\zeta - z|^2}\}, \quad z \in \mathbb{C} \setminus \sigma. \quad (5.3.15)$$

It is easy to see that

$$\lim_{z \to z_0} \int\int_D \frac{dA(\zeta)}{|\zeta - z|^2} = +\infty,$$

where $z_0 \in \partial D$, $z \in \mathbb{C} \setminus D$. Thus from (5.3.15), $P_D(z_0, z_0) = 0$ for $z_0 \in \partial D$, which proves (5.3.12). $\qquad \square$

On the L, $P(z, \overline{S(z)}) = P(z, z) = 0$. But $P(z, \overline{S(z)})$ is a meromorphic function on D. Thus $P(z, \overline{S(z)}) = 0$ for $z \in D$.

Define $\tilde{\sigma}(H) = \{\overline{S(z)} : S'(z) = 0, z \in D\}$. It is a finite set. For $w \notin \tilde{\sigma}(H)$, let

$$E(z, w) = E_D(z, w) \overset{\text{def}}{=} \frac{P_D(z, w)}{(S(z) - \bar{w})(z - \overline{S(w)}) Q_D(z) \overline{Q_D(w)}}, \quad z \in D.$$

For the poles of $S(\cdot)$, $E(\cdot, w)$ is defined by limit. $E(\cdot, \cdot)$ is said to be an eigen kernel of H. This kernel is Hermitian: $\overline{E(z, w)} = E(w, z)$.

Proposition 5.3.4. *Let H be a hyponormal operator on $\mathcal{H}$ associated with a quadrature domain D. Then for $w \in D \setminus \tilde{\sigma}(H)$, $E(\cdot, w) \in \mathcal{H}_g$,*

$$H^* E(\cdot, w) = \bar{w} E(\cdot, w) \quad (5.3.16)$$

and this kernel $E(\cdot, \cdot)$ is reproducing:

$$(f; E(\cdot, w))_D = f(w), \quad \text{for } f \in \mathcal{H}_g, w \in D \setminus \tilde{\sigma}(H). \quad (5.3.17)$$

Proof. It is obvious that $E(z, w)$ is a meromorphic function of $z \in D$, it is analytic on the boundary L of $\sigma(H)$, except possible finite set of points. But at those points, the function $E(\cdot, w)$ is continue. Let us prove that there is no pole of $E(\cdot, w)$ in D. The zero of $Q_D(\cdot)$ is cancelled by the factor $(S(z) - \bar{w})$. Let us prove that the zero of $S(z) - \bar{w}$ is not a pole of $E(\cdot, w)$. Actually, from (5.3.13) and (5.3.14), we have

$$\frac{P_D(z, w)}{Q_D(z)\overline{Q_D(w)}} = 1 - ((z - \Lambda)^{-1}(\bar{w} - \Lambda^*)^{-1}1, 1). \tag{5.3.18}$$

From (5.3.18), we have

$$\frac{P_D(z, \overline{S(z)})}{Q_D(z)\overline{Q_D(\overline{S(z)})}} = 1 - ((z - \Lambda)^{-1}(S(z) - \Lambda^*)^{-1}1, 1), \tag{5.3.19}$$

is a meromorphic function on D. But the boundary value of (5.3.19) at L is 0, since $\overline{S(z)} = z$ for $z \in L$ and (5.3.12). Thus (5.3.19) is identically zero on $D \cup L$. Therefore the simple zero of $S(z) - \bar{w}$ is cancelled by $P_D(z, w)$ of the numerator. Besides, there is no zero of $S(z) - \bar{w}$ with multiplicity greater than 1, since

$$w \notin \{\overline{S(z)} : S'(z) = 0, z \in D\}.$$

Similarly, we may prove that the zero $\overline{S(w)}$ of the factor $z - \overline{S(w)}$ in the denominator of $E(\cdot, w)$ is also cancelled by $P_D(z, w)$, which proves $E(\cdot, w) \in \mathcal{H}_g$.

By (5.3.2), we have

$$((H^* - \bar{w})E(\cdot, w))(z) = \frac{1}{2\pi i} \int_L \frac{(\bar{\zeta} - \bar{w})E(\zeta, w)d\zeta}{\zeta - z}$$
$$= \frac{1}{2\pi i} \int_L \frac{P_D(\zeta, w)d\zeta}{(\zeta - \overline{S(w)})Q_D(\zeta)\overline{Q_D(w)}(\zeta - z)}. \tag{5.3.20}$$

Let us assume that w is close to the boundary L, then $\overline{S(w)} \in \mathbb{C} \setminus \sigma = \bigcup_{j=0}^n D_j$, where $D_j, j = 0, 1, 2, \ldots, n$ are exterior domains of σ and D_0 is unbounded. Let us calculate the residues of the contour integration (5.3.20) at the poles in $\mathbb{C} \setminus \sigma$. The zeros of $Q_D(\zeta)$ are in D, and the zeros of $\zeta - \overline{S(w)}$ are cancelled by

$$\frac{P_D(\overline{S(w)}, w)}{(\overline{S(w)} - z)Q_D(\overline{S(w)})\overline{Q_D(w)}} = 0,$$

by a formula similar to (5.3.19). There is no residue at ∞, since the Laurent expansion near ∞ is

$$\frac{P_D(\zeta, w)}{(\zeta - \overline{S(w)})Q_D(\zeta)\overline{Q_D(w)}(\zeta - z)} = \frac{1}{\overline{Q_D(w)}}\frac{1}{\zeta^2} + \cdots,$$

which proves that $(H^* - \bar{w})E(z,w) = 0$ for w close to the boundary L. However $(H^* - \bar{w})E(z,w)$ is a meromorphic function of $\bar{w} \in \{\bar{z} : z \in D\}$. Thus $E(\cdot,w)$ is an eigenvector of H^* corresponding to the eigenvalue $\bar{w}$ for all $w \notin \tilde{\sigma}(H)$.

To prove (5.3.17), let us prove that

$$(E(\cdot,w),1)_g = \frac{1}{2\pi i} \int_L \bar{\zeta} E(\zeta,w)d\zeta = 1. \tag{5.3.21}$$

Let us calculate

$$\frac{1}{2\pi i} \int_L (S(\zeta) - \bar{w})E(\zeta,w)d\zeta = \frac{1}{2\pi i} \int_L \frac{P_D(\zeta,w)d\zeta}{(\zeta - \overline{S(w)})Q_D(\zeta)\overline{Q_D(w)}}. \tag{5.3.22}$$

Similar to the integral (5.3.20), we only have to examine the Laurent expansion of the function

$$\frac{P_D(\zeta,w)}{(\zeta - \overline{S(w)})Q_D(\zeta)\overline{Q_D(w)}} = \frac{1}{\zeta} + \cdots$$

near infinity. Its residue at ∞ is 1. Thus (5.3.22) equals to 1 which proves (5.3.21). Now for any $\lambda \in \rho(H)$

$$((\lambda - (\cdot))^{-1}, E(\cdot,w)) = ((\lambda - H)^{-1}1, E(\cdot,w)) = (1, (\bar{\lambda} - H^*)^{-1}E(\cdot,w))$$
$$= (1, (\bar{\lambda} - \bar{w})^{-1}E(\cdot,w)) = 1/(\lambda - w) \tag{5.3.23}$$

by (5.3.21). But $\mathcal{H}_g = $ closure of $\bigvee\{(\lambda - H)^{-1}1 : \lambda \in \rho(H)\}$. Thus (5.3.23) implies (5.3.17). □

For the quadrature domain D, there is the quadrature identity

$$\frac{1}{\pi} \iint_D f(\zeta)dA(\zeta) = \sum_{j=1}^{m} \sum_{l=0}^{m_j-1} \alpha_{lj} f^{(l)}(a_j), \tag{5.3.24}$$

where $a_j, j = 1,2,\ldots,m$ are the poles of the Schwatz function $S(\cdot)$ in D and $\alpha_{lj} = c_{(l+1)j}/l!$ where c_{lj} are the coefficients of the Laurent expansion of $S(\cdot)$ at a_j in (5.3.5). We can prove (5.3.24) by

$$\frac{1}{\pi} \iint_D f(\zeta)dA(\zeta) = \frac{1}{2\pi i} \int_L \bar{\zeta} f(\zeta)d\zeta = \frac{1}{2\pi i} \int_L S(\zeta)f(\zeta)d\zeta$$

and the calculus of residues.

Corollary 5.3.5. *Under the condition of Proposition 5.3.4,*

(i)

$$1 = \sum_{j=1}^{m} \sum_{l=0}^{m_j-1} \bar{\alpha}_{lj} \bar{\partial}^l E(z,a_j), \quad z \in D \tag{5.3.25}$$

where $\bar{\partial}^l E(\cdot, w) = \frac{\partial^l}{\partial \bar{w}^l} E(\cdot, w)$.

(ii) *The set*

$$\{\bar{\partial}^l E(\cdot, a_j) : l = 0, \ldots, m_j - 1, j = 1, 2, \ldots, m\} \qquad (5.3.26)$$

is a basis for $K = \bigvee \{H^{*l}[H^*, H]\mathcal{H} : l = 0, 1, 2, \ldots\}$.

(iii) *The denominator* $Q_D(z) = \Pi(z - a_j)^{m_j}$ *of the Schwartz function is the unique polynomial with minimal degree and leading coefficient* 1 *satisfying* $Q_D(H)^* 1 = 0$.

Proof. By the quadrature identity (5.3.24), for every $f \in \mathcal{A}(\sigma)$ we have

$$(f, 1)_D = \sum \alpha_{lj} f^{(l)}(a_j). \qquad (5.3.27)$$

On the other hand, from (5.3.17), it is easy to calculate that

$$(f, \sum \overline{\beta}_{lj} \bar{\partial}^l E(\cdot, w_j))_D = \sum \beta_{lj} f^{(l)}(w_j), \qquad (5.3.28)$$

for every $f \in \mathcal{A}(\sigma)$, $w_j \in D \setminus \tilde{\sigma}(H)$, $\beta_{lj} \in \mathbb{C}$. From (5.3.27) and (5.3.28) it follows (5.3.25). From (5.3.28), it is easy to see that (5.3.26) is a set of linearly independent vectors in $\mathcal{H}$. We have

$$q(H^*)\bar{\partial}^l E(\cdot, w) = \sum_{j=0}^{l} q^{(j)}(\bar{w})\bar{\partial}^{(l-j)} E(\cdot, w)l!/j!(l-j)!, \qquad (5.3.29)$$

for any polynomial q, since (5.3.16). From (5.3.29), we may conclude that $q(H^*)1 = 0$ iff $q^{(j)}(\bar{a}_l) = 0$, for $0 \leq j \leq m_l - 1$, $l = 1, 2, \ldots, m$, which prove statement (iii) in the corollary. $\qquad \square$

Proposition 5.3.6. *Under the same condition of Proposition 5.3.4, the Hilbert space* $\mathcal{H}_g$ *is a reproducing kernel Hilbert space with inner product*

$$(E(\cdot, w), E(\cdot, z))_g = E(z, w), \qquad (5.3.30)$$

and the operator H^* *is diagonalizable by the kernel* $E(\cdot, \cdot)$ *as in* (5.3.16), *where* H *is the multiplication operator* $HE(\cdot, w) = (\cdot)E(\cdot, w)$.

5.4 Mosaic for hyponormal operator associated with a quadrature domain

Let H be a hyponormal operator on $\mathcal{H}$ associated with a quadrature domain D. Denote $P_D(\cdot, \cdot)$ by $P(\cdot, \cdot)$ and $P_{\bar{w}}(z, w) = \frac{\partial}{\partial \bar{w}} P(z, w)$. Let

$$D_0 = \{z \in D : z \in \rho(\Lambda), S(z) \in \rho(\Lambda^*) \text{ and } P_{\bar{w}}(z, \overline{S(z)}) \neq 0\}.$$

Then $D \setminus D_0$ is a finite set. Define an $L(K)$-valued function

$$\mu(z) \overset{\text{def}}{=} (S(z) - \Lambda^*)^{-1} C(z - \Lambda)^{-1}(S(z) - \Lambda^*)^{-1} k(z), \quad z \in D_0,$$

where

$$k(z) \overset{\text{def}}{=} \overline{P_{\bar{w}}(z, \overline{S(z)})}^{-1} Q_D(z) \overline{Q_D(\overline{S(z)})}.$$

Then $\mu(z)$ is analytic on D_0 and rank $\mu(z) = 1$, since rank $C = 1$. This $\mu(\cdot)$ is said to be the mosaic for H. As in Chapter 1, define

$$R(z) = C(z - \Lambda)^{-1} + \Lambda^*, \quad z \in \rho(\Lambda).$$

Lemma 5.4.1. *The mosaic satisfies the following identities:*

$$\mu(z)^2 = \mu(z), \tag{5.4.1}$$

and

$$R(z)\mu(z) = \mu(z)R(z) = S(z)\mu(z), \quad z \in D_0. \tag{5.4.2}$$

Proof. It is obvious that (5.4.1) is equivalent to

$$k(z)C(z - \Lambda)^{-1}(S(z) - \Lambda^*)^{-2}C = C. \tag{5.4.3}$$

The identity (5.3.18) is equivalent to

$$C - C(z - \Lambda)^{-1}(\bar{w} - \Lambda^*)^{-1}C = \frac{P(z,w)}{Q(z)\overline{Q(w)}}C, \tag{5.4.4}$$

where $Q(\cdot) = Q_D(\cdot)$. Therefore

$$C(z - \Lambda)^{-1}(\bar{w} - \Lambda^*)^{-2}C = \left(\frac{P_{\bar{w}}(z,w)}{Q(z)\overline{Q(w)}} - \frac{P(z,w)\overline{Q'(w)}}{Q(z)\overline{Q(w)}^2}\right)C. \tag{5.4.5}$$

Letting $\bar{w} = S(z)$ in (5.4.5) and using $P(z, \overline{S(z)}) = 0$ on D_0, (5.4.5) implies (5.4.3).

From (5.4.4) and $P(z, S(z)) = 0$, it follows

$$C - C(z - \Lambda)^{-1}(S(z) - \Lambda^*)^{-1}C = 0. \tag{5.4.6}$$

Thus, by the direct calculation, we have

$$(R(z) - S(z))\mu(z)$$
$$= (C(z - \Lambda)^{-1}(S(z) - \Lambda^*)^{-1}C - C)(z - \Lambda)^{-1}(S(z) - \Lambda^*)^{-1}k(z) = 0.$$

Similarly, we have $\mu(z)(R(z) - S(z)) = 0$ which proves (5.4.2). $\qquad\square$

For the subnormal operator with finite rank self-commutator on the boundary of spectrum, we have

$$e(du) = \frac{1}{2\pi i}(u - \Lambda)^{-1}\mu(u)du. \tag{5.4.7}$$

For the hyponormal operator H associated with a quadrature domain D, let us define an $L(K)$-valued measure $e(du)$ on ∂D by (5.4.7).

Lemma 5.4.2. *The function* $iP_{\bar{w}}(u, \overline{S(u)})^{-1}du/ds$ *is real for* $u \in \partial D$, *where* ds *is the arc element of the* ∂D, *and* $e(du)/ds$ *is self-adjoint for* $u \in \partial D$.

Proof. From $P(u, \overline{S(u)}) = 0$, it follows that
$$P_z(u, \overline{S(u)}) + P_{\bar{w}}(u, \overline{S(u)})S'(u) = 0,$$
where $P_z(z, w) \stackrel{\text{def}}{=} \frac{\partial}{\partial z}P(z, w)$. From the Hermitian property $\overline{P(z, w)} = P(w, z)$, it follows that
$$\overline{P_z(z, w)} = \overline{P_{\bar{w}}(w, z)}.$$
Therefore $\overline{P_{\bar{w}}(\overline{S(u)}, u)} + P_{\bar{w}}(u, \overline{S(u)})S'(u) = 0$. For $u \in \partial D$,
$$S'(u) = \frac{\overline{du}}{ds} \bigg/ \frac{du}{ds}.$$
Thus on ∂D, $\text{Re}(P_{\bar{w}}(u, u)^{-1}\frac{du}{ds}) = 0$, i.e. $iP_{\bar{w}}(u, \overline{S(u)})^{-1}du/ds$ is real for $u \in \partial D$. Thus $ik(u)\frac{du}{ds}$ is real for $u \in \partial D$. On the other hand
$$(u - \Lambda)^{-1}(\bar{u} - \Lambda^*)^{-1}C(u - \Lambda)^{-1}(\bar{u} - \Lambda^*)^{-1} \geq 0, \quad u \in \partial D.$$
Hence $e(du)/ds$ is real on ∂D. $\qquad\square$

Later, we will prove that $P_{\bar{w}}(u, \overline{S(u)})^{-1}du/ds \geq 0$ on L.

Let $\rho_1 = \rho(H) \setminus \{\overline{S(u)} : S'(u) = 0, \text{ or } P_{\bar{w}}(u, \overline{S(u)}) = 0, u \in D\}$ and $\rho_1^* = \{\bar{w} : w \in \rho_1\}$. Then ρ_1^* is dense in $\rho(H^*)$. The function $P(z, \bar{w})$ is a polynomial of z and w, therefore, from $P(z, \overline{S(z)}) = 0, z \in D$, the analytic continuation of $w = S(z)$, $z \in D$ is an algbraic function and its inverse function $z = S^{-1}(w)$ is also an algbraic function. For every $w \in \rho_1^*$, the variation of the argument of $S(u) - w$ along the boundary $u \in L$ of D is zero, since $S(u) = \bar{u}$, $u \in L$ and w is in ρ_1^*. Therefore the number of the zeros $u = u(w) \in D$ of the function $S(u) - w$ equals the total number (counting multiplicity) of poles of $S(\cdot) - w$ in D. It is also equals the order n of the quadrature domain D or the dimension of K. Since $S'(u) \neq 0$ for $w \in \rho_1^*$, these zeros $u = u(w)$ are simple zeros. Therefore there are a union of domains $\mathcal{D}_1^*$ in the Riemann surface of the function $S^{-1}(w)$ covers ρ_1^*, a projection ψ^* from $\mathcal{D}_1^*$ to ρ_1^* which is n to 1 mapping, and an analytic function $u(\cdot)$ on $\mathcal{D}_1^*$ such that
$$S(u(\zeta)) = \psi^*(\zeta), \quad \zeta \in \mathcal{D}_1^*.$$
Let $\mathcal{D}_1$ be the Schottky dual of $\mathcal{D}_1^*$, which covers ρ_1, ψ be the projection from $\mathcal{D}_1$ to ρ_1. Let $\zeta \to \zeta^*$ be the mapping from $\mathcal{D}_1$ to $\mathcal{D}_1^*$ satisfying
$$\psi^*(\zeta^*) = \overline{\psi(\zeta)}.$$
Thus $\overline{S(u(\zeta^*))} = \psi(\zeta)$. Define an $L(K)$-valued function λ on ρ_1:
$$\lambda(\zeta) \stackrel{\text{def}}{=} -(\psi(\zeta) - \Lambda)\mu(u(\zeta^*))^*(\overline{u(\zeta^*)} - \Lambda^*)^{-1}\overline{S'(u(\zeta^*))}^{-1}. \tag{5.4.8}$$

Lemma 5.4.3. *For $\zeta \in \mathcal{D}_1$ and $\psi(\zeta) \in \rho_1$, $\lambda(\zeta)$ is the projection to the eigenspace of $R(\psi(\zeta))$ corresponding to the simple eigenvalue $\overline{u(\zeta^*)}$, $\lambda(\zeta)$ is idempotent, and*

$$R(z) = \sum_{\psi(\zeta)=z} \overline{u(\zeta^*)}\lambda(\zeta) \tag{5.4.9}$$

for $z \in \rho(\Lambda)$ satisfying $P_{\bar{w}}(z, \overline{S(z)}) \neq 0$.

Proof. First, let us prove that

$$\lambda(\zeta)^2 = \lambda(\zeta). \tag{5.4.10}$$

It is obvious that (5.4.10) is equivalent to

$$\overline{k(w)}C(z - \Lambda)^{-1}(\bar{w} - \Lambda^*)^{-2}C = -\overline{S'(w)}C \tag{5.4.11}$$

where $z = \psi(\zeta)$, $w = u(\zeta^*)$ and hence $\overline{S(w)} = z$. Take the adjoint of the both sides of (5.4.11), we see that (5.4.10) is equivalent to

$$k(w)C(w - \Lambda)^{-2}(S(w) - \Lambda^*)^{-1}C = -S'(w)C. \tag{5.4.12}$$

From (5.4.6) we have

$$C(w - \Lambda)^{-2}(S(w) - \Lambda^*)^{-1}C + C(w - \Lambda)^{-1}(S(w) - \Lambda^*)^{-2}CS'(w) = 0. \tag{5.4.13}$$

From (5.4.13) and (5.4.3), it follows (5.4.11) and hence (5.4.10).

Let us prove that

$$R(\psi(\zeta))\lambda(\zeta) = \overline{u(\zeta^*)}\lambda(\zeta) = \lambda(\zeta)R(\psi(\zeta)). \tag{5.4.14}$$

Let $\psi(\zeta) = z$, $u(\zeta^*) = w$, then

$$(R(z) - \bar{w})\lambda(\zeta) = -(C - (\bar{w} - \Lambda^*)(z - \Lambda))\mu(w)^*(\bar{w} - \Lambda^*)^{-1}\overline{S'(w)}^{-1}$$

$$= -(\overline{w} - \Lambda^*)(R(w)^* - z)\mu(w)^*(\bar{w} - \Lambda^*)^{-1}\overline{S'(w)}^{-1} = 0$$

by (5.4.2). Similarly, we may prove that $\lambda(\zeta)(R(z) - \bar{w}) = 0$, which proves (5.4.14).

From (5.4.14) and that $\{\overline{u(\zeta^*)} : \psi(\zeta) = z\}$ contains n numbers, it gives the Jordan decomposition (5.4.9) of $R(z)$. $\qquad\square$

The meromorphic function $\lambda(\cdot)$ is also a sort of mosaic for H.

Define $v(z) = (z - \Lambda)^{-1}\mu(z)$, $z \in D_0$.

Lemma 5.4.4. *The functions $v(\cdot)$ and $\mu(\cdot)$ extend meromorphic functions on D with poles only in $\sigma_p(\Lambda) \cup \{z \in D : P_{\bar{w}}(z, \overline{S(z)}) = 0\}$. The order of the pole $a_k \in \sigma_p(\Lambda)$ of the functions $v(\cdot)$ and $\mu(\cdot)$ is less than or equal to the order m_k of the Schwartz function $S(\cdot)$ at a_k.*

Proof. First for $u \in \rho(\Lambda)$, let us examine the function of $(\bar{u} - \Lambda^*)^{-1}1$. Notice that for $u \in \rho(H)$, we have

$$(\bar{u} - H^*)^{-1}\bar{\partial}^l E(\cdot, w) = \bar{\partial}^l((\bar{u} - \bar{w})^{-1}E(\cdot, w))$$

$$= \sum_{j=0}^{l}(\bar{u} - \bar{w})^{-(j+1)}\bar{\partial}^{(l-j)}E(\cdot, w)\frac{l!}{j!(l-j)!},$$

where $\bar{\partial} = \partial_{\bar{w}}$. Therefore for $u \in \rho(\Lambda)$,

$$(\bar{u} - \Lambda^*)^{-1}\bar{\partial}^l E(\cdot, z_k) = l! E(\cdot, u)(\bar{u} - \bar{a}_k)^{-(l+1)}$$

$$+ \text{ an analytic function of } (\cdot) \text{ and } \bar{u}. \tag{5.4.15}$$

From (5.3.25) and (5.4.15), it follows that

$$(\bar{u} - \Lambda^*)^{-1}1 = E(\cdot, u)\overline{S(u)} + h(\cdot, u)$$

where $h(\cdot, u)$ is an analytic function of $(\cdot)$ and $\bar{u}$, since

$$S(u) - \sum_{j=1}^{k}\sum_{i=1}^{m_j}\frac{c_{ij}}{(u - a_j)^i}$$

is an analytic function of $u \in D$ by (5.3.4). From (5.4.3), we have

$$k(u) = ((S(u) - \Lambda^*)^{-2}1, (\bar{u} - \Lambda^*)^{-1}1)^{-1}.$$

At the neighborhood of a_k, $k(u) = S(u)(1 + (u - a_k)f_0(u))$, where $f_0(u)$ is an analytic function. For $x, y \in K$, $y = \sum y_{l,k}\bar{\partial}^l E(\cdot, a_k)$,

$$(v(u)x, y) = k(u)((S(u) - \Lambda^*)^{-1}1, (\bar{u} - \Lambda^*)^{-1}y)((S(u) - \Lambda^*)^{-1}x, (\bar{u} - \Lambda^*)^{-1}1).$$

$$((S(u) - \Lambda^*)^{-1}x, (\bar{u} - \Lambda^*)^{-1}1)$$
$$= (x, E(\cdot, u))(1 + (u - a_k)f(u, x)),$$

$$((S(u) - \Lambda^*)^{-1}1, (\bar{u} - \Lambda^*)^{-1}\bar{\partial}^l E(\cdot, a_k))$$
$$= S(u)^{-1}l!(u - a_k)^{-(l+1)}(1 + (u - a_k)f_l(u)),$$

and

$$((S(u) - \Lambda^*)^{-1}1, (\bar{u} - \Lambda^*)^{-1}y)$$
$$= S(u)^{-1}\sum \bar{y}_{l,k}l!(u - a_k)^{-l+1}(1 + (u - a_k)f_l(u)),$$

where $f_l(u)$ and $f(u, x)$ are analytic functions of u at a neighborhood of a_k. Therefore $v(u)$ may have a pole at a_k of order less than or equal to m_k.

Next, if $S(z_k) = \bar{a}_k \in \sigma_p(\Lambda^*)$ for some $z_k \in D$. For simplicity of notation, we only calculate the case that all $a_j \in \sigma_p(\Lambda)$ are simple eigenvalues. Then

$$\mu(z)E(\cdot, a_j) = k(z) \sum_{l,m} \frac{\bar{\alpha}_l \alpha_m E(\cdot, z_l) E(z_m, z_l)}{(S(z) - \bar{z}_l)(S(z) - \bar{a}_j)(z - z_m)},$$

where $1 = \sum \bar{\alpha}_l E(\cdot, z_l)$ and

$$k(z) = \left(\sum_{l,m} \frac{\bar{\alpha}_l \alpha_m E(z_m, z_l)}{(S(z) - \bar{z}_l)^2 (z - z_m)} \right)^{-1}.$$

Therefore

$$\mu(z)E(\cdot, a_i) = E(\cdot, a_i)(S(z) - \bar{a}_j)(S(z) - \bar{a}_i)^{-1}(1 + O(|S(z) - \bar{a}_j|)),$$

as $z \to a_j$. Therefore z_k is a removable singularity for $\mu(\cdot)$. If $z_k \notin \sigma_p(\Lambda)$, then z_k is also a removable singularity for $v(\cdot)$.

Thus the only singularities of $v(\cdot)$ must be $\{z \in D \backslash \sigma(\Lambda) : P_{\bar{w}}(z, \overline{S(z)}) = 0\}$ and they must be poles, since $P_{\bar{w}}(z, \bar{w})$ is a polynomial of z and w, which proves the lemma. $\qquad\square$

Let the order of the pole a_k of $\mu(\cdot)$ be p_k. Let $\{b_k\}$ be the zeros $P_{\bar{w}}(z, \overline{S(z)})$ of order q_k. If $f(\cdot)$ is an analytic function on a neighborhood of $D \cup L$ with zeros at a_k of order $\geq p_k$ and zeros at b_k of order $\geq q_k$. Then

$$\mu(z) = \frac{1}{f(z)} \int_L \frac{(u - \Lambda)e(du)f(u)}{u - z}.$$

5.5 The inner product on an invariant subspace

Let us continue the study in §5.4. For sufficiently small positive number ϵ, let L_ϵ be the contour in $\sigma(H)$ consisting of the arcs of L which are not in the disks $\{z : |z - b_k| < \epsilon\}$ and the arcs in the circles $\{z : |z - b_k| = \epsilon\} \cap D$ such that for every analytic function f on some neighborhood of $\sigma(H)$,

$$\frac{1}{2\pi i} \int_{L_\epsilon} \frac{f(u)}{u - z} du = f(z), \quad z \in D_\epsilon,$$

where $D_\epsilon = D \backslash \cup \{z : |z - b_k| \leq \epsilon\}$ and that all the poles in D of $v(\cdot)$ are in D_ϵ. Let $\text{Res}(f, a)$ denote the residue of the analytic function f at its isolated singularity a. Let

$$\{u_k\} \stackrel{\text{def}}{=} \{u \in D \backslash \sigma(\Lambda) : P_{\bar{w}}(u, \overline{S(u)}) = 0\}.$$

Let P_K be the projection from $\mathcal{H}$ to K.

Lemma 5.5.1. *Let $f(\cdot)$ be an analytic function on a neighborhood of $\sigma(H)$. Then for $w \in \rho(H)$.*

$$\frac{1}{2\pi i} \int_{L_e} f(u)(\bar{w} - S(u))^{-1} v(u) du - \sum_j \text{Res}(f(u)(\bar{w} - S(u))^{-1} v(u), u_j)$$
$$= P_K(\bar{w} - H^*)^{-1} f(H)|_K. \quad (5.5.1)$$

Proof. By (3.3.7) and (3.3.17), we have

$$P_K(\bar{w} - H^*)^{-1}(z - H)^{-1}|_K = Q(z, w)^{-1} = (z - R(w)^*)^{-1}(\bar{w} - \Lambda^*)^{-1}. \quad (5.5.2)$$

For $w \in \rho_1$, by (5.4.8) and (5.4.9), we have

$$(z - R(w)^*)^{-1} = \sum_{\psi(\zeta)=w} (z - u(\zeta^*))^{-1} \lambda(\zeta)^*$$
$$= \sum_{\psi(\zeta)=w} (z - u(\zeta^*))^{-1} v(u(\zeta)^*)(\bar{w} - \Lambda^*) S'(u(\zeta^*))^{-1}. \quad (5.5.3)$$

Therefore

$$(z - R(w)^*)^{-1}(\bar{w} - \Lambda^*)^{-1} = \sum_{\psi(\zeta)=w} (z - u(\zeta^*))^{-1} v(u(\zeta)^*) S'(u(\zeta^*))^{-1}. \quad (5.5.4)$$

On the other hand, the poles of $(\bar{w} - S(u))^{-1}$ are those $u(\zeta^*)$, satisfying $\psi(\zeta) = w$ with

$$\text{Res}(\frac{1}{\bar{w} - S(u)}, u(\zeta^*)) = -S'(u(\zeta^*))^{-1}.$$

Besides, according to Lemma 5.4.4, a_j is a removable singularity of $(\bar{w} - S(\cdot))^{-1} v(\cdot)$. Therefore the left hand side equals to the right hand side of (5.5.1), which proves (5.5.1). $\qquad \square$

Let $F_H \overset{\text{def}}{=} \{u \in D \cap \rho(\Lambda) : P_{\bar{w}}(u, \overline{S(u)}) = 0 \text{ and } \overline{S(u)} \in D\}$. For every pair of analytic functions f and h on D, define

$$B_H(f, h) = \sum_{u \in F_H} \text{Res}(f(\cdot)\overline{h(\overline{S(\cdot)})} v(\cdot), u). \quad (5.5.5)$$

Corollary 5.5.2. *Let f and h be analytic functions on some neighborhood of $\sigma(H)$ satisfying the condition that*

$$\int_{\partial\sigma(H)} |f(u)h(u) P_{\bar{w}}(u, u)^{-1} du| < +\infty, \quad (5.5.6)$$

then

$$\int_{\partial\sigma(H)} f(u)\overline{h(u)}e(du) = P_K h(H)^* f(H)|_K + B_H(f, h) \tag{5.5.7}$$

(where $e(\cdot)$ is defined in (5.4.7)) and

$$B_H(f, h) = \overline{B_H(h, f)}. \tag{5.5.8}$$

Proof. The identity (5.5.7) may be obtained from (5.5.1) by the Cauchy formula for the analytic function $h(\cdot)$, and then letting $\epsilon \to 0$. Notice that F_H is possible only a part of $\{b_j\}$, since at the point b_j, if $\overline{S(b_j)} \notin D$ then

$$\mathrm{Res}\Big(\frac{1}{2\pi i} \int \frac{h(w)dw}{w - \overline{S(\cdot)}}, b_j\Big) = 0.$$

The identity (5.5.8) comes from the fact that the left hand side of (5.5.7) is Hermitian by Lemma 5.4.2, and the other term in the right hand side of (5.5.7) is also Hermitian, by direct calculation. $\qquad\square$

Let $q(\cdot) = q_H(\cdot)$ be a polynomial with minimal degree and leading co-efficient 1 satisfying

$$\int_{\partial\sigma(H)} |q(u)|^2 |P_{\tilde{w}}(u, \overline{S(u)})^{-1}| |du| < +\infty.$$

Let

$$q_H(u) = \Pi_{j=1}^t (u - v_j)^{l_j}, \tag{5.5.9}$$

where $\{v_j : j = 1, 2, \ldots, t\}$ is the set of different zeros of $q_H(\cdot)$. Let

$$G_H = \bigvee \{\bar{\partial}^l E(\cdot, v_j) : l = 0, 1, \ldots, l_j - 1, \ j = 1, 2, \ldots, t\}.$$

This is a subspace of $\mathcal{H}_g$. Then by (5.3.30),

$$(\bar{\partial}^l E(\cdot, v_i), \bar{\partial}^m E(\cdot, v_i))_D = \partial^m \bar{\partial}^l E(v_j, v_i), \tag{5.5.10}$$

where $\partial^m \bar{\partial}^l E(z, w) = \frac{\partial^m}{\partial z^m} \frac{\partial^l}{\partial \bar{w}^m} E(z, w)$. Define a measure

$$\Theta(dz) = \frac{1}{2\pi i} |Q_D(z)|^2 P_{\tilde{w}}(z, z)^{-1} dz$$

on $\partial\sigma(H)$. In the next theorem, we will prove that

$$\Theta(dz) = \frac{1}{2\pi} |Q_D(z)|^2 |P_{\tilde{w}}(z, z)^{-1}| |dz|.$$

Let $H^2(D, \Theta)$ be the Hilbert completion of all rational functions f with possible poles in $\rho(H)$ with respect to the following inner product

$$(f, g)_{H^2(D,\Theta)} \overset{\text{def}}{=} \int_{\partial\sigma} f(z)\overline{g(z)}\Theta(dz). \tag{5.5.11}$$

It is easy to see that

$$(v(z)1, 1) = k(z), \quad z \in \sigma(H),$$

by $((z - \Lambda)^{-1}(S(z) - \Lambda^*)^{-1}1, 1) = 1$ for $z \in \sigma(H)$, from (5.4.4). Therefore

$$\Theta(dz) = (e(dz)1, 1),$$

where $e(dz)$ is define in (5.4.7). Define

$$< f, h >_D \overset{\text{def}}{=} (B_H(f, h)1, 1)_D = \sum_{u \in F_H} \text{Res}(f(\cdot), \overline{h(\overline{S(\cdot)})}k(\cdot), u). \quad (5.5.12)$$

Now let us give a more concrete form for $< f, h >_D$.

Lemma 5.5.3. *If $u \in F_H$ then $\overline{S(u)} \in F_H$. The set $F_H = F_H^{(1)} \cup F_H^{(2)}$ where*

$$F_H^{(1)} = \{u_1, \dots, u_{p_1} : u_j \in F_H \text{ and } \overline{S(u_j)} = u_j\}$$

and

$$F_H^{(2)} = \{w_j : j = 1, 2, \dots, 2p_2, \ w_j \in F_H, \overline{S(w_{2j-1})} = w_{2j}\}.$$

(Some of these two set may be empty.) The hermitian form

$$< f, h >_D = B_H^{(1)}(f, h) + B_H^{(2)}(f, h),$$

where

$$B_H^{(1)}(f, h) = \sum_{j=1}^{p_1} \sum_{l=0}^{s_j} f^{(l)}(u_j) \overline{h^{(s_j-l)}}(u_j) \gamma_{l,j}^{(1)}, \quad (5.5.13)$$

where $\gamma_{l,j}^{(1)} = \gamma_{s_j-l,j}^{(1)}$ are real numbers, and

$$B_H^{(2)}(f, h) = \sum_{j=1}^{p_2} \sum_{l=0}^{t_j} (f^{(l)}(w_{2j-1}) \overline{h^{(t_j-l)}(w_{2j})} \gamma_{l,j}^{(2)}$$

$$+ \overline{h^{(l)}(w_{2j-1})} f^{(t_j-l)}(w_{2j}) \overline{\gamma_{l,j}^{(2)}}). \quad (5.5.14)$$

Proof. From (5.5.8), it is obvious that $< f, h >_D$ is Hermitian, i.e.

$$< f, h >_D = \overline{< h, f >_D}. \quad (5.5.15)$$

First, let us prove that $u \in F_H$ implies $\overline{S(u)} \in F_H$. Suppose on contrary, there is a $u_0 \in F_H$ satisfying $\overline{S(u_0)} \notin F_H$. Suppose the order of the pole u_0 of the function $v(\cdot)$ is k and the coefficient of $((\cdot) - u_0)^{-k}$ in the singular part of $v(\cdot)$ be γ. Suppose J is the maximum of all the orders of the poles of $v(\cdot)$ in F_H. Choose the function f satisfying the condition that

$$f^{(l)}(u_0) = 0, \quad l = 0, 1, 2, \dots, k - 2,$$

and $f^{(k-1)}(u_0) \neq 0$. Choose another analytic function h satisfying the condition that

$$h^{(l)}(u) = 0, \quad l = 0, 1, \ldots, J - 1, \quad u \in F_H,$$

and $h(\overline{S(u_0)}) \neq 0$. Then

$$< f, h >_D = f^{(k-1)}(u_0)\overline{h(\overline{S(u_0)})}\gamma/(k-1)! \neq 0. \tag{5.5.16}$$

But $< h, f >_D = 0$ which contradicts (5.5.15). Thus $u \in F_H$ implies $\overline{S(u)} \in F_H$.

Suppose that $u_0 \in F_H$ and $u_1 = \overline{S(u_0)} \in F_H$ but $u_1 \neq u_0$. We have to prove that

$$\overline{S(u_1)} = u_0. \tag{5.5.17}$$

Suppose on the contrary that (5.5.17) is false, i.e. $u_2 \neq u_0$ where $u_2 = \overline{S(u_1)} \in F_H$. We choose the function f as above. But we choose a different h satisfying

$$h^{(l)}(u) = 0, \quad l = 0, \ldots, J - 1, \quad u \in F_H \setminus \{u_1\},$$

$h^{(l)}(u_1) = 0, l = 1, \ldots, J - 1$ and $h(u_1) \neq 0$. Then we still have (5.5.16). But $< h, f >_D = 0$, if we require that $f^{(k'-1)}(u_2) = 0$, where k' is the order of the pole u_1 of the function $v(\cdot)$, which contradicts (5.5.15).

The rest of the lemma can be proves by (5.5.15) as well. We omit the details. $\qquad\square$

Theorem 5.5.4. *Let H be a pure hyponormal operator associated with a quadrature domain D. Then H is unitarily equivalent to an operator, denoted still by H, on a Hilbert space*

$$\mathcal{H}_g = G_H \oplus H^2(D, \Theta), \tag{5.5.18}$$

where the inner product in G_H is defined by (5.5.12), and

$$(f, h)_D = (f, h)_{H^2(D,\Theta)} - < f, h >_D, \quad f, h \in H^2(D, \Theta). \tag{5.5.19}$$

The operator H is a multiplication operator

$$(Hf)(u) = uf(u), \quad u \in D, \ f \in \mathcal{H}_g, \tag{5.5.20}$$

and

$$(H^*f)(u) = \frac{1}{2\pi i} \int_{\partial D} \frac{\bar{\zeta}f(\zeta)d\zeta}{\zeta - u}, \quad f \in \mathcal{A}(\sigma(H)).$$

Proof. Let $R(\sigma(H))$ be the space of rational functions with poles in $\rho(H)$ and $\mathfrak{X} = \bigvee\{q(\cdot)f(\cdot) : f \in R(\sigma(H))\}$. For $f, h \in \mathfrak{X}$, the condition (5.5.6) is satisfied. From

$$(P_K h(H)^* f(H)|_K 1, 1)_D = (f(H)1, h(H)1)_D = (f(\cdot), h(\cdot))_D$$

and (5.5.7), it follows (5.5.19).

Let

$$p_H(z) = \prod_{j=1}^{p_1} (z - u_j)^{s_j+1} \prod_{j=1}^{p_2} [(z - w_{2j-1})(z - w_{2j})]. \qquad (5.5.21)$$

Then for every $f \in \mathcal{A}(\sigma(H))$,

$$\|q_H p_H f\|_D^2 = \int |q_H(u) p_H(u)|^2 |f(u)|^2 \Theta(du).$$

Thus, $|q_H(u) p_H(u)|^2 \Theta(du)$ is a positive measure and hence $\Theta(du) = |\Theta(du)|$.

It is easy to see that

$$(q_H(\cdot) r(\cdot), \bar{\partial}^l E(\cdot, v_i))_D = \left(\frac{d}{du}\right)^l (q_H(u) r(u))|_{u=v_j} = 0$$

for $f \in R(\sigma(H))$, (i.e. $q_H(\cdot) r(\cdot) \in \mathfrak{X}$) and $0 \le l \le l_j - 1$, where v_j and l_j are the numbers in (5.5.9). Thus G_H is orthogonal to $\mathfrak{X}$ and hence orthogonal to $H^2(D, \Theta)$ too.

Let $\mathcal{P}_H$ be the set of all the polynomials with degree less than or equals to the degree of $q_H(\cdot)$. Then it is easy to see that $\mathcal{P}_H + \mathfrak{X}$ is dense in $\mathcal{H}_g$. Therefore

$$\text{codimension of } H^2(D, \Theta) = \text{dimension of } \mathcal{P}_H$$

$$= \text{degree of } q_H(\cdot) = \text{dimension of } G_H,$$

which proves (5.5.18). □

Corollary 5.5.5. *Under the condition of Theorem 5.5.4, the restriction of the operator H on the invariant subspace, the closure of $q_H p_H \mathcal{H}$ (where q_H, p_H are polynomials in (5.5.9) and (5.5.21) respectively) with*

$$\text{codimension of } H^2(D, \Theta) \le \text{degree of } q_H + \text{degree of } p_H$$

is subnormal.

Proof. By (5.5.19), the inner product on $q_H p_H \mathcal{H}$ is

$$(f, h)_D = \int_{\partial \sigma(H)} f(z) \overline{g(z)} \Theta(dz),$$

since $< f, h >= 0$ for $f, h \in q_H p_H \mathcal{H}$. Therefore the multiplication operator H in (5.5.20) is subnormal. □

5.6 Simply connected quadrature domains

The following lemma is well-known.

Lemma 5.6.1. *Let D be a simply connected quadrature domain with Schwartz function $S(\cdot)$. Let $r(\cdot)$ be any analytic and univalent function on the unit disk $\mathbb{D}$ such that $r(\mathbb{D}) = D$. Then $r(\cdot)$ is a rational function and*

$$S(r(z)) = \overline{r(\frac{1}{\bar{z}})}, \quad for \ z \in \mathbb{D}. \tag{5.6.1}$$

Proof. By a theorem of conformal mapping, $r(\cdot)$ is continuous on $|z| \le 1$. Define

$$r^*(z) = \overline{r(\frac{1}{\bar{z}})}, \quad |z| \ge 1. \tag{5.6.2}$$

Then $r^*(z)$ is analytic function on $|z| > 1$ and continuous on $|z| \ge 1$. We have

$$r^*(z) = S(r(z)), \quad for \ |z| = 1, \tag{5.6.3}$$

since $r^*(z) = \overline{r(z)}$ for $|z| = 1$ and $S(u) = \bar{u}$ for $u \in \partial D$. But $S(r(z))$ is a meromorphic function for $|z| < 1$ and continuous on $1 - \delta \le |z| \le 1$ for some $\delta > 0$. By (5.6.3) and the Painlevi's theorem $S(r(\cdot))$ and $r^*(\cdot)$ are the analytic continuation of each other. Therefore $r^*(\cdot)$ has the analytic continuation on $\mathbb{C} \cup \infty$ with only finite set of poles in $\mathbb{D}$. Thus $r^*(\cdot)$ is a meromorphic function on $\mathbb{C} \cup \infty$, i.e. $r^*(\cdot)$ is a rational function and so is $r(\cdot)$ by (5.6.2). The formulas (5.6.2) and (5.6.3) imply (5.6.1). $\square$

Theorem 5.6.2. *Suppose H is a pure hyponormal operator on $\mathcal{H}_g$ associated with a simply connected quadrature domain D. Then there is a unitary operator V from $\mathcal{H}_g$ onto $\hat{G}_H \oplus \hat{H}^2(\mathbb{T})$, where $\hat{H}^2(\mathbb{T})$ is the Hardy space $H^2(\mathbb{T})$ on the unit disk with a modified inner product*

$$(f, h) = \frac{1}{2\pi} \int_0^{2\pi} f(e^{i\theta})\overline{h(e^{i\theta})}d\theta - [f, h]_D \tag{5.6.4}$$

where $\hat{G}_H = VG_H$, and

$$[f, h]_D = <V^{-1}f, V^{-1}h>_D \tag{5.6.5}$$

where V is the operator

$$(Vf)(\cdot) = f(r(\cdot))\omega(\cdot), \tag{5.6.6}$$

r is the rational function in Lemma 5.6.1, and ω is a rational function with zeros in $\{z : |z| \le 1\}$ and poles in $\{z : |z| > 1\}$ satisfying

$$\overline{\omega(z)\omega(\frac{1}{\bar{z}})} = P_{\bar{w}}(r(z), r(\frac{1}{\bar{z}}))^{-1} r'(z) z Q_D(r(z)) \overline{Q_D(r(\frac{1}{\bar{z}}))}. \tag{5.6.7}$$

Besides,

$$(VHV^{-1}f)(z) = r(z)f(z).$$

Proof. Let

$$F(z) \stackrel{\text{def}}{=} P_{\bar{w}}(r(z), r(\frac{1}{\bar{z}}))^{-1} r'(z) z Q_D(r(z)) \overline{Q_D(r(\frac{1}{\bar{z}}))}.$$

Then $F(\cdot)$ is a rational function and

$$F(e^{i\theta}) = 2\pi\Theta(dr(e^{i\theta}))/d\theta \ge 0.$$

Therefore, there are $a_j, b_j \in \mathbb{C}$ and $\gamma > 0$ such that

$$F(e^{i\theta}) = \gamma\Pi_i|e^{i\theta} - a_i|^2\Pi_j|e^{i\theta} - b_j|^{-2}.$$

Without loss of generality, we may assume that $|a_i| \ge 1$ and $|b_j| > 1$, since it is obvious that if a_i or b_j is zero, then we may cancel the factor $|e^{i\theta} - a_i| = 1$ or $|e^{i\theta} - b_j| = 1$. If $|a_i| < 1$, then we may change the factor $|e^{i\theta} - a_i|$ to $|a_i||e^{i\theta} - 1/\bar{a}_i|$. We can do same thing for the factor $|e^{i\theta} - b_j|$. Define

$$\omega(z) = \gamma^{1/2}\Pi_i(z - a_i)\Pi_j(z - b_j)^{-1}.$$

Then $F(e^{i\theta}) = |\omega(e^{i\theta})|^2$. Thus

$$F(z) = \omega(z)\overline{\omega(\frac{1}{\bar{z}})}, \tag{5.6.8}$$

since both sides of (5.6.8) are meromorphic functions of z and are equal on $|z| = 1$. Let V be the operator from $\mathcal{H}_g$ onto $G_H \oplus \hat{H}^2(\mathbb{T})$ defined by (5.6.6). Then it follows this theorem from Theorem 5.5.4. □

Let us study the cases according to the dim K.

(1) Case of one dimensional K.

If dim $K = 1$, then $K = M$. From §2.1, H must be subnormal, and it is a linear combination of the unilateral shift with multiplicity one and identity, and $\sigma(H)$ is a circular disk.

(2) Case of two dimensional K.

As an exercise of the theory, we calculate this case. In this case, the polynomial

$$P_H(z, \bar{w}) = P_2(z)w^2 + P_1(z)w + P_0(z),$$

where $P_i(\cdot)$, $i = 0, 1$ are polynomials of degree no more than 2, and

$$P_2(z) = (z - \lambda_1)(z - \lambda_2)$$

where λ_1 and λ_2 are the eigenvalues of the 2×2 matrix Λ. Therefore the Schwartz function $S(z)$ of the interior domain D of $\sigma(H)$ is

$$S(z) = (-P_1(z) + \eta(z))/2P_2(z) \tag{5.6.9}$$

where $\eta(z) = (P_1(z)^2 - 4P_0(z)P_2(z))^{1/2}$. The only poles of $S(\cdot)$ must be $\{\lambda_1, \lambda_2\}$. They must be in D, since the dimension of K equals to the number of the poles of $S(\cdot)$ in D, counting the multiplicity.

Lemma 5.6.3. *Suppose H is a pure hyponormal operator associated with a quadrature domain D. If the dimension of $K_H = 2$ (or the order of D equals 2), then D must be simply connected.*

Proof. The domain D is a quadrature domain which is finitely connected and bounded by piecewise smooth closed curves. Suppose on contrary that D is n-connected with $n > 1$, and D_1 is a bounded component of $\mathbb{C} \setminus D$. Let γ be the common boundary curve of D_1 and D with counter-clockwise orientation. The Schwartz function $S(\cdot)$ as shown in (5.6.9) is an algebraic function and $S(z) = \bar{z}$ on γ, a part of ∂D. Thus $S(\cdot)$ is single valued on the boundary γ of D_1. If there are branch points of $S(\cdot)$ in $D_1 \cup \gamma$, the total order must be even. If there is no branch points in $D_1 \cup \gamma$, then $S(\cdot)$ is a single-valued analytic function on a neighborhood of $D_1 \cup \gamma$, since the poles of $S(\cdot)$ are in D. Then the image of γ by $w = S(z)$ must be counter-clockwise by the property of the conformal mapping. However, $w = \bar{z}$, $z \in \gamma$ is clockwise. This contradiction leads that there must be some branch points in $D_1 \cup \gamma$. Similarly we may discuss the same situation in case of D_1 being unbounded.

Therefore every component must contains at least two branch points of $S(\cdot)$. But from (5.6.9), there are at most four branch points of $S(\cdot)$. Thus $n \leq 2$.

Now let us exclude the case $n = 2$. Suppose $n = 2$. Then the polynomial $\eta(z)^2$ must have two simple zeros in the closure of one component of $\mathbb{C} \setminus D_1$. The rest must be simple poles in the unbounded component of $\mathbb{C} \setminus D$. Therefore there is no zero of $\eta(\cdot)$ in D, and

$$\int_{\partial D} |P_{\bar{w}}(z, \overline{S(z)})^{-1} dz| < \infty,$$

since $P_{\bar{w}}(z, \overline{S(z)}) = \eta(z)$ on ∂D, and the possible zeros of $\eta(z)^2$ are simple. Thus the polynomial $q_H(\cdot) \equiv 1$. On the other hand there is no zero of

$\eta(z) = P_w(z, \overline{S(z)})$ in D. Hence $p_H(\cdot) \equiv 1$. Thus $p_H(\cdot)q_H(\cdot)\mathcal{H} = \mathcal{H}$. By Corollary 5.5.5, H is subnormal and then $\dim K_H = 1$. It leads to a contradiction. Thus D must be simply-connected. $\qquad\qquad\square$

From Lemma 5.6.1, there is a rational function $\omega = r(\cdot)$ which is analytic and univalent on $\mathbb{D} = \{z : |z| < 1\}$ such that $r(\mathbb{D}) = D$. By (5.6.1), there are either two simple poles or one double pole of $r(\cdot)$ on the domain $\{z \in C : |z| > 1\}$. We may choose $r(\cdot)$ such that $r(0)$ is a pole of $S(\cdot)$ in D. Then it is not difficult to calculate that $r(\cdot)$ must be one of the following two types.

Case 1. $\lambda_1 = \lambda_2$, one double pole of $S(\cdot)$.

From (5.6.1), $r(\frac{1}{\bar{z}})$ only has a double pole at $z = 0$, since $r(0) = \lambda_1$. Therefore $r(\cdot)$ is a polynomial

$$r(z) = c_1(z + \lambda z^2) + \lambda_1,$$

where $c_1 \neq 0$, and $0 < |\lambda| \leq \frac{1}{2}$, since $r(\cdot)$ is univalent on $\mathbb{D}$. It is easy to see that $H = c_1 H_\lambda + \lambda_1$, where H_λ is a pure hypomormal operator associated with $D_\lambda = \{(w - \lambda_1)/c_1 : w \in D\}$. Let $S_\lambda(\cdot)$ be the Schwartz function of D_λ and $w = r_\lambda(z)$ be the rational mapping from $\mathbb{D}$ to D_λ. Then

$$r_\lambda(z) = z + \lambda z^2, \quad z \in \mathbb{D} \qquad\qquad (5.6.10)$$

and $S_\lambda(r_\lambda(z)) = \overline{S_\lambda(r_\lambda(\frac{1}{\bar{z}}))}$. Therefore

$$S_\lambda(u) = (\bar{\lambda} + (2|\lambda|^2 + 1)u + (u + \bar{\lambda})(1 + 4\lambda u)^{1/2})/2u^2,$$

where we choose the branch of $(1 + 4\lambda u)^{1/2} = 1 + 2\lambda u + O(|u|^2)$ as $u \to 0$. It is easy to see that the polynomial $P_{H_\lambda}(\cdot, \cdot)$ corresponding to the operator H_λ is

$$P(z, \bar{w}) = z^2 w^2 - (2|\lambda|^2 + 1)zw - \bar{\lambda}w - \lambda z + |\lambda|^4 - |\lambda|^2,$$

since the corresponding $P_1(z) = -(\bar{\lambda} + (2|\lambda|^2 + 1)z)$, $P_2(z) = z^2$,

$$\eta(z) = (z + \bar{\lambda})(1 + 4\lambda z)^{1/2}$$

and hence $P_0(z) = -\lambda z + |\lambda|^4 - |\lambda|^2$. Therefore

$$P_{\bar{w}}(u, \overline{S(u)}) = \eta(u) = (u + \bar{\lambda})(1 + 4\lambda u)^{1/2}.$$

The only zero of $P_{\bar{w}}(z, \overline{S(z)})$ in D_λ is $u_1 = -\bar{\lambda}$, since $-1/4\lambda$ is not in

$$D_\lambda = \{z + \lambda z^2 : z \in \mathbb{D}\},$$

by $|\lambda| \leq \frac{1}{2}$. Besides $\overline{S(-\bar{\lambda})} = -\lambda$. Therefore $F_H = \{u_1\}$. On the other hand

$$\frac{dr(z)}{P_{\bar{w}}(r(z), \overline{S(r(z))})} = \frac{dz}{z + \lambda z^2 + \bar{\lambda}},$$

and $e^{i\theta} + \lambda e^{2i\theta} + \bar{\lambda} = e^{i\theta}(1 + 2\text{Re}(e^{i\theta}\lambda)) \neq 0$, if $|\lambda| \leq \frac{1}{2}$. Now let us consider the case $|\lambda| < 1/2$. Then

$$\int_{\partial D_\lambda} |P_{\bar{w}}(u, S(u))^{-1} du| < +\infty$$

and $q_{H_\lambda}(\cdot) \equiv 1$. By Theorem 5.6.2, $\hat{G}_{H_\lambda}$ is $\{0\}$. The function ω defined by (5.6.7) is

$$\omega(z) = (1 + \lambda z)^2 |v_\lambda \lambda|^{-1/2} (z - 1/\bar{v}_\lambda)^{-1}$$

where $v_\lambda = (-1 + (1 - 4|\lambda|^2)^{1/2})/2\lambda$ which is the unique solution of $r(v_\lambda) = u_1$ in $\mathbb{D}$. Notice that in the calculation of $\omega(z)$, we have cancelled a factor z^2 from $\overline{Q_D(r(\frac{1}{\bar{z}}))}^{-1}$. Besides

$$[f, h]_D = \text{Res}(\hat{f}(\cdot)\overline{\hat{h}(\overline{S(\cdot)})}k(\cdot), u_1) = \hat{f}(u_1)\overline{\hat{h}(u_1)}u_1^2 S(u_1)^2 (1 + 4\lambda u_1)^{-1/2}$$

where $\hat{f}(u) = f(r^{-1}(u))\omega(r^{-1}(u))^{-1}$, $\hat{h}(u) = h(r^{-1}(u))\omega(r^{-1}(u))^{-1}$. Thus

$$[f, h]_D = k_\lambda f(v_\lambda)\overline{h(v_\lambda)},$$

where

$$k_\lambda = |1 + \lambda v_\lambda|^{-4} |v_\lambda| |\lambda|^3 (1 - 4|\lambda|^2)^{-1}.$$

Therefore

$$(f, h)_{H_\lambda} = \frac{1}{2\pi} \int_0^{2\pi} f(e^{i\theta})\overline{h(e^{i\theta})}d\theta - k_\lambda f(v_\lambda)\overline{h(v_\lambda)}. \qquad (5.6.11)$$

The operator H_λ is defined as

$$(H_\lambda f)(z) = (z + \lambda z^2)f(z), \quad z \in \mathbb{D}, \ f \in \mathcal{H}_\lambda. \qquad (5.6.12)$$

If $|\lambda| = \frac{1}{2}$, then the only zero of $P_{\bar{w}}(u, \overline{S(u)})$ in $\sigma(H_\lambda)$ is $u = -\bar{\lambda}$ which is on the $\partial\sigma(H_\lambda)$. The subspace G_{H_λ} in the Theorem 5.5.4 is

$$G_{H_\lambda} = \{cE(\cdot, -\bar{\lambda}) : c \in \mathbb{C}\},$$

where

$$E(u, -\bar{\lambda}) = \frac{4}{(1 + (1 + 4\lambda u)^{1/2})^2},$$

with norm $E(-\bar{\lambda}, -\bar{\lambda}) = 4$. Therefore $E(\cdot, -\bar{\lambda})/2$ is a unit vector in G_{H_λ}. The function

$$\omega(z) = \sqrt{2}(1 + \lambda z)^2 (z + 2\bar{\lambda})^{-1}.$$

Therefore the unit vector in $\hat{G}_{H_\lambda}$ is

$$\frac{\sqrt{2}}{z + 2\bar{\lambda}}. \qquad (5.6.13)$$

Thus for $|\lambda| = \frac{1}{2}$, H_λ can be represented as an operator defined in (5.6.12) on a Hilbert space

$$\mathcal{H}_\lambda = \hat{G}_{H_\lambda} \oplus H^2(\mathbb{T}) \qquad (5.6.14)$$

where $H^2(\mathbb{T})$ is the Hardy space on unit disk and $\hat{G}_{H_\lambda}$ is a one-dimensional Hilbert space with unit vector (5.6.13).

Case 2. There are two simple poles of $S(\cdot)$ in D.

By a translation, we may assume that 0 is a pole of $S(\cdot)$. Choose r such that $r(0) = 0$. Therefore

$$r(z) = c_1 \frac{z(z-b)}{z-a}.$$

Thus we may choose $c_1 \neq 0$, and c_2 such that $H = c_1 H_{a,b} + c_2$ where $H_{a,b}$ is associated with a quadrature domain of degree 2 of which $S(\cdot)$ has different poles and

$$r(z) = z(z-b)/(z-a). \qquad (5.6.15)$$

In this case, there should be a z_0 such that $z_0 \in D$ and $r(1/\bar{z}_0) = \overline{S(r(z_0))} = \infty$. This z_0 must be $1/\bar{a}$. Therefore $|a| > 1$. Besides, from $r(0) = r(b) = 0$, it follows that $|b| > 1$.

Let us prove that the mapping (5.6.15) is univalent on $\mathbb{D}$ iff

$$|a||\bar{a}b - 1| \geq |a||a - b| + |a|^2 - 1. \qquad (5.6.16)$$

It is obvious that the function (5.6.15) is univalent, iff

$$z_1 z_2 - a(z_1 + z_2) + ab \neq 0 \qquad (5.6.17)$$

for $|z_i| < 1$, $i = 1, 2$. For fixed $z_2 \in \mathbb{D}$, (5.6.17) is equivalent to

$$0 \notin \{(z_1 - a)z_2 + a(b - z_1) : z_1 \in \mathbb{D}\},$$

i.e. $|z_1 - a| \leq |a||b - z_1|$ for $|z_1| < 1$. It is equivalent to

$$|z_1 - a(b\bar{a} - 1)(|a|^2 - 1)^{-1}| \geq |a||a - b|(|a|^2 - 1)^{-1}, \quad \text{for } |z_1| < 1, \qquad (5.6.18)$$

which is equivalent to (5.6.16).

In this case, the two poles of $S(\cdot)$ are $0 = r(0)$ and $(1 - \bar{a}b)(\bar{a}(1 - |a|^2))^{-1} = r(1/\bar{a})$. Therefore the polynomial $P_2(\cdot)$ in (5.6.9) is

$$P_2(u) = u(u - (1 - \bar{a}b)(\bar{a}(1 - |a|^2))^{-1}).$$

By means of $S(r(z)) = \overline{r(1/\bar{z})}$, we may calculate the functions in (5.6.9), we have

$$P_1(u) = ((1 - a\bar{b})\bar{a}u^2 - u(|\bar{a}b - 1|^2 + 2\mathrm{Re}(\bar{a}(a - b))) + b(\bar{a}b - 1))|a|^{-2}(|a|^2 - 1)^{-1}$$

and

$$\eta(u) = (1 - a\bar{b} + \bar{a}(a\bar{b} - 1)u)(u^2 + (2b - 4a)u + b^2)^{1/2}|a|^{-2}(|a|^2 - 1)^{-1},$$

where $(u^2 + (2b - 4a)u + b^2)^{1/2} \to b$ as $u \to 0$. As before, we have

$$P_{\bar{w}}(u, \overline{S(u)}) = \eta(u).$$

But for $u = r(z)$,

$$(u^2 + (2b - 4a)u + b^2)^{1/2} = -r'(z)(z - a) \neq 0, \quad z \in \mathbb{D}.$$

The polynomial $u^2 + (2b - 4a)u + b^2$ can not have a double zero. So it only can have simple zero on ∂D. Therefore the only zero of $P_{\bar{w}}(u, \overline{S(u)})$ is a simple zero:

$$u_1 = (\bar{a}b - 1)(a\bar{b} - 1)^{-1}\bar{a}^{-1}.$$

Let us examine the necessary and sufficient condition for that $u_1 \in F_H$.

If $u_1 \in F_H$, by Lemma 5.5.3, then $\overline{S(u_1)} = u_1$, since F_H contains no more than one point. Let $v_1 \in \mathbb{D}$ satisfying $r(v_1) = u_1$. Let $w = (b + u_1)/2$ and

$$\rho = ((b + u_1)^2/4 - au_1)^{1/2}/w.$$

Then $v_1 = w(1 - \rho)$. From $r(1/\bar{v}_1) = \overline{S(u_1)} = u_1$ and $|1/v_1| > 1$, we have

$$1/\bar{v}_1 = w(1 + \rho).$$

Thus $|w|^2(1 + \bar{\rho})(1 - \rho) = 1$. Therefore ρ is real and $|w|^2(1 - \rho^2) = 1$. Then

$$(1 - \rho^2)w^2 = au_1. \tag{5.6.19}$$

However $w^2 = (|ab|^2 - 1)^2/(2\bar{a}(a\bar{b} - 1))^2$. Therefore (5.6.19) becomes $2|a|^2|a\bar{b} - 1|^2(|ab|^2 - 1)^{-2} = 1 - \rho^2$. Thus we have a necessary condition

$$|ab|^2 - 1 > 2|a||a\bar{b} - 1|. \tag{5.6.20}$$

By the same calculation, we know that (5.6.20) is also the sufficient condition for $u_1 \in F_H$. In this case, rename the v_1 by $v_{a,b}$

$$v_{a,b} \stackrel{\text{def}}{=} (|ab|^2 - 1 - ((|ab|^2 - 1)^2 - 4|a|^2|a\bar{b} - 1|^2)^{1/2})(2\bar{a}(a\bar{b} - 1))^{-1}. \tag{5.6.21}$$

Now, let us find $\omega(\cdot)$ by (5.6.7). Let $v_2 \stackrel{\text{def}}{=} a(\bar{a}b - 1)(|a|^2 - 1)^{-1}$. If $|v_2| \geq 1$, then

$$\omega(z) = |av_1^{-1}v_2|^{-1/2}(z - b)(z - v_2)(z - 1/\bar{v}_{a,b})^{-1}(z - a)^{-1}.$$

If $|v_2| < 1$, then

$$\omega(z) = |v_2v_1^{-1}|^{1/2}(z - b)(z - 1/\bar{v}_2)(z - 1/\bar{v}_{a,b})(z - a)^{-1}.$$

Let

$$k_{a;b} \overset{\text{def}}{=} v_{a,b}^{-1} \overline{\omega(v_{a,b})}^{-1} \lim_{z \to v_{a,b}} \overline{(z - 1/\bar{v}_{a,b})\omega(z)}. \qquad (5.6.22)$$

Then $H_{a,b}$ is a multiplication operator defined by

$$(H_{a,b}f)(z) = z(z - b)(z - a)^{-1}f(z), \quad f \in \mathcal{H}_{a,b}, \qquad (5.6.23)$$

where $\mathcal{H}_{a,b}$ is the Hardy space $H^2(\mathbb{T})$ with a modified scalar product

$$(f, h)_{\mathcal{H}_{a,b}} = \frac{1}{2\pi} \int_0^{2\pi} f(e^{i\theta})\overline{h(e^{i\theta})}d\theta - k_{a,b}f(v_{a,b})\overline{h(v_{a,b})}, \qquad (5.6.24)$$

where $v_{a,b}$ and $k_{a,b}$ are defined by (5.6.21) and (5.6.22) respectively.

The third case is $u_1 \in \partial D$, i.e. $|v_{a,b}| = 1$, i.e. $\rho = 0$, $|w| = 1$. Therefore

$$|ab|^2 - 1 = 2|a||a\bar{b} - 1|. \qquad (5.6.25)$$

In this case $H_{a,b}$ is still a multiplication operator on

$$\mathcal{H}_{a,b} = \hat{G}_{H_{a,b}} \oplus H^2(\mathbb{T}), \qquad (5.6.26)$$

where $H^2(\mathbb{T})$ is the Hardy space and $\hat{G}_{H_{a,b}}$ is one-dimensional with unit vector

$$E(u_1, u_1)^{-1/2}E(r(\cdot), u_1)\omega(\cdot).$$

Now, we have to exclude the case

$$|ab|^2 - 1 < 2|a||a\bar{b} - 1|. \qquad (5.6.27)$$

Suppose there are a and b satisfying $|a| > 1$, $|b| > 1$, and (5.6.25). Define

$$y_j = (|ab|^2 - 1 + (-1)^j(4|a|^2|a\bar{b} - 1|^2 - (|ab|^2 - 1)^2))(2\bar{a}(a\bar{b} - 1))^{-1},$$

for $j = 1, 2$. It is easy to see that $|y_j| = 1$, $y_1 \neq y_2$ and

$$|y_j - a(b\bar{a} - 1)(|a|^2 - 1)^{-1}| = |a||a - b|(|a|^2 - 1)^{-1}.$$

Then $z_1 = (y_1 + y_2)/2$ satisfies $|z_1| < 1$, but

$$|z_1 - a(b\bar{a} - 1)(|a|^2 - 1)^{-1}| < |a||a - b|(|a|^2 - 1)^{-1},$$

it contradicts to (5.6.18). Thus the case (5.6.27) cannot happen.

In conclusion, we have the following:

Lemma 5.6.4. *For every quadrature domain D of order 2, there exist c_1, $c_2 \in \mathbb{C}$, $c_1 \neq 0$, such that*

$$D = \{c_1 r(z) + c_0 : z \in \mathbb{D}\}$$

where $r(z)$ is either the function in (5.6.11) with $|\lambda| \leq \frac{1}{2}$ or the function in (5.6.15) with $a, b \in \mathbb{C}$ satisfying $|a| > 1$, $|b| > 1$ and (5.6.16).

Proposition 5.6.5. *Suppose H is a pure hyponormal operator associated with a quadrature domain satisfying* $\dim K_H = 2$. *Then there are* $c_1, c_0 \in \mathbb{C}$, $c_1 \neq 0$ *such that H is either unitary equivalent to* $c_1 H_\lambda + c_0$ *on a Hilbert space* $\hat{G}_{H_\lambda} \oplus H^2(\mathbb{T})$ *as in* (5.6.14) *with inner product in* (5.6.12) *where* H_λ *is defined in* (5.6.13), *or unitarily equivalent to* $c_1 H_{a,b} + c_0$ *on the Hilbert space* $\mathcal{H}_{a,b}$ *where* $H_{a,b}$ *is defined by* (5.6.23) *and* $\mathcal{H}_{a,b}$ *is the* $H^2(\mathbb{T})$ *with modified inner product* (5.6.24) *in the case of* (5.6.20), *or is the Hilbert space in* (5.6.26) *in the case of* (5.6.25).

Theorem 5.6.6. *Let H be a pure hyponormal Hilbert space $\mathcal{H}$ associated with a quadrature domain D. If* $\dim K_H = 2$, *then there is a point* $a \in \sigma(H)$ *such that the subspace*

$$\mathcal{H}_1 = \text{closure}(H - a)\mathcal{H}$$

is with codimension one and the restriction $H_1 \overset{def}{=} H|_{\mathcal{H}_1}$ *is subnormal. In this case* $H_1 = r(U_+)$ *where* $r(\cdot)$ *is the conformal mapping from unit disk to the interior of* $\sigma(H)$ *and* U_+ *is the unilateral shift with multiplicity one.*

Proof. This is only a corollary of the Proposition 5.6.5 and can be proved by checking the structures of $\mathcal{H}_\lambda$ and $\mathcal{H}_{a,b}$. ☐

5.7   Operator of finite type

Let T be a pure operator on a Hilbert space $\mathcal{H}$. Let M_T and $K_T \overset{def}{=} \mathcal{H}_0$ be the subspaces defined in §1.2. If $\dim K_T < +\infty$, then T is said to be *of finite type*. By Lemma 1.3.1, pure subnormal operator with finite rank self-commutator is of finite type. By Lemma 5.3.2, a pure hyponormal operator associated with a quadrature domain is of finite type.

Proposition 5.7.1. *Let T be a pure operator on $\mathcal{H}$ with rank* $[T^*, T] = m < +\infty$. *If there is an invariant subspace $\mathcal{H}_1$ of T with codimension* $n < +\infty$ *such that the restriction of T on $\mathcal{H}_1$ is subnormal. Then T is of finite type with*

$$\dim K_T \leq m + 2n. \tag{5.7.1}$$

If $m = n = 1$, *then* $\dim K_T \leq 2$.

Proof. Let $\mathcal{H}_2 = \mathcal{H} \ominus \mathcal{H}_1$. Then T can be written in a matrix form

$$T = \begin{pmatrix} S & B \\ 0 & A \end{pmatrix} \tag{5.7.2}$$

with respect to the decomposition $\mathcal{H} = \mathcal{H}_1 \oplus \mathcal{H}_2$, where S is subnormal, $\dim \mathcal{H}_2 = n$, $A \in L(\mathcal{H}_2)$ and B is an operator from $\mathcal{H}_2$ to $\mathcal{H}_1$. There exists an orthonormal set of eigenvectors w_j, $j = 1, 2, \ldots, m$ of the self adjoint operator $[T^*, T]$ corresponding to eigenvalues λ_j, $j = 1, 2, \ldots, m$ such that

$$[T^*, T]x = \sum_{j=1}^{m} (x, w_j) \lambda_j w_j, \quad x \in \mathcal{H}. \tag{5.7.3}$$

By the decomposition (5.7.2), it is easy to see that

$$[S^*, S]x = BB^*x + \sum_{j=1}^{m} (x, g_j) \lambda_j g_j, \quad x \in \mathcal{H}_1, \tag{5.7.4}$$

$$S^*Bz = BA^*z + \sum_{j=1}^{m} (z, \eta_j) \lambda_j g_j, \quad z \in \mathcal{H}_2, \tag{5.7.5}$$

and

$$BB^*z + [A^*, A]z = \sum_{j=1}^{m} (z, \eta_j) \lambda_j \eta_j, \quad z \in \mathcal{H}_2, \tag{5.7.6}$$

where $w_j = g_j + \eta_j$, $g_j \in \mathcal{H}_1$ and $\eta_j \in \mathcal{H}_2$. Besides,

$$T^*(x + z) = S^*x + B^*x + A^*z, \quad x \in \mathcal{H}_1, \; z \in \mathcal{H}_2. \tag{5.7.7}$$

From (5.7.3), $M_T = \bigvee \{w_j, j = 1, 2, \ldots, m\}$. From (5.7.4),

$$M_S = BB^*\mathcal{H}_1 + \bigvee \{g_j : j = 1, 2, \ldots, m\}. \tag{5.7.8}$$

Thus $\dim M_S \leq m + n$. By Lemma 1.3.1,

$$S^*M_S \subset M_S. \tag{5.7.9}$$

Define

$$L = M_T + BB^*\mathcal{H}_1 + \mathcal{H}_2. \tag{5.7.10}$$

Let us show that L is an invariant subspace of T^*. From (5.7.8) and (5.7.9), for every g_j, there are $b_j \in \mathcal{H}_1$ and $a_{kj} \in \mathbb{C}$, $k = 1, 2, \ldots, m$ such that

$$S^*g_j = \sum_{k} a_{kj} g_k + BB^*b_j, \quad j = 1, 2, \ldots, m.$$

Therefore, from (5.7.7), it follows

$$T^*w_j = \sum a_{kj} g_k + BB^*b_j + B^*g_j + A^*\eta_j \tag{5.7.11}$$

$$= \sum a_{kj} w_k + BB^*b_j + (B^*g_j + A^*\eta_j - \sum a_{kj}\eta_k) \in L.$$

For $x \in \mathcal{H}_1$, there are $c_j(x) \in \mathcal{H}_1$ and $b_{kj}(x) \in \mathbb{C}$ such that

$$
\begin{aligned}
T^* B B^* x &= \sum b_{kj}(x) g_k + B B^* c_j(x) + B^* B B^* x \\
&= \sum b_{kj}(x) w_k + B B^* c_j(x) - \sum b_{kj}(x) \eta_k \in L. \quad (5.7.12)
\end{aligned}
$$

Besides $T^* \mathcal{H}_2 = A^* \mathcal{H}_2 \subset \mathcal{H}_2$. Thus (5.7.11) and (5.7.12) imply that $T^* L \subset L$. Therefore $L \supset K_T$, since $M_T \subset L$ which proves (5.7.1).

Now suppose $m = n = 1$. Let η be a unit vector in $\mathcal{H}_2$, there exist $g \in \mathcal{H}_1$, $c \in \mathbb{C}$ such that

$$
[T^*, T](x + z\eta) = j((x, g) + z\bar{c})(g + c\eta), \quad \text{for } x \in \mathcal{H}_1 \text{ and } z \in \mathbb{C}, \quad (5.7.13)
$$

where $j = \pm 1$, since $\mathrm{rank}[T^*, T] = 1$. Let $T\eta = f + b\eta$, $f \in \mathcal{H}_1$, $b \in \mathbb{C}$. Then $B\eta = f$ and $A = b$. From (5.7.2) and (5.7.13), it is easy to see that

$$
[S^*, S]x = (x, f)f + (x, g)g, \quad x \in \mathcal{H}_1 \quad (5.7.14)
$$

$$
S^* f = \bar{b} f + \bar{c} g, \quad (5.7.15)
$$

$$
\|f\| = |c|, \quad (5.7.16)
$$

and

$$
T^*(x + z\eta) = S^* x + ((x, f) + \bar{b}z)\eta, \quad \text{for } x \in \mathcal{H}_1 \text{ and } z \in \mathbb{C}. \quad (5.7.17)
$$

From (5.7.14), it follows that

$$
M_S = \bigvee \{f, g\}. \quad (5.7.18)
$$

From (5.7.13), $M_T = (g + c\eta)\mathbb{C}$. From (5.7.17)

$$
T^*(g + z\eta) = S^* g + ((g, f) + \bar{b}z)\eta, \quad z \in \mathbb{C}. \quad (5.7.19)
$$

From (5.7.18) and $K_S = M_S$, there are $a_1, a_2 \in \mathbb{C}$ such that

$$
S^* g = a_1 f + a_2 g. \quad (5.7.20)
$$

If $a_1 = 0$, then $\bigvee \{g, \eta\}$ is an invariant subspace of T^*. But $M_T \subset \bigvee \{g, \eta\}$. Thus $\dim K_T \leq 2$.

If $a_1 \neq 0$, then from (5.7.17) and (5.7.20), we have

$$
T^*(g + c\eta) = a_1(f + z_1\eta) + a_2(g + c\eta),
$$

where $z_1 = ((g, f) + (\bar{b} - a_2)c)/a_1$. From (5.7.15), (5.7.16) and (5.7.17), we have

$$
T^*(f + z_1\eta) = \bar{b}(f + z_1\eta) + \bar{c}(g + c\eta).
$$

Therefore $\bigvee \{(f + z_1\eta), (g + c\eta)\}$ is an invariant subspace of T^*. Besides this subspace contains M_T as a subspace. Therefore $\dim K_T \leq 2$, which proves the proposition. $\qquad \square$

Lemma 5.7.2. *Let H be a pure hyponormal operator with rank one self commutator on a Hilbert space $\mathcal{H}$. Let $k \in [H^*, H]\mathcal{H}$, $k \neq 0$. Then the following two assertions are equivalent.*

a) There is a non-zero polynomial $P(\cdot)$ satisfying $P(H)^*[H^*, H] = 0$, and

b) there is a non-zero polynomial $P(\cdot)$ such that

$$\overline{P(z)}P(w)(((H - z)^{-1})^*k, ((H - w)^{-1})^*k) \tag{5.7.21}$$

is a polynomial of $\bar{z}$ and w for z and w in the unbounded component of $\rho(H)$.

Proof. Let us point out that $P(H)^*[H^*, H] = 0$ is equivalent to $P(H)^*k = 0$. Suppose that a) is satisfied, then

$$(P(z)(H - z)^{-1})^*k = ((P(z) - P(H))(H - z)^{-1})^*k + (H - z)^{-1*}P(H)^*k$$
$$= V(z, H)^*k,$$

where $V(z, H) = (P(z) - P(H))(z - H)^{-1}$ is a polynomial of z and H, which proves b).

Suppose b) is satisfied. Then for fixed w in the unbounded component of $\rho(H)$,

$$((P(H)(H - z)^{-1})^*k, (P(w)(H - w)^{-1})^*k) = \overline{P(z)}P(w)((H - z)^{-1*}k,$$
$$(H - w)^{-1*}k) + (V(z, H)^*k, P(w)^*(H - w)^{-1*}k) \tag{5.7.22}$$

is a polynomial of z for z in the unbounded component of $\sigma(H)$. But it is obvious that the left hand side of (5.7.22) approaches to zero as $z \to \infty$. Therefore (5.7.22) must equal to zero for any fixed w, satisfying $|w| > \|H\|$. Expanding (5.7.22) as Laurent series for $|z| > \|H\|$ and $|w| > \|H\|$, we have

$$(P(H)^*k, H^m H^{*n}k) = 0, \quad m, n = 0, 1, 2, \ldots \tag{5.7.23}$$

From Proposition 1.2.1, $\mathcal{H} = \mathrm{cl} \vee \{H^m H^{*n}k : m, n = 0, 1, 2, \ldots\}$, since H is pure. Thus (5.7.23) implies a), which proves the lemma. $\qquad \square$

Notice that a) is equivalent to that H is of finite type, since if $P(H)^*[H^*, H] = 0$ then $H^{*n}[H^*, H]k = -\sum_{j=0}^{n-1} p_j H^{*j}[H^*, H]k$, where $p(\lambda) = \lambda^n + \sum_{j=0}^{n-1} p_j \lambda_j$. Thus

$$K_H \subset \vee\{H^{*j}M_H : j = 0, \ldots, n - 1\}.$$

On the other hand, if $\dim K_H = n < +\infty$, then

$$H^{*n}k, H^{*n-1}k, \ldots, H^*k, k$$

must be linearly dependent. Thus there is a non-zero polynomial $P(\cdot)$ such that

$$P(H)^*k = 0,$$

which implies a).

Theorem 5.7.3. *Let H be a pure hyponormal operator on a Hilbert space $\mathcal{H}$ with rank one self-commutator and of finite type. Assume that the Pincus principal function $g(z) = 1$, for $z \in \sigma(H)$. Then H is associated with a quadrature domain.*

Proof. By Lemma 5.7.2, there is a polynomial $P(\cdot)$ such that (5.7.21) is a polynomial of z and w, for z and w in the unbounded component of $\rho(H)$. By Lemma 5.2 of [M. Putinar [6]] (in that Lemma, it needs Lemma 5.7.2), the interior of $\sigma(H)$ is a finitely connected domain D with a system of piecewise smooth Jordan curves as its boundary. Let L be the union of the boundary curves of D with positive orientation.

Let k be the vector in Lemma 5.7.2. Then $P(H)^*k = 0$. But in Lemma 5.7.2, we may assume that all the zeros of $P(\cdot)$ lie in the set $\sigma(H)$, since for any point $\lambda \in \rho(H)$, $(H - \lambda)^*$ is invertible. If $P(\cdot)$ contains a factor $(\cdot) - \lambda$, $\lambda \in \rho(H)$, then

$$(P(H)(H - \lambda)^{-1})^*k = (H^* - \bar{\lambda})^{-1}P(H)^*k = 0.$$

Therefore we may cancel the factor $(\cdot) - \lambda$ from $P(\cdot)$. Besides, we may assume that k is the vector in §5.2. Then by (5.2.6) and (5.2.19), we have

$$((H - z)^{-1}k, k) = (((\cdot) - z)^{-1}, 1)_g = \frac{1}{\pi} \iint_{\sigma(H)} \frac{dA(\zeta)}{\zeta - z}, \quad \text{for } z \in \rho(H).$$

Thus the function

$$P(z)\frac{1}{\pi} \iint_{\sigma(H)} \frac{dA(\zeta)}{\zeta - z} = P(z)(k, (H^* - \bar{z})^{-1}k)$$

$$= (k, ((P(z) - P(H))(H - z)^{-1})^*k)$$

$$= (V(z, H)k, k), \quad z \in \rho(H) \qquad (5.7.24)$$

since $P(H)^*(H^* - \bar{z})^{-1}k = 0$, where $V(z, H) = (P(z) - P(H))(H - z)^{-1}$ is a polynomial of z and H. Therefore the right hand side of (5.7.24) is a polynomial of z. Thus there is a rational function $r(\cdot)$ on $\mathbb{C}$ with poles on $\sigma(H)$ such that

$$\frac{1}{\pi} \iint_{\sigma(H)} \frac{dA(\zeta)}{\zeta - z} = r(z), \quad z \in \rho(H). \qquad (5.7.25)$$

From (5.7.25), it is easy to see that $\lim_{z \to \lambda} r(z)$ is finite for $\lambda \in L$. Therefore the poles of $r(\cdot)$ are in D. By Green's formula

$$r(z) = \frac{1}{2\pi i} \int_L \frac{\bar{\zeta} d\zeta}{\zeta - z}, \quad z \in \rho(H).$$

Define

$$u(z) = \frac{1}{2\pi i} \int_L \frac{\bar{\zeta} d\zeta}{\zeta - z}, \quad z \in D$$

then $u(\cdot)$ is an analytic function on D. By the Plemelj's formula, at the smooth point λ of L

$$\lim_{z \to \lambda} (u(z) - r(z)) = \bar{\lambda}.$$

Define $S(z) = u(z) - r(z)$. Then $S(\cdot)$ is a meromorphic function with boundary value $S(\lambda) = \bar{\lambda}$, $\lambda \in L$. Thus D is a quadrature domain with Schwarz function $S(\cdot)$. $\qquad \square$

Proposition 5.7.4. *Let T be a pure operator on the Hilbert space $\mathcal{H}$ of finite type with $\dim K_T = 1$. Then T is a linear combination of a unilateral shift with multiplicity one and the identity.*

Proof. In this case K_T must be M_T. Thus Proposition 5.7.4 is a corollary of Theorem 2.1.1. $\qquad \square$

5.8 The reproducing kernels for operators of finite type

Let A be an operator on a Hilbert space $\mathcal{H}$. Let

$$K = K_A \overset{\text{def}}{=} \text{cl} \bigvee_{n=0}^{\infty} A^{*n}[A^*, A]\mathcal{H},$$

$$C = C_A \overset{\text{def}}{=} [A^*, A]|_{K_A},$$

$$\Lambda = \Lambda_A \overset{\text{def}}{=} (A^*|_{K_A})^*,$$

and

$$R(\lambda) \overset{\text{def}}{=} C(\lambda - \Lambda)^{-1} + \Lambda^*, \quad \lambda \in \rho(\Lambda).$$

Let P_K be the projection from $\mathcal{H}$ to K. For $v \in \mathcal{H}$, let $\rho_v(A)$ be the set of all $\lambda \in \mathbb{C}$ satisfying the condition that there exists a unique vector f such that

$$(\lambda - A)f = v.$$

Let us denote this f by $\mathcal{F}(\lambda, v)$, i.e

$$(\lambda - A)\mathcal{F}(\lambda, v) = v. \tag{5.8.1}$$

Lemma 5.8.1. *Let $v \in K$, and $\lambda \in \rho_v(A) \cap \rho_{R(\lambda)v} \cap \rho(\Lambda)$. Then*

$$A^* \mathcal{F}(\lambda, v) = \mathcal{F}(\lambda, R(\lambda)v). \tag{5.8.2}$$

Proof. It is easy to see that

$$(\lambda - A)A^* \mathcal{F}(\lambda, v) = [A^*, A]\mathcal{F}(\lambda, v) + A^*(\lambda - A)\mathcal{F}(\lambda, v)$$
$$= CP_K \mathcal{F}(\lambda, v) + \Lambda^* v.$$

On the other hand, for all $\beta \in K$, we have

$$(P_K \mathcal{F}(\lambda, v), (\bar{\lambda} - \Lambda^*)\beta) = (\mathcal{F}(\lambda, v), (\bar{\lambda} - A^*)\beta) = ((\lambda - A)\mathcal{F}(\lambda, v), \beta)$$
$$= (v, \beta).$$

Thus

$$(\lambda - \Lambda)P_K \mathcal{F}(\lambda, v) = v \tag{5.8.3}$$

or $P_K \mathcal{F}(\lambda, v) = (\lambda - \Lambda)^{-1}$, which proves that

$$(\lambda - A)A^* \mathcal{F}(\lambda, v) = R(\lambda)v,$$

and so the lemma. $\square$

Lemma 5.8.2. *Let $\lambda \in \rho(\Lambda) \cap \rho_v(A)$ and $v \in K$. Then $\mathcal{F}(\lambda, v)$ is an eigenvector of A^* corresponding to the eigenvalue z, iff v is an eigenvector of $R(\lambda)$ corresponding to $\bar{z}$, or $(\lambda - \Lambda)^{-1}v$ is an eigenvector of $R(z)^*$ corresponding to the eigenvalue λ, in the case of $z \in \rho(\Lambda)$.*

Proof. From (5.8.2), $(\bar{z} - A^*)\mathcal{F}(\lambda, v) = 0$, iff $(\bar{z} - R(\lambda))v = 0$. On the other hand, for $\lambda, z \in \rho(\Lambda)$,

$$\bar{z} - R(\lambda) = (\bar{z} - \Lambda^*)(\lambda - R(z)^*)(\lambda - \Lambda)^{-1}.$$

Thus $(\bar{z} - R(\lambda))v = 0$, iff $(\lambda - R(z)^*)(\lambda - \Lambda)^{-1}v = 0$, which proves the lemma. $\square$

For the operator A of finite type, as in (5.3.8), let

$$Q(z, w) = (\bar{w} - \Lambda^*)(z - \Lambda) - C,$$

and

$$P(z, w) \overset{\text{def}}{=} \det Q(z, w).$$

Then $P(z, w)$ is a polynomial of z and $\bar{w}$ with leading term $z^n \bar{w}^n$, where $n = \dim K$. Evidently, there is a decomposition of

$$P(z, w) = P(z)\overline{P(w)}\Pi_{k=1}^{l} P_k(z, w)^{t_k}, \tag{5.8.4}$$

where $P(z)$ is a polynomial of z with leading term z^p, and $P_k(z, w) = \overline{P_k(w, z)}$ is an irreducible polynomial of z and $\bar{w}$ with leading term $z^{m_k} \bar{w}^{m_k}$ such that the equation $P_k(z, w) = 0$ has no solution of type $w \equiv$ constant. Besides, $P_k(z, w) \neq P_{k'}(z, w)$ for $k \neq k'$. Let Σ_k be the Riemann surface of the algebraic function $f_k(\cdot)$ defined by

$$P_k(z, \overline{f_k(z)}) = 0.$$

Let $\Sigma = \cup_{k=1}^l \Sigma_k$, and

$$n_A = \sum_{k=1}^l m_k l_k. \tag{5.8.5}$$

Then there are an analytic function $\psi(\cdot)$ and a meromorphic function $S(\cdot)$ on Σ satisfying (i) the restriction of the mapping $z = \psi(\zeta)$ from Σ_k to $\mathbb{C}$ is "m_k to 1", except at the branch points of the algebraic function $f_k(\cdot)$, and (ii)

$$P(\psi(\zeta), \overline{S(\zeta)}) = 0, \text{ for } \zeta \in \Sigma_k. \tag{5.8.6}$$

If $\lambda \in \rho(\Lambda)$, then $P(\lambda, z) = \det(\lambda - \Lambda) \det(\bar{z} - R(\lambda))$. Thus $\bar{z} \in \sigma_p(R(\Lambda))$, iff

$$P(z, \lambda) = \overline{P(\lambda, z)} = 0.$$

Then for $\lambda \in \rho(\Lambda) \setminus \{\lambda : P(\lambda) = 0\}$, $\bar{z} \in \sigma_p(R(\Lambda))$ iff there is a $\zeta \in \Sigma$ such that

$$z = \psi(\zeta) \text{ and } \lambda = \overline{S(\zeta)}. \tag{5.8.7}$$

Let $\mathcal{D}_0$ be the subset of all pints $\zeta \in \Sigma$ satisfying the condition that $\overline{S(\zeta)} \in \rho_v(A) \cap \rho(\Lambda) \cap \{\lambda : P(\lambda) \neq 0\}$ for some eigenvector v of $R(\overline{S(\zeta)})$ corresponding to the eigenvalue $\overline{\psi(\zeta)}$. For $\zeta \in \mathcal{D}_0$, let

$$K_\zeta \stackrel{\text{def}}{=} \{(\overline{S(\zeta)} - \Lambda)^{-1} v \in K :$$
$$R(\overline{S(\zeta)})v = \overline{\psi(\zeta)}v, \overline{S(\zeta)} \in \rho_v(A) \cap \rho(\Lambda) \cap \{\lambda : P(\lambda) \neq 0\}\}.$$

For $\zeta \in \mathcal{D}_0$, $\alpha \in K_\zeta$, let

$$E_{\zeta,\alpha} \stackrel{\text{def}}{=} \mathcal{F}(\overline{S(\zeta)}, (\overline{S(\zeta)} - \Lambda)\alpha).$$

Lemma 5.8.3. *For* $\zeta \in \mathcal{D}_0$, $\alpha \in K_\zeta$,

$$A^* E_{\zeta,\alpha} = \overline{\psi(\zeta)} E_{\zeta,\alpha}.$$

Proof. Let $\lambda = \overline{S(\zeta)}$, $z = \psi(\zeta)$ and $v = (\lambda - \Lambda)\alpha$. Then $R(\lambda)v = \bar{z}v$. By Lemma 5.8.1, we have $A^* \mathcal{F}(\lambda, v) = \mathcal{F}(\lambda, R(\lambda)v) = \bar{z}\mathcal{F}(\lambda, v)$. $\qquad\square$

Notice that if A is subnormal with finite rank of self-commutator, then

$$E_{\zeta,\alpha} = \nu(\zeta)^*\gamma,$$

for some $\gamma \in M$, as in §2.3.

Lemma 5.8.4. *For $\xi, \zeta \in \mathcal{D}_0$, $\alpha \in K_\zeta$ and $\beta \in K_\xi$,*

$$(E_{\xi,\beta}, E_{\zeta,\alpha}) = (\frac{(S(\zeta) - \Lambda^*)(\overline{S(\xi)} - \Lambda) - C}{(\psi(\zeta) - \overline{S(\xi)})(\overline{\psi(\xi)} - S(\zeta))}\beta, \alpha). \qquad (5.8.8)$$

Proof. For the simplicity of notation, let $\lambda = \overline{S(\zeta)}$, $\mu = \overline{S(\xi)}$, $z = \psi(\zeta)$, $w = \psi(\xi)$, $h = (\lambda - \Lambda)\alpha$ and $k = (\mu - \Lambda)\beta$. Then

$$E_{\zeta,\alpha} = \mathcal{F}(\lambda, h) \quad \text{and} \quad E_{\xi,\beta} = \mathcal{F}(\mu, k).$$

Besides, $R(\lambda)h = \bar{z}h$ and $R(\mu)k = \bar{w}k$. Therefore

$$\begin{aligned}
(\mu - z)(\bar{\lambda} - \bar{w})(E_{\xi,\beta}, E_{\zeta,\alpha}) &= (\mu - z)((\bar{\lambda} - A^*)E_{\xi,\beta}, \mathcal{F}(\lambda, h)) \\
&= (\mu - z)(E_{\xi,\beta}, h) \\
&= (E_{\xi,\beta}, (\bar{\mu} - R(\lambda))h). \qquad (5.8.9)
\end{aligned}$$

On the other hand, from (5.8.3), we have $P_K E_{\xi,\beta} = P_K \mathcal{F}(\mu, k) = (\mu - \Lambda)^{-1}k$. Therefore

$$\begin{aligned}
(E_{\xi,\beta}, (\bar{\mu} - R(\lambda))h) &= ((\mu - \Lambda)^{-1}k, (\bar{\mu} - \Lambda^*)(I - (\bar{\mu} - \Lambda^*)^{-1}C(\lambda - \Lambda)^{-1})h) \\
&= ((I - (\bar{\lambda} - \Lambda^*)^{-1}C(\mu - \Lambda)^{-1})k, h). \qquad (5.8.10)
\end{aligned}$$

From (5.8.9) and (5.8.10), it follows (5.8.8). $\qquad \square$

Let $\mathcal{H}_1 \stackrel{\text{def}}{=} \text{cl} \bigvee \{E_{\zeta,\alpha} : \alpha \in K_\zeta, \zeta \in \mathcal{D}_0\}$.

Lemma 5.8.5. *If A is a pure operator of finite type and K is the span of all eigenvectors of Λ^*, then $K \subset \mathcal{H}_1$.*

Proof. Without loss of generality, we may assume $0 \in \sigma(\Lambda)$. We only have to show that if $e \neq 0$ and e is an eigenvector of Λ^* then $e \in \mathcal{H}_1$.

Let δ be a sufficiently small positive number. Choose a number θ, $0 < \theta \leq 2\pi$ such that on

$$\Omega = \{z \in \mathbb{C} : |z| < \delta, \theta \leq \arg z < \theta + 2\pi\}$$

there exists an analytic function $w(z)$ satisfying

$$\det(R(\frac{1}{z}) - w(z)) = 0, \quad \text{and} \quad \lim_{z \to 0} w(z) = 0,$$

where $R(\frac{1}{z}) = C(1 - z\Lambda)^{-1}z + \Lambda^*$, and eigenvectors $g(z)$, $z \in \Omega$ satisfying

$$R(\frac{1}{z})g(z) = w(z)g(z),$$

and $\lim_{z\to 0} g(z) = e$. Let

$$f(z) = (1 - zA)^{-1}g(z).$$

By Lemma 5.8.1, $f(z)$ is an eigenvector of A^*. There must be a $\zeta \in \mathcal{D}_0$, which depends on z, such that $\overline{S(\zeta)} = z^{-1}$ and $\overline{\psi(\zeta)} = w(z)$. Then $f(z) = E_{\zeta,\alpha} \in \mathcal{H}_1$, where $\alpha = z^{-1}(\overline{S(\zeta)} - \Lambda)^{-1}g(z)$. Therefore

$$e = \lim_{z\to 0} g(z) = \lim_{z\to 0} f(z) \in \mathcal{H}_1,$$

which proves the lemma. $\qquad\qquad\square$

Proposition 5.8.6. *Suppose A is a pure operator of finite type on a Hilbert space $\mathcal{H}$, and K is the span of all eigenvectors of Λ^*. Then*

$$\mathrm{cl} \bigvee \{E_{\zeta,\alpha} : \zeta \in \mathcal{D}_0, \alpha \in K_\zeta\} = \mathcal{H}.$$

Proof. It is obvious that $\mathcal{H}_1$ is invariant with respect to A^*, by Lemma 5.8.3. On the other hand,

$$AE_{\zeta,\alpha} = (\Lambda - \overline{S(\zeta)})\alpha + \overline{S(\zeta)}E_{\zeta,\alpha}$$

which is in $\mathcal{H}_1$ by Lemma 5.8.5. Hence $\mathcal{H}_1$ reduces A and containing K. Therefore $\mathcal{H}_1 = \mathcal{H}$ since A is pure, which proves the proposition. $\qquad\square$

For $\zeta \in \mathcal{D}_0$, let

$$\nu(\zeta) \overset{\text{def}}{=} \left((\overline{S(\zeta)} - \Lambda)^{-1}\frac{1}{2\pi i}\int_{\gamma_{\zeta,\epsilon}} (\lambda - R(\psi(\zeta)))^{-1}d\lambda\right)^*,$$

where $\gamma_{\zeta,\epsilon}$ is a circle centered at $\overline{\Psi(\zeta)}$ with radius $\epsilon > 0$ which is small enough such that there is no eigenvalue of $R(\overline{S(\zeta)})$ in $\{z \in \mathbb{C} : 0 < |z - \overline{\Psi(\zeta)}| \le \epsilon\}$. Besides, the orientation of $\gamma_{\zeta,\epsilon}$ is positive. Then $\nu(\zeta)K = K_\zeta$ for almost all $\zeta \in \mathcal{D}_0$. Thus $\mathcal{H} = \mathrm{cl} \bigvee \{E_{\zeta,\nu(\zeta)^*\alpha} : \alpha \in K, \zeta \in \mathcal{D}_0\}$.

As in (2.3.4), for $\zeta, \xi \in \mathcal{D}_0$, let

$$E(\zeta,\xi) \overset{\text{def}}{=} \nu(\zeta)Q(\overline{S(\xi)}, \overline{S(\zeta)})\nu(\xi)^*(\psi(\zeta) - \overline{S(\xi)})^{-1}(\overline{\psi(\xi)} - S(\zeta))^{-1}.$$

From (5.8.8), we have

$$(E_{\xi,\nu(\xi)^*\beta}, E_{\zeta,\nu(\zeta)^*\alpha}) = (E(\zeta,\xi)\beta, \alpha).$$

Define a positive semi-definite kernel on $\mathcal{D}_0 \times K$ as following

$$K((\xi,\beta),(\zeta,\alpha)) \overset{\text{def}}{=} (E(\zeta,\xi)\beta, \alpha).$$

Let $K((\xi, \alpha), \cdot)$ be the function on $\mathcal{D}_0 \times K$ defined as

$$(\zeta, \beta) \mapsto K((\xi, \alpha), (\zeta, \beta)).$$

Let $\hat{\mathcal{H}}_1 \stackrel{\text{def}}{=} \bigvee \{K((\xi, \alpha), \cdot) : (\xi, \alpha) \in \mathcal{D}_0 \times K\}$. Define an inner product on $\hat{\mathcal{H}}_1$ as

$$(K((\xi, \alpha), \cdot), K((\zeta, \beta), \cdot)) \stackrel{\text{def}}{=} K((\xi, \alpha), (\zeta, \beta)).$$

Define a linear operator V from $\mathcal{H}_1$ to $\hat{\mathcal{H}}_1$

$$V E_{\xi, \nu(\xi)^* \alpha} = K((\xi, \alpha), \cdot). \tag{5.8.11}$$

Then by the same way of proving Theorem 2.3.3 and previous lemmas and proposition 5.8.6, we have the following.

Theorem 5.8.7. *Let A be a pure operator of finite type on $\mathcal{H}$. Let V be the extension of (5.8.11) from $\mathcal{H}_1$ to $\mathcal{H}$, and K is the span of all eigenvectors of Λ^*. Then V is a unitary operator from $\mathcal{H}$ to the reproducing kernel Hilbert space $\hat{\mathcal{H}}$, the completion of $\hat{\mathcal{H}}_1$, satisfying*

$$(Vx)(\zeta, \beta) = (x, E_{\zeta, \nu(\zeta)^* \beta}), \quad z \in \mathcal{H},$$

$$(VAV^{-1}f)(\zeta, \beta) = \psi(\zeta)f(\zeta, \beta), \quad f \in \hat{\mathcal{H}},$$

and

$$VA^*V^{-1}K((\xi, \alpha), \cdot) = \overline{\psi(\xi)}K((\xi, \alpha), \cdot), \quad (\xi, \alpha) \in \mathcal{D}_0 \times K.$$

5.9   Trace formulas for some operators of finite type

Let A be an operator of finite type as defined in §5.8. In order to establish a trace formula of commutators, we have to establish a formula similar to (1.6.8).

Lemma 5.9.1. *Let A be an operator on a Hilbert space satisfying $\dim K_A < +\infty$. Let C and Λ be the operators defined in §5.8. Then*

$$\operatorname{tr}(\bar{\nu} - A^*)^{-1}[(\bar{\xi} - A^*)^{-1}, (\lambda - A)^{-1}]$$
$$= (\bar{\nu} - \bar{\xi})\frac{d}{d\bar{\xi}}q_\lambda(\xi) - (\bar{\nu} - \bar{\xi})^{-2}(q_\lambda(\nu) - q_\lambda(\xi)), \tag{5.9.1}$$

where $q_\lambda(\xi) = \operatorname{tr}(\lambda - R(\xi)^*)^{-1}$, and $\lambda, \xi, \nu \in \rho(A)$.

Proof. From

$$[(\bar{\xi} - A^*)^{-1}, (\lambda - A)^{-1}] = (\xi - A^*)^{-1}(\lambda - A)^{-1}|_K C P_K (\lambda - A)^{-1}(\bar{\xi} - A^*)^{-1} \tag{5.9.2}$$

for $\lambda, \xi \in \rho(A)$, we have for $\lambda, \xi, \nu \in \rho(A)$,

$$\operatorname{tr}(\bar{\nu} - A^*)^{-1}[(\bar{\xi} - A^*)^{-1}, (\lambda - A)^{-1}]$$
$$= \operatorname{tr} C P_K (\lambda - A)^{-1}(\bar{\nu} - A^*)^{-1}(\bar{\xi} - A^*)^{-2}(\lambda - A)^{-1}|_K. \tag{5.9.3}$$

From (5.9.2) again, we have

$$(\lambda - A)^{-1}(\bar{\xi} - A^*)^{-1}(\lambda - A)^{-1} - (\lambda - A)^{-2}(\bar{\mu} - A^*)^{-1}$$
$$= (\lambda - A)^{-1}[(\bar{\xi} - A^*)^{-1}, (\lambda - A)^{-1}]$$
$$= (\lambda - A)^{-1}(\bar{\xi} - A^*)^{-1}(\lambda - A)^{-1}|_K C P_K (\lambda - A)^{-1}(\bar{\xi} - A^*)^{-1}.$$

As in (3.3.11), we have $P_K (\lambda - A)^{-1} = (\lambda - \Lambda)^{-1} P_K$ for $\lambda \in \rho(A)$. Therefore

$$P_K (\lambda - A)^{-1}(\bar{\xi} - A^*)^{-1}(\lambda - A)^{-1}|_K (I - C(\lambda - \Lambda)^{-1}(\bar{\xi} - \Lambda^*)^{-1})$$
$$= (\lambda - \Lambda)^{-2}(\bar{\xi} - \Lambda^*)^{-1}.$$

Thus

$$P_K (\lambda - A)^{-1}(\bar{\mu} - A^*)^{-1}(\lambda - A)^{-1}|_K = (\lambda - \Lambda)^{-1} Q(\lambda, \xi)^{-1}, \tag{5.9.4}$$

since $Q(\lambda, \xi)$ is invertible for $\lambda, \xi \in \rho(A)$ (cf. Lemma 3.3.1). On the other hand it is easy to see that

$$\operatorname{tr}(\lambda - \Lambda)^{-1} Q(\lambda, \xi)^{-1} C = -\operatorname{tr}(\lambda - \Lambda)^{-1}$$
$$+ \operatorname{tr}(\lambda - \Lambda)^{-1} Q(\lambda, \xi)^{-1}(\bar{\xi} - \Lambda^*)(\lambda - \Lambda)$$
$$= -\operatorname{tr}(\lambda - \Lambda)^{-1} + \operatorname{tr}(Q(\lambda, \xi)^{-1}(\bar{\xi} - \Lambda^*))$$
$$= -\operatorname{tr}(\lambda - \Lambda)^{-1} + \operatorname{tr}(\lambda - R(\xi)^*)^{-1}. \tag{5.9.5}$$

From (5.9.3), (5.9.4) and (5.9.5), it follows (5.9.1). $\qquad\square$

Next, let us consider the Riemann surfaces $\sum$, functions $\psi(\cdot)$, $S(\cdot)$ etc. as in §5.8. Suppose $\mathcal{D}_{j,k}$ is a domain in $\sum_j$ satisfying the conditions that (i) $\{\psi(\zeta), \zeta \in \mathcal{D}_{j,k}\}$ is bounded and (ii)

$$S(\zeta) = \overline{\psi(\zeta)}, \quad \zeta \in \partial \mathcal{D}_{j,k}.$$

Then this $\mathcal{D}_{j,k}$ is said to be a quadrature domain associated with the operator A. In this case, the restriction of $S(\cdot)$ on $\mathcal{D}_{j,k}$ is the Schwartz function of $\mathcal{D}_{j,k}$. Let $\mathcal{D}$ be the union of all quadrature domains associated with the operator A. The union $\mathcal{D}$ is said to be *complete*, if in every component of $\rho(A)$, there is a point w in $\mathbb{C} \setminus \psi(\mathcal{D} \cup \partial \mathcal{D})$ such that there exists n_A zeros(counting multiplicity) in $\mathcal{D}$ of the function $S(\cdot) - \bar{w}$, where n_A is the number in (5.8.5). A zero ξ of $S(\cdot) - \bar{w}$ is said to be of multiplicity $l \geq 1$, if

$$S(\zeta) - \bar{w} = \sum_{j=l}^{\infty} a_j(\psi(\zeta) - \psi(\xi))^j, \quad a_l \neq 0,$$

for ζ in a neighborhood of ξ. Let $\mathcal{F}$ be the family of all pure operator A of finite type satisfies the condition that $\mathcal{D}$ is complete. All hyponormal operators associated with quadrature domain belongs to $\mathcal{F}$. There are a lot of examples of non-hyponormal operators in $\mathcal{F}$.

For $\xi \in \rho(\Lambda)$, define $m(\lambda, \xi)$ as the multiplicity of the eigenvalue λ of $R(\xi)^*$. For $\zeta \in \mathcal{D}$, let $n(\zeta)$ be the multiplicity of the zero ζ of the function $S(\cdot) - S(\zeta)$.

Lemma 5.9.2. *Let $A \in \mathcal{F}$. Let $\rho(A) = \cup_j G_j$, where G_j is a component of $\rho(A)$. Then for every G_j there exists an open set $O_j, G_j \supset O_j \neq \emptyset$, such that for every $\xi \in O_j$ and*

$$\lambda \in \sigma_p(R(\xi)^*) \cap \{z : P(z) \neq 0\},$$

there is some $\zeta \in \mathcal{D}$ satisfying $\psi(\zeta) = \lambda$, $S(\zeta) = \bar{\xi}$ and

$$m(\lambda, \xi) = \sum \{n(\zeta) : \zeta \in \mathcal{D}, \psi(\zeta) = \lambda, S(\zeta) = \bar{\xi}\}.$$

Proof. Since $\mathcal{D}$ is complete, there exists a point w in G_j such that there are n_A zeros in $\mathcal{D}$ of the function $S(\cdot) - \bar{w}$. By Rouché theorem, it is easy to prove that there is a neighborhood O_j of w, such that the function $S(\cdot) - \bar{\xi}$, for $\xi \in O_j$ still has n_A zeros in $\mathcal{D}$.

Suppose $\xi \in O_j$, then $\xi \in \rho(\Lambda)$. We have

$$\det(\lambda - R(\xi)^*) = P(\lambda, \xi) \det(\bar{\xi} - \Lambda^*)^{-1}. \tag{5.9.6}$$

Therefore $\lambda \in \sigma_p(R(\xi)^*)$, iff $P(\lambda, \xi) = 0$ and $m(\lambda, \xi)$ is just the multiplicity of the zero λ of the polynomial $P(\cdot, \xi) = 0$ for the fixed ξ.

For $\xi \in \rho(\Lambda)$, let $\{\lambda_1(\xi), \ldots, \lambda_{i_\xi}(\xi)\}$ be the set of different zeros of the polynomial

$$\prod_{k=1}^{l} P_k(z, \xi)^{t_k}$$

in (5.8.4). We may choose O_j such that for $\xi \in O_j$, $P(\lambda_\nu(\xi)) \neq 0$, $\nu = 1, 2, \ldots, i_\xi$. Then for $\xi \in O_j$,

$$\sum_{\nu=1}^{i_\xi} m(\lambda_\nu, \xi) \leq n_A.$$

Let $m_\nu(\xi)$ be the sum of the multiplicities $n(\zeta)$ of the zero $\zeta \in \mathcal{D}$ of the function $S(\zeta) - \bar{\xi}$ satisfying $\psi(\zeta) = \lambda_\nu(\xi)$. Then

$$m_\nu(\xi) \leq m(\lambda_\nu, \xi).$$

But $\sum m_\nu(\xi) = n_A$. Therefore $m_\nu(\xi) = m(\lambda_\nu(\xi), \xi)$, which proves the lemma. $\qquad\square$

For a pure operator A of finite type, let $L = L_A$ be the union of $\psi(\partial \mathcal{D}_j)$, where $\mathcal{D}_j$ is a component of $\mathcal{D}$. Let $\partial \mathcal{D}_j$ be positively oriented, and $\psi(\partial \mathcal{D}_j)$ keeps the orientation $\partial \mathcal{D}_j$ through the projection ψ. It is easy to see that

$$L \subset \{z \in \mathbb{C} : P(z, z) = 0\}.$$

From Lemma 3.3.1, $Q(z, z)$ is invertible for $z \in \rho(A)$. Therefore

$$L \subset \sigma(A). \tag{5.9.7}$$

Let $R(\sigma)$ be the algebra of functions generated by functions $f(\cdot)$ analytic on a neighborhood of the compact σ and $\overline{f(\cdot)}$. For $f = \sum_j \bar{h}_j f_j \in R(\sigma(A))$, where f_j, h_j are analytic on a neighborhood of $\sigma(A)$, define $f(A) = \sum_j h_j(A)^* f_j(A)$.

Theorem 5.9.3. *Let $A \in \mathcal{F}$. Then for $f, h \in R(\sigma(A))$*

$$\mathrm{tr}[f(A), h(A)] = \frac{1}{2\pi i} \int_L m(z) f(z) dh(z), \tag{5.9.8}$$

where $m(z)$ is the cardinal number of the set $\{\zeta \in \partial \mathcal{D} : \psi(\zeta) = z\}$.

Proof. For $\lambda \in (\cup_j G_j) \cap (\mathbb{C} \setminus \psi(\mathcal{D} \cup \partial \mathcal{D}))$ and $\xi \in \cup_j O_j$, where O's are the neighborhoods in the Lemma 5.9.2, let us calculate

$$F(\lambda, \xi) \overset{\text{def}}{=} \frac{1}{2\pi i} \int_{\partial \mathcal{D}} \frac{dS(\zeta)}{(\lambda - \psi(\zeta))(S(\zeta) - \bar{\xi})}.$$

Let $\{\zeta_k(\xi)\}$ be the set of the different zeros of the function $S(\cdot) - \bar{\xi}$ in $\mathcal{D}$. Then it is easy to calculate that

$$F(\lambda, \xi) = \sum_k (\lambda - \psi(\zeta_k(\xi)))^{-1} n(\zeta_k(\xi)) + F(\lambda),$$

where $F(\lambda)$ contains some contributions from the poles of $S(\cdot)$, but it is independent of ξ. Let $\sigma_p(R(\xi)^*) \cap \{z : P(z) \neq 0\} = \{\xi_1, \dots, \xi_\nu\}$. Then by Lemma 5.9.2,

$$\sum_k (\lambda - \psi(\zeta_k(\xi)))^{-1} n(\zeta_k(\xi)) = \sum_p (\lambda - \xi_p)^{-1} m(\xi_p, \xi).$$

Therefore

$$F(\lambda, \xi) = \text{tr}(\lambda - R(\xi)^*)^{-1} - \sum_{j=1}^{k} \frac{m_j}{\lambda - \lambda_j} + F(\lambda),$$

where $\{\lambda_j\}$ is the set of all zeros of the polynomial $P(\cdot)$ in (5.8.4) with multiplicities $\{m_j\}$, respectively.

Thus by Lemma 5.9.1, for $\lambda \in (\cup_j G_j) \cap (\mathbb{C} \setminus \psi(\mathcal{D} \cup \partial\mathcal{D}))$ and $\xi, \nu \in O_j$, we have

$$\text{tr}\,(\bar{\nu} - A)^{*-1}[(\bar{\xi} - A^*)^{-1}, (\lambda - A)^{-1}]$$
$$= (\bar{\nu} - \bar{\xi})^{-1}\frac{d}{d\bar{\xi}}F(\lambda, \xi) - (\bar{\nu} - \bar{\xi})^{-2}(F(\lambda, \nu) - F(\lambda, \xi))$$
$$= -\frac{1}{2\pi i}\int_{\partial\mathcal{D}} \frac{dS(\zeta)}{(\lambda - \psi(\zeta))(S(\zeta) - \bar{\xi})^2(S(\zeta) - \bar{\nu})}$$
$$= \frac{1}{2\pi i}\int_L \frac{m(z)d\bar{z}}{(\lambda - z)(\bar{z} - \bar{\xi})^2(\bar{\nu} - \bar{z})}.$$

By Cauchy's formula, for any analytic functions $f(\cdot), h(\cdot), u(\cdot)$ in $\mathcal{A}(\sigma(A))$, we have

$$\text{tr}(u(A)^*[h(A)^*, f(A)]) = \frac{1}{2\pi i}\int_L m(z)f(z)\overline{u(z)}\overline{dh(z)}. \tag{5.9.9}$$

From (5.9.9) we may prove (5.9.8) in a standard way as shown in §1.6. □

Corollary 5.9.4. *Let $A \in \mathcal{F}$, then the Pincus principal function*

$$g(z) = \frac{1}{2\pi i}\int_L \frac{m(w)dw}{w - z}, \quad z \in \text{interior of } \sigma(A). \tag{5.9.10}$$

Proof. The Pincus principal function $g(\cdot)$ is determined by the trace formula

$$i\,\text{tr}[(\bar{\xi} - A^*)^{-1}, (\lambda - A)^{-1}] = \frac{1}{\pi}\iint \frac{g(z)dA(z)}{(\bar{\xi} - \bar{z})^2(\lambda - z)^2}, \tag{5.9.11}$$

for $\lambda, \xi \in \rho(A)$, where $dA(z)$ is the Lebesgue planar measure. From (5.9.8) it is easy to see that the function defined by (5.9.11) satisfies (5.9.10), which proves the corollary. □

Appendix A

The Singular Integral Model, Mosaic and Trace Formula of Hyponormal Operator

Let H be an operator on a Hilbert space $\mathcal{H}$. Let $[H^*, H] \stackrel{\text{def}}{=} H^*H - HH^*$. If

$$([H^*, H]x, x) \geq 0, \quad \text{for } x \in \mathcal{H}. \tag{A.1}$$

Then H is said to be *hyponormal*. The formula (A.1) is equivalent to $\|H^*x\| \leq \|Hx\|$, for $x \in \mathcal{H}$.

First, let us introduce the singular integral model of hyponormal operators.

Let $L^2(\mathbb{R})$ be the Hilbert space of all measurable, square integrable functions in $\mathbb{R}$. Let P_+ be the singular integral operator for $f \in \mathrm{L}^2(\mathbb{R})$,

$$(P_+ f)(x) = l.i.m._{\epsilon \to 0^+} \frac{1}{2\pi} \int \frac{f(t)dt}{t - (x + i\epsilon)}, \tag{A.2}$$

i.e.

$$\lim_{\epsilon \to 0^+} \left\| P_+ f - \frac{1}{2\pi} \int \frac{f(t)dt}{t - ((\cdot) + i\epsilon)} \right\| = 0.$$

Actually, P_+ is a projection operator (Cauchy projection). Let $\mathcal{D}$ be an auxiliary separable Hilbert space. Let $L^2(\mathbb{R}, \mathcal{D})$ be the space of all $\mathcal{D}$-valued, measurable, and square integrable functions, with inner product

$$(f, h) \stackrel{\text{def}}{=} \int_{\mathbb{R}} (f(t), h(t))_{\mathcal{D}} dt,$$

then $L^2(\mathbb{R}, \mathcal{D})$ is a Hilbert space. By (A.2), we may define the same projection operator P_+ on $L^2(\mathbb{R}, \mathcal{D})$. Let σ be a compact set in $\mathbb{R}$. Let $R(\cdot)$ be a $L(\mathcal{D})$-valued, bounded, measurable function on σ satisfying the condition that $R(t)$ is a projection operator on $\mathcal{D}$, for every $t \in \sigma$. Let $L^2(\sigma, R(\cdot))$ be the subspace of all functions $f(\cdot)$ in $L^2(\mathbb{R}, \mathcal{D})$ satisfying the condition that $f(t) = 0$ for $t \in \mathbb{R} \setminus \sigma$ and $f(t) \in R(t)\mathcal{D}$ for $t \in \sigma$. In the following, we need the concept of a pure operator which is defined in §1.2.

Theorem A.1. Let H be a pure hyponormal operator on $\mathcal{H}$. Let $H = u + iv$, where u and v are self-adjoint. Let $\sigma = \sigma(u)$. Then there are Hilbert space $\mathcal{D}$, projection-valued measurable function $R(t), t \in \mathcal{D}$ and bounded measurable, self-adjoint operator-valued functions $\alpha(\cdot)$ and $\beta(\cdot)$ on σ satisfying conditions

$$\alpha(\cdot)R(\cdot) = R(\cdot)\alpha(\cdot) = \alpha(\cdot), \beta(\cdot)R(\cdot) = R(\cdot)\beta(\cdot) = \beta(\cdot),$$

and $\alpha(\cdot) \geq 0$, there is a unitary operator V from $\mathcal{H}$ onto $L^2(\sigma, R(\cdot))$ such that

$$(VuV^{-1}f)(x) = xf(x), \ x \in \sigma, \tag{A.3}$$

and

$$(VvV^{-1}f)(x) = \beta(x)f(x) + \alpha(x)(P_+\alpha(\cdot)f(\cdot))(x), \ x \in \sigma. \tag{A.4}$$

The operators defined by (A.3) and (A.4) are said to be the singular integral model of the hyponormal operator $H = u + iv$.

Theorem A.2. Under condition of Theorem A.1, There is a unique $L(\mathcal{D})$-valued measurable function $B(z), z \in \mathbb{C}$ satisfying

$$0 \leq B(z) \leq I, \ z \in \sigma(H), \ B(z) = 0, \ z \in \rho(H)$$

and

$$I + \alpha(x)(\beta(x) - l)^{-1}\alpha(x) = \exp \int_{\mathbb{R}} \frac{B(x + iy)dy}{y - l}, \ I_m l \neq 0.$$

The function $B(\cdot)$ defined in Theorem A.2 is said to be the *mosaic* of the hyponormal operator H. It is a complete unitary invariant for H. Let us define the function

$$g(z) = g_H(z) \overset{\text{def}}{=} \operatorname{tr} B(z).$$

Then $g(z) = 0$ for $z \in \rho(H)$ and if $B(z)$ is not in trace class then $g(z) = +\infty$. This function is said to be the *Pincus principal function* of H. (In some literatures, $-g(\cdot)$ is defined as the Pincus principal function.)

For a compact set $\sigma \subset \mathbb{C}$, let $\mathcal{R}(\sigma)$ be the function algebra defined in §1.6. For a function of $f \in \mathcal{R}(\sigma(H))$, let $f(H)$ be the operator defined in §1.6.

Theorem A.3. Let H be a pure, hyponormal operator on a Hilbert space $\mathcal{H}$. Suppose $[H^*, H] \in L^1(\mathcal{H})$, i.e. $\operatorname{tr}([H^*, H]) < \infty$. Then for any pair of functions $f, h \in \mathcal{R}(\sigma(H))$,

$$\operatorname{tr}[f(H), h(H)] = \frac{1}{2\pi i} \int \int_{\sigma(H)} g(z)df(z) \wedge dh(z).$$

Appendix B

Quadrature Domain

Let us introduce the concept of quadrature domain in $\mathbb{C}$, originally studied by D. Aharonow and H. S. Shapiro. Let D be a bounded domain in $\mathbb{C}$. If there is a finite set $\{a_j, j = 1, \ldots, n\} \subset D$ and a set of constants $\{c_{jk} : k = 0, \ldots, m_j, j = 1, \ldots, n\}$ such that for every function f which is analytic on a neighborhood of $D \cup \partial D$, the following quadrature identity

$$\frac{1}{\pi} \int \int_D f(z) dA(z) = \sum_{j=1}^{n} \sum_{k=0}^{m_j} c_{jk} f^{(k)}(a_j) \tag{B.1}$$

holds good, then D is said to be a *quadrature domain*.

It is easy to see that if D is a quadrature domain, then the interior of $D \cup \partial D$ is a quadrature domain. Therefore, we always assume that the quadrature domain satisfies the following condition

$$D = \text{interior of } D \cup \partial D.$$

For example, if the quadrature domain D is with some isolated boundary points, then we always add those points into D.

Theorem B.1. A planar bounded domain D is a quadrature domain iff D satisfies the following conditions: (1) D is finitely connected, (ii) ∂D is the union of finite set of piecewise smooth Jordan curves, and (iii) there exists a Schwarz function $S(\cdot)$ for D, i.e. there exists a meromophic function on D which also can be extended as a continuous function on a neighborhood of ∂D satisfying

$$S(z) = \bar{z}, \ z \in \partial D. \tag{B.2}$$

By the Green's formula and (B.2), if $f(\cdot)$ is any analytic function on a neighborhood of $D \cup \partial D$, then

$$\frac{1}{\pi} \int \int_D f(z) dA(z) = \frac{1}{2\pi i} \int_{\partial D} f(z) \bar{z} dz$$

$$= \frac{1}{2\pi i} \int_{\partial D} f(z) S(z) dz.$$

From (B.1), it is easy to see that $\{a_j\}$ is the set of all poles of $S(\cdot)$ in D and the singular part of $S(\cdot)$ at a_j is

$$\sum_{k=0}^{m_j} \frac{c_{jk}}{k!(z - a_j)^{k+1}}.$$

Bibliography

Guidance to the bibliography

Chapter 1

Sec. 1.1: A standard book for the theory of subnormal operator is J. B. Conway [1]. A lot of basic theorems about subnormal operators can be found there. Theorem 1.1.1 and Theorem 1.1.2 are also there.

Sec. 1.2: Most part of this section comes from M. Putinar [5].

Sec. 1.3, 1.4 and 1.5 are in D. Xia [5] and [6]. The readers may also read J. Gleason and C. R. Rosentrater [1] and [2].

Sec. 1.6: This is a special case of Section 5 in D. Xia [7].

Sec. 1.7: This is in D. Xia [20].

Chapter 2

Sec 2.1: B. B. Morrel [1] firstly proved the Corollary 2.1.2. Theorem 2.1.1 is in D. Xia [13]. Corollary 2.1.2 can also be proved by analytic models as in J. Gleason and C. R. Rosentrater [1]. The model of pure subnormal operators with rank two self-commutator can be found in D. Xia [4] and J. Gleason and C. R. Rosentrater [1].

Sec. 2.2 and 2.3: Most of the materials in these two sections can be found in D. Xia [12], [13], [16]. There are several interesting papers related these two sections: J. E. McCarthy and L. Yang [1], R. Olin, J. Thomson and T. Trents [1], and D. V. Yakobovich [1] and [2].
About quadrature domain, also see B. Gustafsson [1], B. Gustafsson and H. S. Shapiro [1], and M. Sakai [1].

About Plemelj's formula see I. I. Priralov [1].

Sec. 2.4: Theorem 2.4.3 is a special case of a theorem in J. D. Pincus and D. Xia [3]. This is also related to principal current of R. W. Carey [1] and R. W. Carey and J. D. Pincus [5]. The rest also see R. W. Carey and J. D. Pincus [3], J. W. Helton and R. How [1], [2] and J. Gleason and C. R. Rosentrater [1].

Chapter 3

Sec. 3.1: Theorem 3.1.1 is a special case of B. Fuglede [1] and C. R. Putnam [1]. The proof of this theorem is from M. Rosenblum [1]. Theorem 3.1.4 is in M. Putinar [1].

About the theory of subnormal tuples of operators and its related topics, there are several interesting papers, such as A. Athavale [1]-[7], A. Athavale and S. Pederson [1], J. B. Conway [2] and [3], R. E. Curto [1], R. E. Curto, H. Lee and W. Y. Lee [1], R. E. Curto and W. Y. Lee [1], R. E. Curto, W. Y. Lee and J. Yoon [1], [2], R. E. Curto, P. Muhly and J. Xia [1], R. E. Curto, Y. T. Poon and J. Yoon [1], and R. E. Curto and J. Yoon [1].

Sec. 3.2 comes from J. Eschmeier and M. Putinar [1]. A related paper is J. Gleason [1].

Sec. 3.3 and Sec. 3.6 are in D. Xia [18].

Sec. 3.4 and Sec. 3.5 are in D. XIa [6].

Sec. 3.7 comes from D. Xia [7].

Chapter 4

This chapter mostly comes from D. Xia [17], except Lemma 4.1.1 and Theorem 4.4.1 are in J. D. Pincus and D. Xia [3], see also J. D. Pincus and D. Zheng [1], see also D. Xia [21].

Chapter 5

Sec. 5.1 and Sec. 5.2 are from J. D. Pincus, D. Xia and J. Xia [1], [2].

Sec. 5.3 mostly comes from M. Putinar [5], [6] and M. Putinar and B. Gustafsson [1], [2], except the formula for the kernel $E(\cdot, \cdot)$, Proposition 5.3.4, Corollary 5.3.5 and Corollary 5.3.6 are in D. Xia [13].

Sec. 5.4, 5.5, 5.6 come from D. Xia [16], also see D. V. Yakobovich [4].

In Sec 5.7, part of Proposition 5.7.1 is in D. Xia [16]. Lemma 5.7.2 and Theorem 5.7.3 are in M. Putinar [5].

Sec. 5.8 comes from D. Xia [13] and Sec. 5.9 comes from D. Xia [15].

D. Aharonov and H.S. Shapiro

[1] *Domains on which analytic functions satisfy quadrature identities*, J. Anal. Math. **30** (1996), 39-73.

M. B. Abrahamse and R. G. Douglas

[1] *A class of subnormal operators related to multiply-connected domains*, Adv. Math. **19** (1976), 106-148.

A. Athavale

[1] *On the joint hyponormality of operators*, Proc. Amer. Math. Soc. **103** (1988), 417-423.
[2] *Subnormal tuples quasi-similar to the Szego tuple*, Michigan Math. J. (1988), 409-412.
[3] *On the class of subnormal tuples*, Integr. Equat. Oper. Theory **12** (1989), 305-323.
[4] *On the intertwining of joint isometries*, J. Oper. Theory **23** (1990), 339-350.
[5] *Relating the normal extension and the regular unitary dilation of a subnormal tuple of contractions*, Acta Math. (Szeged) **56** (1992), 121-124.
[6] *A note on joint subnormality and spherical dilations*, Rev. Roum. Math. Pure Appl. **37** (1992), 429-499.
[7] *On a minimal normal dilation problem,* Pro. Amer. Math. Soc. **121** (1994), 343-350.

A. Athavale and S. Pederson

[1] *Moment problems and subnormality*, Jour. Math. Anal. Appl. **146** (1990), 434-441.

R.W. Carey

[1] *A unitary invariant for pairs of self-adjiont operators*, J. Reine Angew. Math. **283** (1976), 294-312.

[2] *Some remarks on principle currents and index for single and several commuting operators*, Surveys of Some Reccent Resuls in Operator Theory, Pitman Res. Notes Math. Ser. no. **171**, Longman Sci.Tech., Harlow, 1988, 133-155.

R. W. Carey and J. D. Pincus

[1] *Contruction of semi-normal operators with prescribed mosaics*, Indiana Univ. Math. J. **23** (1974), 1155-1165.
[2] *Commutators, symbols and determining functions*, J. Funct. Anal. **19** (1975), 50-80.
[3] *Mosaics, principal functions and mean motion in Von Neumann Algebra*, Acta Math. **138** (1977), 153-218.
[4] *An integrality theorem for subnormal operators*, Integr. Equat. Oper. Theory, **4** (1981) 10-44.
[5] *Principal currents*, Integr. Equat. Oper. Theory, **8** (1985), 615-640.

K. Clancey

[1] *Seminormal Operators*, Lect. Notes Math. V. **742**, Springer-Verlag, Berlin, Hedelbuberg, N.Y., 1979.
[2] *A kernel for operators with one-dimensional self-commutators.* Integr. Equat. Oper. Theory, **7** (1984), 441-458.
[3] *Toeplitz model for operators with one dimensional self-commutators*, Oper. Theory: Advan. Appl. V. **11**, Birkhäuser Verlag, Basel, 1983, 81-107.

J. B. Conway

[1] *Theory of Subnormal Operators*, Math. Sur. Mono. V. **36** Amer. Math. Soc. 1991, 1-435.
[2] *Towards a functional calculus for subnormal tuples: the minimal normal extension*, Trans. Amer. Math. Soc. **329** (1991), 543-577, the minimal extension and approximation in several complex variables, Proc. Symp. Pure. Math. 51, Part I, Amer. Math. Soc. Providence (1990).
[3] *Functons of one complex Variable I*, 1991, Springer Verlag, New York, 1-316.

R. E. Curto

[1] *Joint hyponormality: A bridge between hyponormality and subnormality,* Proc. Symp Pure Math. **51** (1990) part 2, 69-89.

R. E. Curto and H. Lee and W. Y. Lee

[1] *Subnormality and 2-hyponormality for Toeplitz operators,* Integr. Equat. Oper. Theory, **44** (2002), 138-148.

R. E. Curto and W. Y. Lee

[1] *Subnormality and k-hyponormality of Toeplitz operators: A brief survey and open questions,* in Operator Theory and Banach Algebra, The Theta Fundation, Bucharest, 2003, 73-81.

R. E. Curto, W. Y. Lee and J. Yoon

[1] *Hyponormality and subnormality for powers of commuting pairs of subnormal Operators,* J. Funct. Anal. **245** (2007), 390-412.
[2] *Which 2-hyponormal 2-variable weighted shifts are subnormal?* Linear Algebra Appl. **429** (2008), 2227-2238.

R. E. Curto, P. Muhly and J. Xia

[1] *Hyponormal pairs of commuting operators,* Oper. Theory, Adv. Appl. **35** (1988), 1-22.

R. E. Curto and Y. T. Poon and J. Yoon

[1] *Subnormality of Bergman-like weighted shift,* J. Math. Ana. Appl. **308** (2005), 334-342.

R. E. Curto and Salinas

[1] *Spectral properties of cyclic subnormal m-tuplcs,* (preprint).

R. E. Curto and J.Yoon

[1] *Jointly hyponormal pairs of commuting subnormal operators need not be jointly subnormal,* Trans. Amer. Math. Soc. **358** (2006), 5139-5159.

M. A. Dritschel and S. McCullouch

[1] *Model theory for hyponormal contractions*, Integr. Equat. Oper. Theory, **36** (2000), 182-192.

J. Eschmeier and M. Putinar

[1] *Some remarks on spherical isometries, in "Systems, Approximations, Singular Integral Operators, and Related Topics"* (A. A. Borichev and N. N. Nikolskii, eds), Birkhäuser, Basel et al. 2001, 271-292.

N. S. Feldman

[1] *The Berger-Shaw theorem for cyclic subnormal operators*, Indiana Univ. Math. J. **46** (1997), 741-751.
[2] *Pure subnormal operatore have cyclic adjoints*, J. Funct. Anal. **162** (1999), 379-399.
[3] *Subnormal operators, self-commutators and pseudocontinuations*, Integr. Equat. Oper. Theory, **37** (2000), 402-432.

J. Gleason

[1] *Matrix Construction of subnormal tuples of finite type*, J. Math. Anal. Appl. **284** (2), 593-602.

J. Gleason and C. R. Rosentrater

[1] *The mosaic and principal function of a subnormal operator*, Integr. Equat. Oper. Theory, **55** (2006), 69-82.
[2] *Xias analytic model of a subnormal operator and its applications*, Rocky Mountain J. Math. **38** (2008), 849-889.

J. W. Helton and R. Howe

[1] *Integral operators, commutator trace, index and homology*, Proc. Conf. Oper. Theory, Lecture Notes in Math. V **345**, Springer, 1973.
[2] *Trace of commutators of integral operators*, Acta Math. **135** (1975), 271-305.

B. Fuglede

[1] *A commutativity theorem for normal operators*, Pro. Nat. Acad. U. S. A., **36** (1950), 35.

B. Gustafsson

[1] *Quadrature identities and the Schottky double*, Acta Appl. Math. **1** (1983), 209-240.

[2] *Singularity and special points on quadrature domains from an algebraic geometric point of view*, J. dAnal. Math. **51** (1988), 91-117.

B. Gustafsson and H. S. Shapiro

[1] *What is a quadrature domaim? In Quadrature Domains and Their Applications*, Oper. Theory, Adv. Appl. **156** (2005) 1-26.

M. Martin and M. Putinar

[1] *Lectures on Hyponormal Operators*, Oper. Theory Adv. Appl. **39** (1990), Birkhäuser Verlag, Bassel-Boston-New York.

M. Martin and N. Salinas

[1] *Weitzenbock type formulas and joint seminomality*, Cont. Math. **212** (1998), 157-165.

J. E. McCarthy and L. Yang

[1] *Subnormal operators and quadrature domains*, Advances in Math. **127** (1997), 52-72.

B. B. Morrel

[1] *A decomposition for some operators*, Indiana Univ. Math. J. **23** (1973), 497-511.

R. Olin, J. Thomson and T. Trent

[1] *Subnormal operators with finte rank of self-commutators*, to appear in Trans. Amer. Math. Soc.

J. D. Pincus

[1] *Commutators and systems of singular integral equations I*, Acta Math. **121** (1968), 219-249.

[2] *The spectrum of semi-normal operators*, Proc. Nat. Acad. Sci. **68** (1971), 1681-1685.

J. D. Pincus and D. Xia

[1] *Mosaic and principal functions of hyponormal and semi-hyponormal operators*, Integr. Equat. Oper. Theory, **4** (1981), 134-150.

[2] *Toeplitz type operators, determining functions, principal functions and trace formulas*, J. Funct. Anal. **88** (1990), 1-63.

[3] *A trace formula for subnormal tuple of operators*, Integr. Equat. Oper. Theory, **14** (1991), 390-398.

J. D. Pinus, D. Xia and J. Xia

[1] *The analytic model of a hyponormal operator with rank one self-commutator Integr.* Equat. Oper. Theory, **8** (1984), 879-893.

[2] *Note on "The analytic model of a hyponormal operator with finite rank self-commutator"*, Integr. Equat. Oper. Theory **7** (1984), 893-894.

J. D. Pincus and D. Zheng

[1] *A remark on the spectral multiplicity of normal extension of commuting subnormal operator tuple*, Integr. Equat. Oper. Theory, **16** (1993), 390-398.

I. I. Privalov

[1] *Boundary Behavior of Analytic Functions*, German translation in Berlin, Deutsch Verlag der Wissenschaften, 1968, Moskow-Lenngrrad:GITTL, 1950.

[2] *Spectral inclusion of subnormal n-tuples*, Proc. Amer. Math. Soc. **90** (1984), 405-406.

[3] *A two moment problem*, J. Funct. Anal. **80** (1988), 1-8.

[4] *The L-problem of moments in two dimensions*, J. Funct. Anal. **94** (1990), 288-307.

[5] *Algebraic operators with rank one self-commutator*, in "Linear and Complex Analysis Problem Book 3" (v. P. Havin and N. K. Nikolskii, eds) Lecture Notes

[6] *Linear analysis of quadrature domains*, Ark. Mat. **33** (1995), 357-376.

[7] *Linear analysis of quadrature domains, III*, J. Math. Anal. Appl. **239** (1999), 101-117.

M. Putinar and B. Gustafsson

[1] *Linear analysis of quadrature domains*, II, Israel J. Math. **119** (2000), 187-216.
[2] *Linear analysis of quadrature domains*, IV, (preprint).

C. R. Putnam

[1] *On normal operators in Hilbert space*, Amer. J. Math. **73** (1951), 357-362.
[2] *Commutation Properties of Hilbert Space Operators and Related Topics*, Springer, Berlin-Heidelberg-New York, 1967.

M. Rosenblum

[1] *On a theorem of Fuglede and Putnam*, J. London Math. Soc. (1958), 376-377.

M. Sakai

[1] *Quadrature Domains*, Lect. Notes Math. V. **934**, Springer Verlag, Berlin-Heidelberg, 1982.

S. A. Steward and D. Xia

[1] *A class of subnormal operators with finite rank of self-commutators*, Integr. Equat. Oper. Theory, **44** (2002), 370-382.

B. Sz-Nagy and C. Foias

[1] *Harmonic Analysis of Operators on Hilbert Space*, Budapest-Amsterdam, 1970.

J. L. Taylor

[1] *A joint spectrum for several commuting operators*, J. Funct. Anal. **6** (1970) 172-191.

J. E. Thomson

[1] *Approximation in the mean by polynomials*, Ann. Math. **133** (1991), 477-507.

D. Xia

[1] *Spectral Theory of Hyponormal Operators*, Sci. Tech. Press, Shanghai 1981, Birkhäuser Verlag, Basel-Boston-Stuttgart, 1983, 1-241.

[2] *On the analytic model of a class of hyponormal operators*, Integr. Equat. Oper. Theory, **6** (1983), 134-157.

[3] *On the kernels associated with a class of hyponormal operators*, Integr. Equat. Oper. Theory, **6** (1983), 444-452.

[4] *The analytic model of a subnormal operator*, Integ. Equat. Oper. Theory, **10** (1987), 258-289.

[5] *Analytic theory of subnormal operators*, Integr. Equat. Oper. Theory, **10** (1987), 880-903. Errata: Integr. Equat. Oper. Theory 12 (1989), 898-899.

[6] *Analytic theory of n-tuple of operators*, Pro. Symp. Pure Math. **51** (1990), part 1, 617-640.

[7] *On some class of hyponormal tuples of commuting operators*, Oper. Theory, Adv. Appl. **48** (1990), 423-448.

[8] *Trace formulas for almost commuting operators, cyclic cohomology and subnormal operators*, Integr. Equat. Oper. Theory, **14** (1991), 276-198.

[9] *Complete unitary invariant for some subnormal operators*, Integr. Equat. Oper. Theory, **15** (1992), 154-166.

[10] *Trace formula for a sclass of subnormal tuple of operators*, Integr. Equat. Oper. Theory, **17** (1993) 417-439.

[11] *Trace formula and complete unitary invariants for some k-tuple of commuting operators*, Cont. Math. **185** (1995), 367-380.

[12] *n pure subnormal operators with finite rank self-commutators and related operator tuples*, Integr. Equat, Oper. Theory, **24** (1996), 107-125.

[13] *Hyponormal operators with finite rank self-commutators and quadrature domains*, J. Math. Ana. Appl. **203** (1996), 540-559.

[14] *On a class of operators with finite rank of self-commutators,* Integr. Equat. Oper. Theory, **33** (1999)489-502.

[15] *Trace formula for some operators related to quadrature domains in the Riemann surfaces*, Integr. Equat. Oper. Theory, **47** (2003), 123-130.

[16] *Hyponormal operators with rank one self-commutators*, Integr. Equat. Oper. Theory, **48** (2004), 115-135.

[17] *On a class of operators of finite type*, Integr. Equat. Oper. Theory, **54** (2006), 131-150.

[18] *Right spectrum and trace formula of subnormal tuple of operators of finite type*, Integr. Equat. Oper. Theory, **55** (2006), 439-452.

[19] *Operator identities of subnormal tuples of operators.* Oper. Theory. Adv. Appl. **218** (2012) 613-638.

[20] *A note on the complete unitary invariant for some subnormal operators,* Ineegr. Equat. Oper. Theory, **77** (2013), 291-298.

[21] *Commutator formulas for some subnormal tuples of operators,* (preprint).

D. V. Yakobovich

[1] *Dual analytic model of semi-normal operators,* Integr. Equat. Oper. Theory **23** (1995), 353-371.

[2] *Subnormal operators of finite type I, Xias model and real algebraic curves,* Revista Matem. Iber. **14** (1998), 95-115.

[3] *Subnormal operators of finite type II, Structure theorems,* Revista Matem. Iber. **14** (1998), 623-689.

[4] *A note on hyponormal operators associated with quadratur domains,* Oper. Theory, Adv. Apll. **123** (2001), 512-525.

[5] *Real separated algebraic curves, quadrature domains, Ahlfors type functions and operator theory,* J. Funct. Ana. **236** (2006), 25-58.

Acknowledgment

Many thanks to Dr. Haichao Wang and Dr. Lujun Wang for typing this. Many thanks to Mr. Shujun Liu for his typing while the author makes some changes of the original draft, and changing the style of the original draft to fit the publisher's need. The author sincerely thanks the editor Mdm Kwong Lai Fun for her great help in the publication of this monograph.

Index

Printed in the United States
By Bookmasters